# Recent Advances and Contribution of Synthetic Aperture Radar (SAR) Applications for Agricultural Monitoring

Editors

**Olaniyi Ajadi**
**David McAlpin**

MDPI • Basel • Beijing • Wuhan • Barcelona • Belgrade • Manchester • Tokyo • Cluj • Tianjin

MDPI

*Editors*
Olaniyi Ajadi
University of Alaska Fairbanks
USA

David McAlpin
University of Alaska
USA

*Editorial Office*
MDPI
St. Alban-Anlage 66
4052 Basel, Switzerland

This is a reprint of articles from the Special Issue published online in the open access journal *Remote Sensing* (ISSN 2072-4292) (available at: https://www.mdpi.com/journal/remotesensing/special_issues/SAR-in-agriculture).

For citation purposes, cite each article independently as indicated on the article page online and as indicated below:

LastName, A.A.; LastName, B.B.; LastName, C.C. Article Title. *Journal Name* **Year**, *Volume Number*, Page Range.

**ISBN 978-3-0365-7358-8 (Hbk)**
**ISBN 978-3-0365-7359-5 (PDF)**

# Recent Advances and Contribution of Synthetic Aperture Radar (SAR) Applications for Agricultural Monitoring

# Contents

# Preface to "Recent Advances and Contribution of Synthetic Aperture Radar (SAR) Applications for Agricultural Monitoring"

Over the last two decades, synthetic aperture radar (SAR) sensors have become recognized as valuable data sources for agricultural monitoring because of their strong penetrating power and high-resolution remote sensing ability. The C-band Sentinel-1 mission, the X-band TerraSAR-X mission, the commercial-grade Capella's satellites, and the upcoming NASA-ISRO synthetic aperture radar (NISAR) mission will introduce a new era for SAR-based agricultural monitoring. The SAR sensors' capabilities for remotely sensing the Earth's surface have been utilized in numerous studies (e.g., [1]). Compared to other sources, SAR's combination of weather and illumination independence with large area coverage at fine spatial resolution (3 m to 20 m) enables frequent and detailed observations [2,3].

This Special Issue covers the state-of-the-art research on SAR for predictive agricultural monitoring using publicly available and commercial datasets. Possessing detailed knowledge of the crop type and distribution across landscapes is important for managing existing lands and new agricultural expansion in a sustainable manner. The research conducted for this Special Issue demonstrated the use of SAR polarization and microwave-derived indices for routine cropland monitoring. In particular, the research conducted by [4] pointed out that high-frequency Earth observation (EO) data was effective in monitoring crop development and crop identification. The authors in [4] highlighted that the ratio of the two Sentinel-1 polarizations (VH/VV) showed great potential when analyzing crop phenology because of its independence from soil moisture changes. Additionally, several state-of-the-art research studies have pointed out that an increase in accuracy was observed when combining data collected at different polarizations in contrast to using a single polarization observation [5–8]. In [9], the authors derived a target characterization parameter for dual-pol SAR data. The parameter utilized the combination of the 2D Barakat degree of polarization and the elements of the covariance matrix [9]. At each crop phenological stage, the authors observed a high sensitivity to the morphological changes. Similarly, in [10], the authors confirmed that when monitoring the biophysical parameters of agricultural fields, the polarimetric decomposition parameters of the dual-pol Sentinel-1 data were valuable.

Altogether, the results obtained in this Special Issue have adequately demonstrated that the relationships between the time series of SAR observations at different polarizations and crop phenology can considerably enhance the capabilities of radar sensors in retrieving parameters related to agricultural crop monitoring.

**References**

1. Hollinger, J.; Lo, R.; Poe, G.; Savage, R.; Pierce, J. *Special Sensor Microwave/Imager User's Guide*; Naval Research Laboratory: Washington, DC, USA, 1987.
2. Ajadi, O.A.; Barr, J.; Liang, S.Z.; Ferreira, R.; Kumpatla, S.P.; Patel, R.; Swatantran, A. Large-scale crop type and crop area mapping across Brazil using synthetic aperture radar and optical imagery. *Int. J. Appl. Earth Obs. Geoinf.* **2021**, *97*, 102294.
3. Ajadi, O.A.; Liao, H.; Jaacks, J.; Delos Santos, A.; Kumpatla, S.P.; Patel, R.; Swatantran, A. Landscape-Scale Crop Lodging Assessment across Iowa and Illinois Using Synthetic Aperture Radar (SAR) Images. *Remote Sens.* **2020**, *12*, 3885. https://doi.org/10.3390/rs12233885.

4. Vavlas, N.-C.; Waine, T.W.; Meersmans, J.; Burgess, P.J.; Fontanelli, G.; Richter, G.M. Deriving Wheat Crop Productivity Indicators Using Sentinel-1 Time Series. *Remote Sens.* **2020**, *12*, 2385. https://doi.org/10.3390/rs12152385.
5. Kim, Y.; Jackson, T.; Bindlish, R.; Lee, H.; Hong, S. Radar Vegetation Index for Estimating the Vegetation Water Content of Rice and Soybean. *IEEE Geosci. Remote Sens. Lett.* **2012**, *9*, 564–568.
6. Bazzi, H.; Baghdadi, N.; El Hajj, M.; Zribi, M.; Minh, D.H.T.; Ndikumana, E.; Courault, D.; Belhouchette, H. Mapping Paddy Rice Using Sentinel-1 SAR Time Series in Camargue, France. *Remote Sens.* **2019**, *11*, 887. https://doi.org/10.3390/rs11070887.
7. Vreugdenhil, M.; Wagner, W.; Bauer-Marschallinger, B.; Pfeil, I.; Teubner, I.; Rüdiger, C.; Strauss, P. Sensitivity of Sentinel-1 Backscatter to Vegetation Dynamics: An Austrian Case Study. *Remote Sens.* **2018**, *10*, 1396. https://doi.org/10.3390/rs10091396.
8. Khabbazan, S.; Vermunt, P.; Steele-Dunne, S.; Ratering Arntz, L.; Marinetti, C.; van der Valk, D.; Iannini, L.; Molijn, R.; Westerdijk, K.; van der Sande, C. Crop Monitoring Using Sentinel-1 Data: A Case Study from The Netherlands. *Remote Sens.* **2019**, *11*, 1887. https://doi.org/10.3390/rs11161887.
9. Dey, S.; Bhogapurapu, N.; Homayouni, S.; Bhattacharya, A.; McNairn, H. Unsupervised Classification of Crop Growth Stages with Scattering Parameters from Dual-Pol Sentinel-1 SAR Data. *Remote Sens.* **2021**, *13*, 4412. https://doi.org/10.3390/rs13214412.
10. Harfenmeister, K.; Itzerott, S.; Weltzien, C.; Spengler, D. Agricultural Monitoring Using Polarimetric Decomposition Parameters of Sentinel-1 Data. *Remote Sens.* **2021**, *13*, 575. https://doi.org/10.3390/rs13040575.

**Olaniyi Ajadi and David McAlpin**
*Editors*

remote sensing

MDPI

Article

# Synergistic Use of Radar and Optical Satellite Data for Improved Monsoon Cropland Mapping in India

**Abdul Qadir** [1,*] **and Pinki Mondal** [1,2]

1 Department of Geography and Spatial Sciences, University of Delaware, Newark, DE 19716, USA; mondalp@udel.edu
2 Department of Plant and Soil Sciences, University of Delaware, Newark, DE 19716, USA
* Correspondence: abdulqa@udel.edu

Received: 26 November 2019; Accepted: 1 February 2020; Published: 5 February 2020

**Abstract:** Monsoon crops play a critical role in Indian agriculture, hence, monitoring these crops is vital for supporting economic growth and food security for the country. However, monitoring these crops is challenging due to limited availability of optical satellite data due to cloud cover during crop growth stages, landscape heterogeneity, and small field sizes. In this paper, our objective is to develop a robust methodology for high-resolution (10 m) monsoon cropland mapping appropriate for different agro-ecological regions (AER) in India. We adapted a synergistic approach of combining Sentinel-1 Synthetic Aperture Radar (SAR) data with Normalized Difference Vegetation Index (NDVI) derived from Sentinel-2 optical data using the Google Earth Engine platform. We developed a new technique, Radar Optical cross Masking (ROM), for separating cropland from non-cropland by masking out forest, plantation, and other non-dynamic features. The methodology was tested for five different AERs in India, representing a wide diversity in agriculture, soil, and climatic variations. Our findings indicate that the overall accuracy obtained by using the SAR-only approach is 90%, whereas that of the combined approach is 93%. Our proposed methodology is particularly effective in regions with cropland mixed with tree plantation/mixed forest, typical of smallholder dominated tropical countries. The proposed agriculture mask, ROM, has high potential to support the global agriculture monitoring missions of Geo Global Agriculture Monitoring (GEOGLAM) and Sentinel-2 for Agriculture (S2Agri) project for constructing a dynamic monsoon cropland mask.

**Keywords:** monsoon cropland; Sentinel; smallholders; Google Earth Engine; SAR; India

## 1. Introduction

India is a primarily agrarian economy with 17% of the national Gross Domestic Product (GDP) contributed by agriculture and approximately 50% of the population supported by agricultural activities [1]. Prior studies have shown a direct impact of monsoon (wet season during June–November) crop production on Indian economic growth and national food security [2,3]. Variations in inter-annual and inter-seasonal distribution of monsoon rainfall also affect monsoon crop production [4]. Accurate information on monsoon cropland can thus help researchers in monitoring the trend in annual crop production and in finding the gap in overall food production [5]. Timely statistics for monsoon cropland are also essential for the policy-makers to decide the crop prices and to provide compensation to the farmers in case of crop failure, as well as for development of agricultural economy and farmers' well-being as a whole [6].

About 56% of the total cultivated area in India is utilized for rainfed monsoon crops. The importance of the rainfed monsoon crop can be gauged from the fact that it contributes to 40% of the country's food production [7]. During the rainfed monsoon crop growing season, both water intensive and dryland crops are grown throughout the country. Water intensive crops, such as rice and sugarcane, are grown in low lying areas and/or areas which receive sufficient rainfall above 1200 mm [8]. Dryland crops are grown in the region where monsoon rainfall is low and erratic (ranging between 500 mm and1200 mm). Dryland crops are also important for the economy as most of the coarse grains, pulses, and cotton are grown on these lands [9]. Traditionally, monsoon crop statistics are documented by different government agencies through agriculture census across India [10,11]. However, these crop statistics are typically aspatial and are collected by sample surveys at either the national or state level [12]. These surveys are expensive, time consuming, and labor intensive. Moreover, the accuracy of the resulting dataset is commonly affected by human and statistical biases [13]. Timely and largely bias-free monsoon cropland statistics as a whole, instead of focusing only on major crops at a spatial scale finer than the district level, could considerably improve targeted intervention for providing government welfare schemes at the village or farm level [14].

Historically, data collected by earth observation satellites have been used for generating reliable crop statistics in many countries and are widely used for operational crop monitoring [15–18]. These satellite-derived data products are particularly important as they can link cropping activities to the environmental factors such as soil, topography, and weather variability [19]. However, most of the prior case studies at the national or global scale are implemented using coarser satellite data such as Moderate Resolution Imaging Spectroradiometer (MODIS) or Advanced Very-High Resolution Radiometer (AVHRR) [20–23]. Methodologies involving such coarser data, when applied to small-scale agriculture (farm sizes less than 2 hectares), common among transitioning economies, result in mixed pixel issues where one aggregated grid-cell value is assigned to many fields with varying cropping practices [24]. The studies on small farms with improved methodology using moderate–high resolution satellite data such as Landsat and Sentinel-2 mainly focus on winter crops when sufficient cloud free optical data are available during the crop-growing season [25,26]. However, optical satellite data are insufficient for operational monsoon cropland mapping as the wet (monsoon) season coincides with the crop growing duration, thus, providing an insufficient number of images for mapping monsoon cropland over a large scale [27]. Previous studies have focused on examining crop phenology over large area using satellite data with high temporal resolution in order to overcome cloud coverage issues [20,28]. Shang et al. [29] applied refined cloud detection approach using MODIS data to separate cloudy scenes from the clear ones to model vegetation growth across India. Similarly, Chakraborty et al. [30] measured the crop phenology trend of the monsoon cropland vegetation and its relationship with rainfall using the Global Inventory Modelling and Mapping Studies (GIMMS) datasets for all of India. Yet, the high revisit frequency of optical satellites such as Landsat (16-day) and Sentinel-2 (5-day) is not enough for effective mapping of monsoon cropland over large area. During the monsoon season, especially for the first 2–3 months, it is unlikely to obtain any cloud-free quality optical image mosaic in tropical or sub-tropical countries [31]. For example, in certain regions in central India, where farmers primarily plant soybean and black gram, the crops are harvested early in the season, even before the cloud free optical satellite images are available. Even when optical satellite data are available during the peak growth stages of the crops such as rice, the spectral signatures of the crops are often mixed with that of plantation, grassland, or forested regions [32,33], thus making it challenging to segregate croplands with monsoon crops from other vegetation covers.

The hindrance in using optical satellite data for intra-seasonal monsoon cropland monitoring over large region requires the remote sensing community to develop new methods, especially for countries with heterogeneous landscapes, such as India. These methods should take into account the variations in cropping practices across different agro-ecological regions (AER). The Synthetic Aperture Radar (SAR) data provide an alternative to optical data for monsoon cropland mapping, as SAR data are not affected by cloud cover [34,35]. In contrast to optical, SAR (also referred to as radar) sensors are active systems with their own source of energy, transmitting microwaves, and receiving the reflected echoes from objects on the earth's surface [36–38]. The SAR sensor transmits longer wavelengths that can easily penetrate through the clouds, thus making it particularly useful on the cloudy days when optical sensors fail to capture reflected sunlight. Moreover, the information collected by radar differs from that obtained from optical sensors due to differences in interaction with the ground objects. The backscattering intensities recorded by SAR sensors are influenced by geometric and dielectric properties of the crops, whereas the interaction with optical sensor are influenced by the chlorophyll and water content. Because of this difference in interaction with crops, data from SAR and optical systems provide complementary information [39–41]. SAR data also have different polarization components (VH, VV, HH, and HV) that interact with crops differently, thus increasing the information content provided [42,43]. Besides, the spatial resolution of SAR sensors such as Sentinel-1 or Radarsat-2, are better suited for small-scale cropland mapping [44].

Multi-temporal SAR images improve the crop classification accuracy and capture the variation in growth process [32,43]. Skakun et al. [45] has shown how multi-temporal SAR images can effectively produce the equivalent classification accuracy as optical images during the cloudy seasons. Studies have also shown that multi-temporal SAR images (>10 scenes) can increase the classification accuracy obtained from optical images (2 or 3 scenes) by 5% [45–47]. Skriver et al. [47] have shown that multi-temporal, multi-polarization SAR images perform better compared to single date multi-polarization or multi-date single polarization SAR images. Hence, the temporal information from SAR, combined with multi-polarization, provides better information of crop conditions. Yet, previous studies utilizing SAR have been confined to examining croplands dominated by specific water intensive monsoon crops such as rice or jute which are easier to detect due to their distinct backscattering signatures compared to dryland monsoon crops [48,49]. Hence, these SAR-based methods need to be evaluated or revised in the context of diverse cropping practices, especially for rainfed monsoon crops grown in dryland regions.

Previously, combinations of SAR and optical data have been extensively used for land cover classification during cloud free seasons, either by stacking bands or by fusing SAR and optical data products [50,51]. Studies have also shown that integrated use of SAR and optical data can significantly improve the crop classification and yield accuracy focused in many regions of the world [45,52]. Recently, studies have emerged to combine SAR and optical data for predicting yields of several monsoon crops including rice, corn, soybean, and cotton [53–55]. These studies either focused on water intensive monsoon crops or over small regions where obtaining a few optical image snapshots was possible during the monsoon season [56,57]. However, there are no studies performed by using SAR data alone or by integrating SAR and optical data for extracting monsoon cropland over a large area in different agro-ecological regions practicing diverse agriculture systems.

This study intends to fill the gap in monsoon cropland monitoring by combining SAR and optical data and has the following objectives:

(1) Evaluating Sentinel-1 (S1) SAR and a combination of SAR and Sentinel-2 (S2) optical data in terms of providing greater accuracy for monsoon cropland mapping.
(2) Developing a high resolution, all weather applicable non-crop mask for segregating monsoon cropland from other land use/land cover (LULC) features with similar signatures (plantation and forest).

## 2. Materials and Methods

### 2.1. Study Area

The study area comprises of ten sub-regions within five agro-ecological regions (AER) covering the Indian states of Uttar Pradesh, Madhya Pradesh, Chhattisgarh, Maharashtra, Andhra Pradesh, and Karnataka (Figure 1) [58]. The study area covers approximately 604,615 km$^2$ and is surrounded by the alluvial Gangetic plains in the north and the Bay of Bengal in the south. It borders the Western Ghats in the west and Chota Nagpur plateau in the east. The region is mostly undulating with the elevation ranging between 0–1560 m (Figure 1).

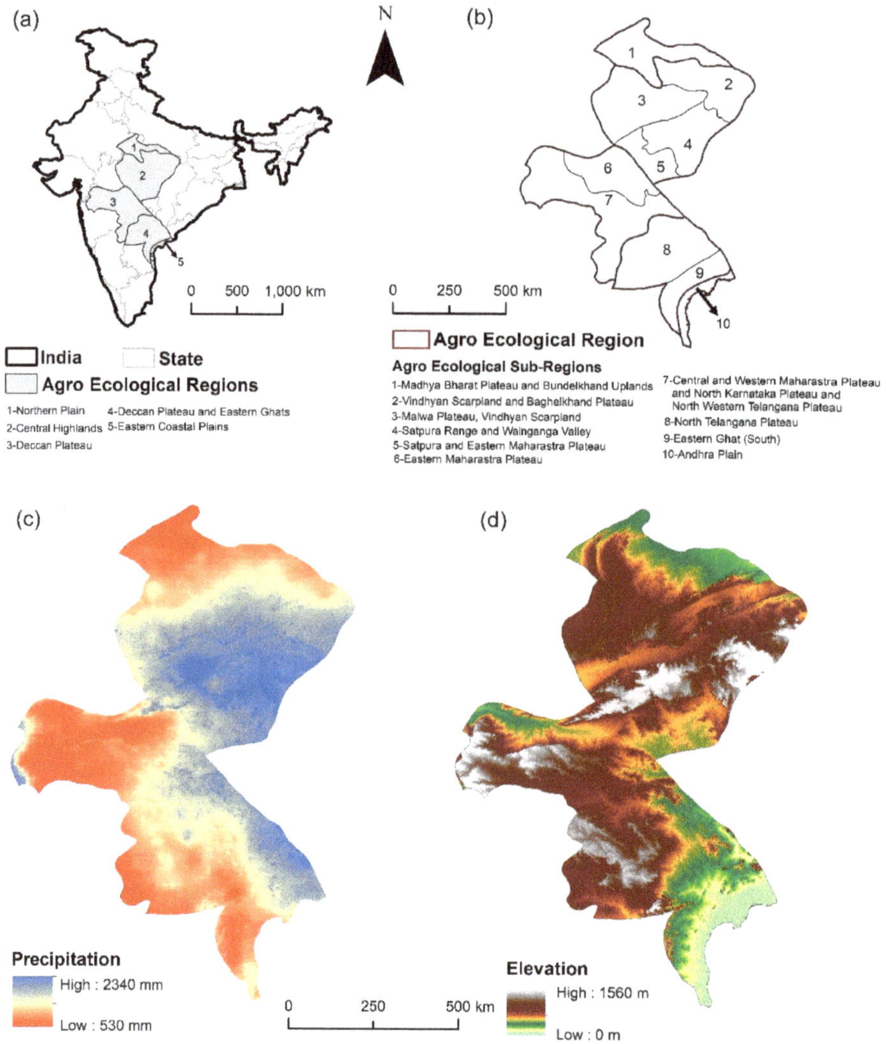


**Figure 1.** Maps of the study area showing: (**a**) Agro-Ecological Regions (AER) selected for this study; (**b**) ten AER sub-regions within five AER; (**c**) spatial variation in annual mean precipitation from the year 2000 to 2018, derived from the Climate Hazards Group InfraRed Precipitation with Station (CHIRPS) data; and (**d**) Digital Elevation Model (DEM) obtained from the Shuttle Radar Topography Mission (SRTM) dataset.

The region has a tropical monsoon climate (Am) as per the Koppen-Geiger climate classification system [59]. The mean annual rainfall ranges approximately from 530 mm to 2300 mm across the region (Figure 1c), as calculated from the Climate Hazards Group InfraRed Precipitation with Station (CHIRPS) data for 2000–2018. Most of the farmers in the study region are smallholders with limited landholdings; they grow crops during three seasons: monsoon (*kharif*) during June-November, winter (*rabi*) during December-April, and summer (*zaid*) during April–June [60]. The major monsoon crops grown in the study region are rice, soybean, black gram (locally known as *Urad*), cotton, maize, and groundnut [61]. The monsoon crop sowing date varies across the study region, starting in the month of June with the onset of monsoon, up to August/September in low-lying regions. The harvesting of the crops widely varies as well and may range from September for soybean and black gram to November for rice. The details regarding the AERs considered for this study and the major monsoon crops grown according to the latest government statistics available are listed in Table 1 [61]. For further analysis, we have combined AER-5 with AER-4, as AER-5 has negligible cropped area to be analyzed as separate unit. Diverse monsoon crops are grown in the AERs comprising of both water intensive monsoon crops and rainfed-dryland crops.

**Table 1.** Agro-Ecological regions and the major crops grown.

| | Agro-Ecological Region | Major Crops |
|---|---|---|
| 1 | Northern Plain | black gram, millet, sesame, rice |
| 2 | Central Highlands | soybean, rice, cotton |
| 3 | Deccan Plateau | cotton, soybean, sorghum |
| 4 | Deccan Plateau and Eastern Ghats, Eastern Coastal Plains | rice, cotton, chili, maize |

### 2.2. Overall Workflow

The flowchart for the methods used in this study is outlined in Figure 2. In the first step, S1 and S2 time series images were loaded on Google Earth Engine (GEE) platform using 'ImageCollection' function [62]. These images were then filtered based on time (June–November 2018) and study region boundary. For S1 SAR, we used images from June to November, but for S2 optical data, we considered images from July to November. The month of June was not considered for S2 optical data as summer crops were still at their peak growth stage in some regions and the land preparation and sowing of monsoon crops were in the initial stages. A 'cropped field' in June would thus be an indication of summer crops, and not monsoon crops. Further image classification was performed on S1 (Figure 2b), and S1+S2 combined (Figure 2c), using pixel-based machine learning classifier (random forest) on GEE. We have used pixel-based classifier instead of object-based classifier for large monsoon cropland mapping, as the latter requires high computation time and has complicated intermediate steps including the segmentation where specific parameter tuning is needed [32,63,64]. Even though object-based classifiers might improve the classification accuracy in some landscapes, this performance improvement is not always evident in complex heterogeneous landscapes such as the one showed in this study. We further performed accuracy assessments for the four AERs (Table 1). Training and testing of the classified images were performed according to the procedure detailed in Figure 2a and Section 2.2.2.

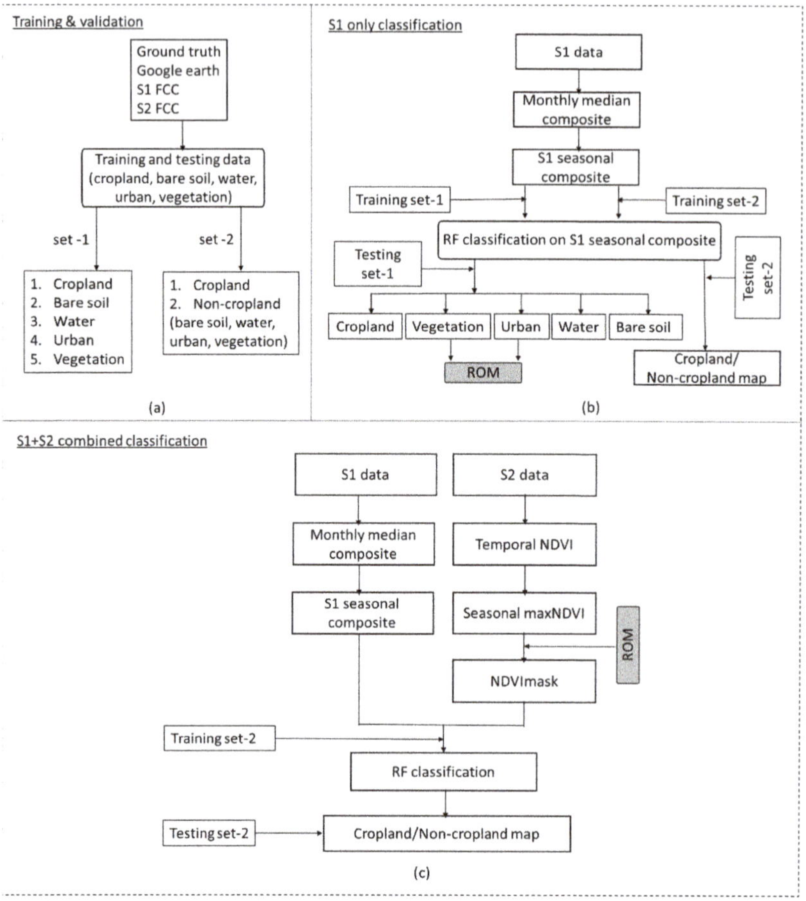


**Figure 2.** Overall workflow followed in this study detailing the steps for: (**a**) Collecting the training and testing points and the classes used for set-1 and set-2; (**b**) Performing the classification using set-1 reference data to obtain Radar Optical cross Masking (ROM) and using set-2 reference data to obtain crop map; and (**c**) For S1+S2 combined classification using set-2 reference data.

### 2.2.1. Satellite Data Pre-Preprocessing

We used freely available S1 SAR data and S2 optical data provided by the European Space Agency (ESA) and images were accessed and processed on the GEE platform [65]. We used S1, C-band, dual-polarization VV (single co-polarization, vertical transmit/vertical receive) and VH (dual-band cross-polarization, vertical transmit/horizontal receive) dataset for the Interferometric Wide Swath (IWS) mode in the descending look angle [66]. The GEE platform provides S1 data pre-processed with thermal noise removal, radiometric calibration and ortho-rectification using the Sentinel-1 toolbox resulting in ground-range detected images with backscattering coefficients in decibel (dB) scale.

Using temporal VH and VV polarization, SAR monthly composite images were created by considering median values. We also used these monthly median composite images to create False Color Composite (FCC) to aid in visual interpretation of the images for training and testing data collection (Figure 4). For optical data, we used the S2 level 1-C, ortho-rectified and geo-referenced top-of-atmosphere (TOA) reflectance data product [67]. The collection contains Multi Spectral Instrument (MSI) bands with a scaling factor of 10,000. To maintain the quality of the data analysis and products during the monsoon season, we considered Sentinel-2 images with cloud cover of 5% or less. On these filtered images, we applied an automated cloud masking algorithm using quality assessment band (band QA60) to mask both opaque and cirrus clouds [68]. The images acquired after the month of November were not considered as we assumed that crops grown after this time are not monsoon crops, based on existing literature [24,60]. A total of 516 S1 images and 1734 S2 images were used for the entire monsoon crop-growing season of 2018.

### 2.2.2. SAR Temporal Backscattering

Temporal backscattering profiles were obtained using C-band VH polarization S1 imagery from monsoon crops (rice and black gram/soybean), bare soil, urban, water and vegetation (forest/plantation/grass) features shown in Figure 3a, similar to what obtained by Singha et al. [31]. The backscattering profiles were generated by taking the mean of 10 sample points for each class spread across the study area. The sample points for each class along with their geolocations are shown in Figure 3b.

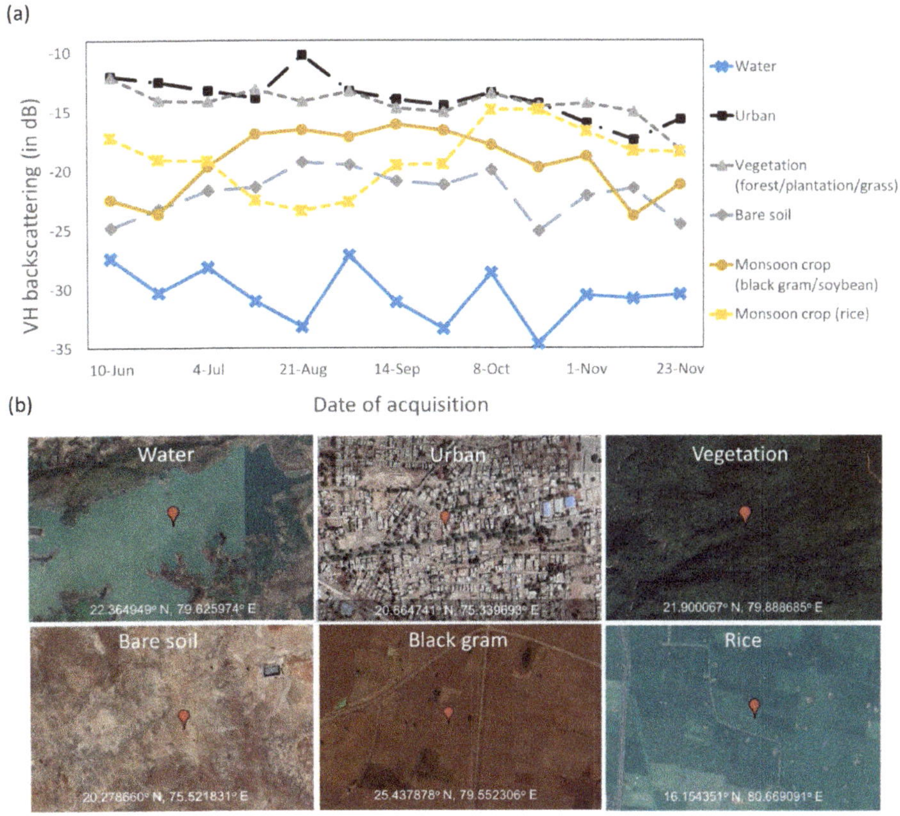


**Figure 3.** (**a**) Sentinel-1 (S1) mean temporal backscattering profile with VH polarization obtained from 10 points each for land cover features, collected from monsoon crops and other land use/cover classes during the monsoon season (June–November 2018). Urban and vegetation class shows constantly high backscattering intensities throughout the monsoon season, water shows very low backscattering intensities and monsoon crops and bare soil has backscattering values between urban/vegetation and water; (**b**) Representative reference points along with its coordinates on the high-resolution google earth imagery.

Vegetation is defined as land surface with plants and includes plantation, grass and forest. The contrasting nature of backscattering from vegetation and monsoon crops forms our basis of utilizing the temporal S1 backscattering signatures for Radar Optical cross Masking (ROM) as discussed in Section 2.2.6 below. As displayed in Figure 3a, temporal SAR backscattering signatures are effective in separating crops (rice, black gram, and soybean) from water, urban and vegetation. However, these signatures are mixed with bare soil during the crop-growing season. So, it becomes difficult to segregate crops from bare soil with very high accuracy using only S1 SAR data. Hindrance in segregating crops from bare soil forms our basis of integrating optical data with SAR as bare soil is very distinct in optical data compared to crops and other vegetation due to its lack of 'greenness' reflected in low Normalized Difference Vegetation Index (NDVI) values [69].

In the study region, the SAR backscattering signatures obtained from vegetation (forest/plantation/grass) and urban class are non-dynamic throughout the monsoon season, and have nearly constant high backscattering values (~−16 dB to −10 dB) compared to other LULC features. During the time of classification, there is high probability of vegetation class being mixed with urban and vice versa. Monsoon crops and bare soil have dynamic backscattering throughout the crop growing season. For crops such as black gram and soybean, the land preparation starts from first week of June and last until mid-July based on the onset of the monsoon. For these monsoon crops, the backscattering is initially low due to land preparation in June and increases with time as the crop grows. For rice, land preparation starts in the July/August when the fields have sufficient amount of water as rice is a water intensive crop [31]. Rice shows very low backscattering during the land preparation/transplanting stage. During the time of maturity, the backscattering increases for both black gram/soybean and rice. The backscattering is high for rice compared to black gram/soybean due to high biomass content resulting in high volume scattering from the rice fields.

For bare soil, the initial backscattering is similar to that obtained from black gram/soybean due to the presence of exposed soil with no crop cover. It can be seen that bare soil signature can get mixed with that from rice in the month of July. Hence, overall it is very difficult to segregate monsoon crops from bare soil with very high accuracy. For water, the backscattering is very low (<−25 dB) throughout the monsoon season, hence it is easily segregated from monsoon crops.

### 2.2.3. Training and Testing the Classifiers

We collected a total of 1500 reference points required for training and testing the classifiers for the five broad land use land cover (LULC) classes: monsoon crop, bare soil, water, vegetation, and urban (Figure 4a). We defined bare soil as any land cover feature, which is devoid of vegetation, either a barren land, fallow land with no crop, or any region with exposed soil. We collected these points through a combination of field visits, high-resolution google earth imagery and visual interpretation of S1 and S2 satellite imagery using the method similar to those explained in [31,70]. Using multiple datasets to generate the training and testing points ensures that only land cover features, which have the high probability of belonging to the actual land cover class on the ground, are selected for training and testing. During field visits, information on land covers along with their geographic coordinates were collected using a handheld Global Positioning System (GPS) device. Field visits were conducted at four agro-ecological sub-regions: (1) Madhya Bharat Plateau and Bundelkhand Uplands, (2) Vindhyan Scarpland and Baghelkhand Plateau, (3) Eastern Ghats (South), and (4) Andhra Plain (Figure 1b). We collected a total of 500 points for monsoon crops, 300 points each for bare soil and vegetation, and 200 points each for water and urban, using stratified random sampling approach. The number of points collected for each land cover was decided based on the relative dominance of these land covers in the study landscape.

First, the field data points were imported on GEE platform and were overlaid on the S1 and S2 FCC images. Using this ground truth data, the extracted features were used to identify similar LULC features in other regions using visual interpretation techniques on FCC of S1 and S2. The extracted features were verified using the high-resolution google earth imagery. Extracting training points for water, forest/plantation and urban is straightforward in S1 as they are not dynamic over time and have very distinct temporal backscattering signatures compared to crops and bare soil as shown in Figure 3 and explained in Section 2.2.2. We only assigned a reference point to a particular LULC, if the corresponding LULC class was confirmed in all three layers (S1 FCC, S2 FCC, and high-resolution google earth imagery). Finally, the field data points and points generated through visual interpretation were merged together to be used for training and testing on GEE platform (Figure 4a). Representative reference points for monsoon crop and bare soil are shown in Figure 4b,c. We randomly identified 70% of the 1500 reference points as 'training points' using the 'randomColumn' function in GEE and used those for training the random forest (RF) classifier. The rest of the reference points (30%) was used as 'testing points', i.e., for post-classification accuracy assessment. To avoid any biases in selecting the training and testing points, we performed the classification and accuracy assessment iteratively for 20 times by randomly dividing training and testing points in 70:30 ratio.

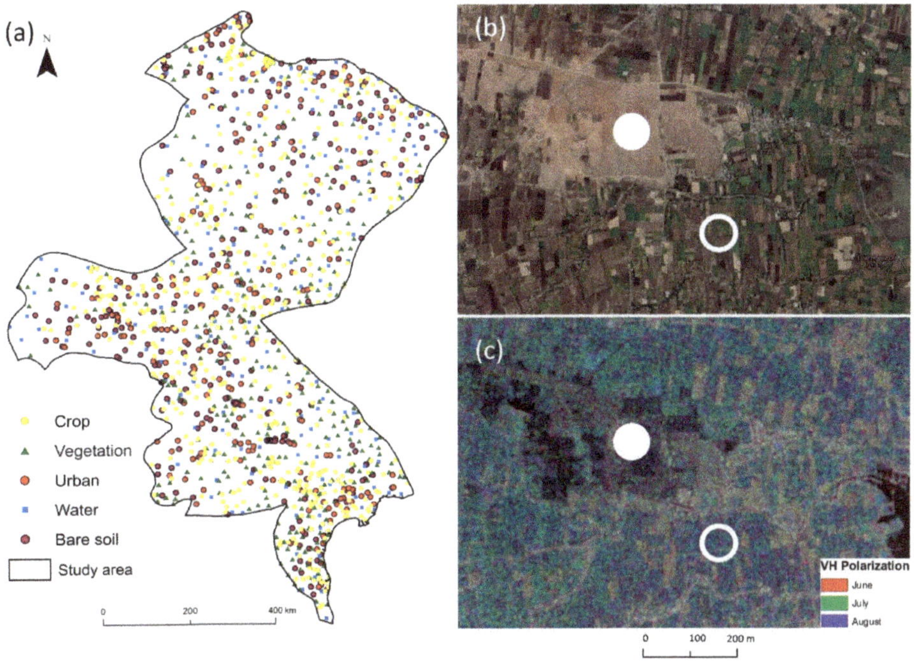


**Figure 4.** (**a**) Spatial distribution of training and testing points across the agro-ecological regions (AER). The five land use/cover classes used for this study are vegetation (forest/plantation/grass), urban, water, bare soil, and monsoon crop; (**b**) Representative reference points on the high resolution google earth imagery for monsoon crop (white hollow circle) and bare soil (white solid circle); (**c**) The same representative reference points as shown in (**b**) confirmed using Sentinel-1 monthly median false color composite imagery (red—June, green—July, and blue—August).

### 2.2.4. Classification Based on Sentinel-1

We considered a monthly composite of SAR data using both VH and VV (VH + VV) polarization, instead of a single date image for S1 classification, as previous studies have shown that multi-temporal SAR data perform far better than single SAR image for crop classification [41,45,47]. Considering multi-temporal SAR data becomes even more important for diversified cropping pattern in India as such data are able to take into account the variation of crops grown in different time of the season. Using the training dataset, the RF classifier was run on monthly median composite of June–November

2018. The RF is an ensemble classifier that follows the decision tree approach in which randomly selected results from multiple decision trees are combined together to obtain highly accurate and stable classification results [65–68]. RF algorithm can handle large quantity of complex multi-source data and is robust against overfitting. For initiating RF classifier, the user must define two parameters, the number of trees to grow and the number of variables used to split each node. In this work, the number of decision trees used are 100 and the variables used to split each node was set to square root of the number of overall variables. Previous studies have shown that RF classifier outperforms other parametric classifiers such as Support Vector Machine (SVM) in obtaining improved classification results [71–74]. RF outperforms SVM in many ways as it can handle large database of temporal images, requires less training time. In addition, the number of user-defined parameters required in RF are less and easier to define compared to SVM [75–77]. RF performs better if there are sufficient amount of training dataset as we have in this study. RF is also successful in discriminating classes with similar characteristics such as crops and vegetation [78]. For this study, S1-based classification was performed using two different output criteria: one with a classified map with five classes (Section 2.2.3) and the other with only two classes – cropland and non-cropland (obtained by combining non-cropland classes, i.e., bare soil, water, vegetation and urban). In addition, we calculated the classification accuracy for each AER separately. The classification and accuracy assessments were performed 20 times using unique set of training and testing data.

#### 2.2.5. Seasonal Normalized Difference Vegetation Index (NDVI)

The Normalized Difference Vegetation Index (NDVI) is a widely used remote sensing measure to assess the health of the vegetation and to differentiate crops, and other vegetation (forest, plantation, grass) from bare soil, water and urban [79]. NDVI is a unitless measure and ranges between −1 and 1. Healthy vegetation typically has higher NDVI values compared to non-vegetated surfaces. For calculating NDVI, we require the red and near-infrared (NIR) reflectance bands (Equation (1)):

$$\text{NDVI} = (\text{NIR} - \text{Red})/(\text{NIR} + \text{Red}) \tag{1}$$

In S2 imagery, we used band 4 and band 8, respectively, for red and NIR in the above equation. We calculated NDVI for all the available cloud free pixels in our image as our main focus was on cropped field identification, especially since previous studies have shown high correlation between NDVI and photosynthetic activities of the cropped fields [80]. To generate the seasonal maximum value of NDVI (maxNDVI), we performed temporal aggregation of NDVI from July to November (Figure 5). Temporal aggregation is an approach to perform pixel-based analysis over a period of time using metrics (i.e., mean, median, maximum etc.) from satellite derived reflectance or satellite-derived indices (e.g., NDVI, Enhanced Vegetation Index (EVI), etc.) [68]. The aggregation addresses the problem of lack of continuity in the optical data due to cloud cover and reduces the volume of data for further processing [68,81]. During monsoon season, optical satellite images in India contain considerable amount of cloud patches, which affects the radiometric quality of the images, thus limiting intra-seasonal crop monitoring capability. We calculated maxNDVI in order to fill this data gap and to capture the crop heterogeneity, i.e., considering all monsoon crops with different intra-seasonal phenology. In this maxNDVI dataset, both crops and vegetation (forest/plantation) have higher values compared to water, urban and bare soil. We further utilized the Otsu's thresholding approach in GEE, to differentiate between the crops/vegetation (forest, plantation) from non-vegetative features with low NDVI values (bare soil, urban and water) [82]. This approach is an automated way of finding an optimal global threshold based on the observed distribution of pixel values. Based on the pixel value distributions for the LULC classes considered in this study, the Otsu's thresholding value for maxNDVI is 0.36.

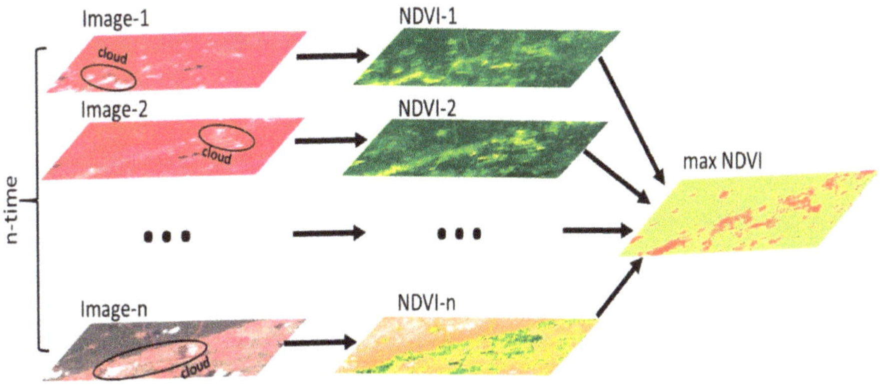


**Figure 5.** Temporal aggregation of normalized difference vegetation index (NDVI) derived from seasonal sentinel-2 (S2) data to obtain the maxNDVI during the monsoon season.

#### 2.2.6. Radar Optical cross Masking (ROM)

It was not possible to differentiate between crop and vegetation within the maxNDVI dataset, which is an important step for a crop mapping procedure involving optical data. Due to non-availability of high-resolution (10m) non-crop mask for the region, we developed a method of masking non-crop vegetation from maxNDVI using classified SAR imagery (Figure 2b). From the S1-derived classified map with five LULC classes, vegetation (forest/plantation/grass) and urban classes were further combined together as a non-dynamic class to obtain the non-crop mask and to segregate crops from vegetation in maxNDVI imagery. We combined vegetation and urban as one class, instead of considering only vegetation, since these two classes have similar backscattering signatures and are difficult to segregate in S1-derived map as described previously (Section 2.2.2, Figure 3). Moreover, combining urban and vegetation to obtain the non-crop mask is less likely to affect our outputs, as urban class is already masked out from the maxNDVI data due to the application of Otsu's thresholding. We coined this technique as Radar Optical cross Masking (ROM) where we used the non-dynamic, non-crop (urban + vegetation) mask to separate vegetation from crops resulting in crop only maxNDVI dataset (NDVImask; Figure 2c).

#### 2.2.7. Classification Based on Combined Sentinel-1 and Sentinel-2

In this step, the S1 data and NDVImask were combined for pixel-based classification to examine if adding NDVImask imagery will result in improved monsoon crop mapping accuracy compared to using S1 only classified map. Combining S1 and NDVImask will also address some of the limitations of using only S1 data for monsoon crop mapping (Section 2.2.2, Figure 3). The monthly median SAR composites from June to November were stacked together with NDVImask data obtained after using ROM. The RF classifier was run on the combined dataset with the number of trees set as 100 and the variables to split each node set to square root of the number of overall variables. The output from this classification is a binary crop/non-crop map using training and testing set-2 (Figure 2c). Similar to the S1-based classification, the combined S1 and S2-based classification and accuracy assessment were repeated for 20 times to avoid any biases in the classification accuracy.

#### 2.2.8. Accuracy Assessment

Both classification outputs obtained from S1 and combined S1 and S2 were evaluated and compared using the standard count-based accuracy assessment methods of overall accuracy (OA) and kappa coefficients obtained from the confusion matrix [83] using 20 different iterations. User's accuracy (UA) and producer's accuracy (PA) were calculated using 30% testing points that were not involved in training the classifiers. UA measures the error of commission, i.e., the proportion of pixels that were incorrectly included in a class that is being evaluated. PA measures the error of omission, i.e., the proportion of pixels in a certain class that is being evaluated that were incorrectly classified in another category, and were omitted from the 'truth' class as identified by the test points.

We further calculated the F-score to determine the degree of discrimination among the five LULC classes obtained from the S1-derived classification and the binary crop vs non-crop classification obtained from S1 only and S1+S2 combined derived classification. The F-score ranges between 0 and 1, with higher values denoting better discriminating power among the classes. The F-score is calculated using Equation (2) mentioned below [84]:

$$\mathrm{F-score} = \left(\frac{(\mathrm{UA} \times \mathrm{PA})}{(\mathrm{UA} + \mathrm{PA})}\right) \times 2 \quad (2)$$

We did not compare the accuracy of the results obtained with the crop estimates provided by government due to non-availability of crop census data for the monsoon crop season 2018–2019.

## 3. Results

### 3.1. Accuracy of S1 Only Classification

The classification accuracy obtained from S1-derived classification using training and testing set-1 is shown in Table 2. The OA obtained after 20 iterations is 80.0%. The kappa coefficient obtained is 0.74 (Table 2). Our results indicate that the water class was easily identified (F-score = 0.96) using SAR data among the five LULC classes. The low F-score obtained for urban (0.64) class indicates that the S1 SAR data has the least discriminative capability to segregate urban from other classes. SAR data was moderately successful in discriminating monsoon crops from other land cover classes (F-score = 0.84).

**Table 2.** Overall Accuracy (OA), Kappa, User's Accuracy (UA), and Producer's Accuracy (PA) for land cover classes obtained from S1 only classification using VH + VV polarization and training and testing set-1.

| Land Cover Type | S1 Only (VH + VV) | | |
|---|---|---|---|
| | UA | PA | F-Score |
| Water | 0.96 | 0.96 | 0.96 |
| Bare soil | 0.79 | 0.8 | 0.79 |
| Urban | 0.78 | 0.54 | 0.64 |
| Vegetation | 0.68 | 0.75 | 0.71 |
| Monsoon crop | 0.81 | 0.87 | 0.84 |
| OA | | 0.80 | |
| Kappa | | 0.74 | |

### 3.2. Accuracy of Binary Crop Maps from S1 Only and Combined S1 and S2

The comparison of accuracy assessment obtained from S1 only and S1+S2 combination using training and testing data set-2 is displayed in Table 3. The crop vs. non-crop overall classification accuracy obtained by S1+S2 combination is approximately 3% higher than obtained by using S1 only dataset. The standard deviation of kappa values while randomly changing the training and testing points for 20 classifier iterations are slightly higher for S1 only compared to S1+S2 classification. The F-score for both S1 and S1+S2 classification shows high discriminative capability (>0.85). Moreover, the F-score for the combined S1+S2 is higher compared to S1 only classification for both crop and non-crop class.

NDVImask image obtained after applying ROM on maxNDVI is shown in Figure 6. It can be seen that non-dynamic forest/plantation regions can be effectively separated from crops using ROM. In the figure, regions with forest and plantation (casuarina and eucalyptus) are masked out using ROM to obtain NDVImask image with crops only. The crop map generated using combined S1 and S2 data is shown in Figure 7. Detailed zoom-in views for selected locations using the combined S1+S2 classification and its comparison with high-resolution imagery are also shown (Figure 8). It can be observed that the combined approach is efficient in differentiating monsoon cropland from plantation (such as mentha/casuarina/eucalyptus) in AER-1 (Figure 8a) and AER-4 (Figure 8c).

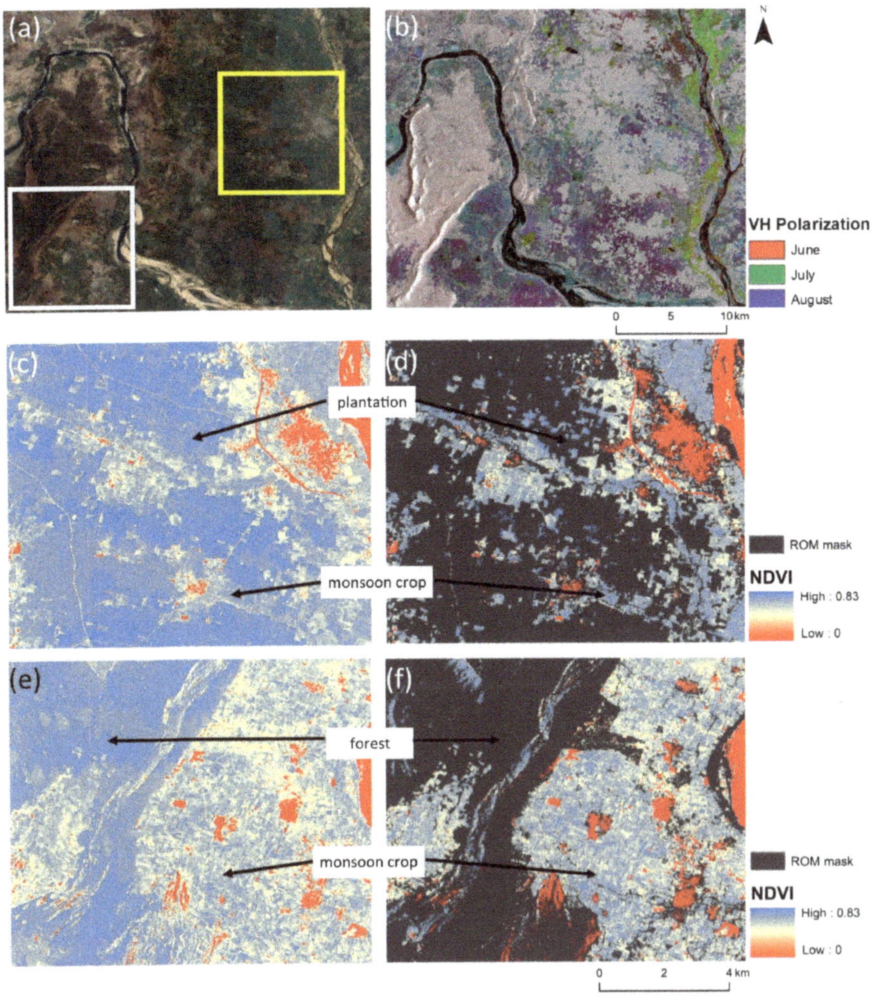


**Figure 6.** Steps for obtaining high-resolution (10m) non-crop mask using the ROM technique: (**a**) High-resolution google earth imagery showing forest class mixed with monsoon crops in white square box and plantation mixed with monsoon crops in yellow square box; (**b**) False Color Composite VH polarization Sentinel 1 (S1) imagery for the same region; (**c**) maxNDVI for plantation region before applying ROM; (**d**) NDVImask obtained after applying ROM for plantation; the plantation regions are masked out from monsoon crop and is shown in the dark grey color; (**e**) maxNDVI for forest region before applying ROM; and (**f**) NDVImask obtained after applying ROM for forest region. It can be observed that regions of hill shadows are not masked completely.

**Figure 7.** Monsoon cropland map obtained using S1+S2 combination and training and testing set-2.

**Table 3.** Overall Accuracy (OA), Kappa coefficient, User's Accuracy (UA), and Producer's Accuracy (PA) for crop vs. non-crop mapping obtained from S1 only and S1+S2 classification using training and testing data set-2.

| | | User's Accuracy | Producer's Accuracy | Overall Accuracy | Kappa | F-Score |
|---|---|---|---|---|---|---|
| S1 Only Classification | cropland | 0.82 | 0.88 | 0.90 + 0.017 | 0.77 + 0.039 | 0.85 |
| | non-cropland | 0.94 | 0.91 | | | 0.92 |
| S1+S2 Classification | cropland | 0.88 | 0.9 | 0.93 + 0.015 | 0.83 + 0.033 | 0.89 |
| | non-cropland | 0.95 | 0.94 | | | 0.95 |

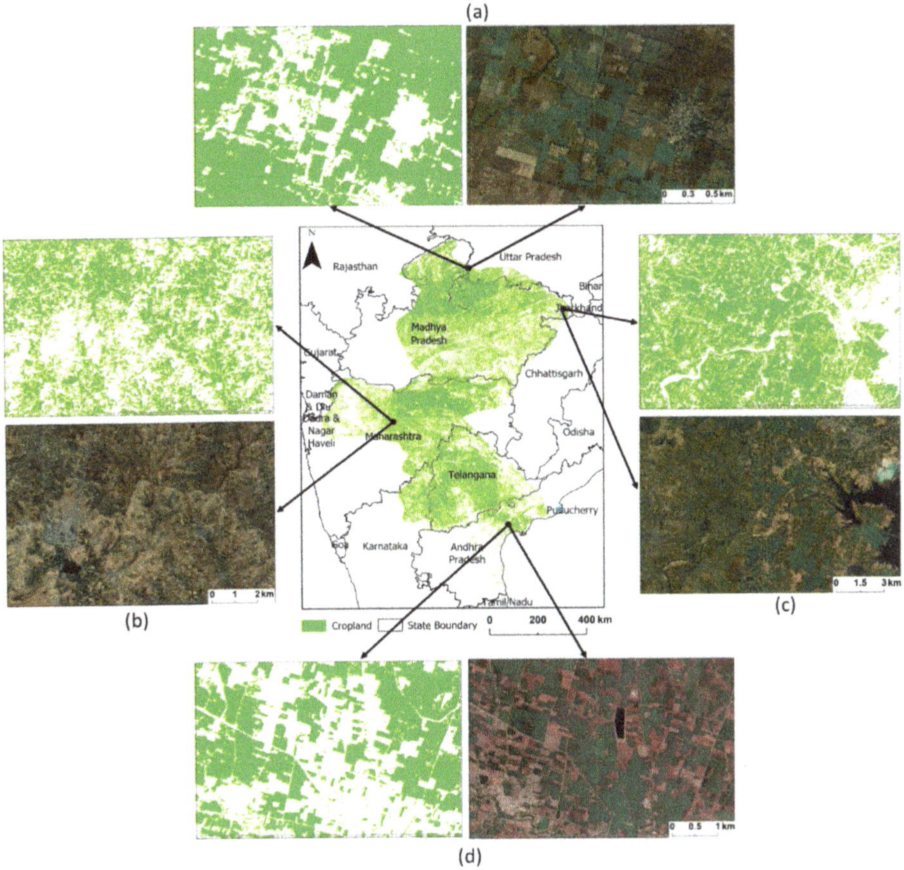


**Figure 8.** Zoom-in view of the monsoon cropland map generated from the combination of S1+S2 for the agro-ecological regions (AER) at various scales and its comparison with high resolution imagery: (**a**) Northern Plain (AER-1); (**b**) Deccan plateau (AER-3); (**c**) Central Highlands (AER-2); and (**d**) Deccan Plateau, Eastern Ghats and Eastern coastal plains (AER-4 and 5).

### 3.3. Accuracy of Binary Crop Maps for Each AER

The comparison of accuracy assessments obtained from S1 only and S1+S2 combination using training and testing data set-2 for the selected AER regions are shown in the Table 4. We found that for all of the AERs, OA obtained from S1+S2 combination outperformed the one obtained from S1 only classification and the improvement varies across the AERs. Also, the OA obtained by the combined S1+S2 was greater than 90% for all of the AERs. The OA difference between S1+S2 and S1 was the lowest for the AER-3 (Table 4), whereas it was greater than 4% for the other three selected AERs. For AER-1, which is dominated by rainfed-dryland crops (90%) with some rice-growing regions (10%), there is a 4% improvement in classification accuracy from S1 to S1+S2. For the S1 dataset, low classification accuracy of AER-2 and AER-4 and 5 compared to the other two AERs is due to the fact that these two regions are dominated by vegetation mixed with crops and have hilly undulating terrain which may have reduced the S1 classification accuracy. AER-2 hosts Vindhyachal and Satpura range whereas AER-4 and 5 are dominated by Eastern Ghats and fragmented vegetation. For both these regions, classification accuracy improved by 5% when S1 data is combined with S2 data. In addition, late maturity crops such as rice or cotton dominate these two regions. Hence, contribution from S2 data towards classification accuracy increases in these regions with increasing availability of cloud free optical data towards the end of the monsoon season. For AER-3, it was observed that combined S1+S2 dataset shows no major improvement over S1 only classification. This region is mainly a plateau with less variation in elevation and negligible forested land. Hence the mixing of crops with natural vegetation is limited resulting in no major inaccuracy in S1-derived classification due to terrain or vegetation.

**Table 4.** Classification accuracy for different AERs obtained from S1 and combined S1+S2 classified maps.

| S1 Classification | | | S1+S2 Classification | | |
|---|---|---|---|---|---|
| | **OA** | **Kappa** | | **OA** | **Kappa** |
| AER-1 | 0.90 | 0.81 | AER-1 | 0.94 | 0.88 |
| AER-2 | 0.89 | 0.76 | AER-2 | 0.94 | 0.86 |
| AER-3 | 0.92 | 0.79 | AER-3 | 0.93 | 0.83 |
| AER-4 and 5 | 0.85 | 0.67 | AER-4 and 5 | 0.90 | 0.77 |

## 4. Discussion

### *4.1. Monsoon Crop Mapping by Combining S1 and NDVImask*

Due to the lack of free time-series of SAR data until recently, previous studies mostly focused on using medium resolution Landsat data (30 m) or MO DIS data (250 m) over large geographic region for monsoon cropland mapping [20–27]. However, using MODIS or Landsat might not be the best approach for monsoon cropland mapping due to frequent cloud cover [24,26]. The spatial resolution of these coarse resolution satellites is not suitable either to capture the small field sizes or mixed agriculture landscapes, thus limiting their usage for preliminary assessment and understanding of croplands over large region. Relying on cloud free optical data alone is not always viable for studying monsoon crops as most of the crops are harvested before cloud free scenes become available in the late monsoon season. Using SAR data during the monsoon season can address this issue. However, SAR data suffers from speckle effects, which makes it difficult to use SAR data alone for generating reliable crop statistics across large regions [65]. Both optical and SAR sensors have limitations for monsoon crop study, but a synergistic approach of combining these data can improve the crop mapping for small-scale farmers at high resolution (Figures 7 and 8). The technique used in this study differ from other published literature as we propose a new way of pixel-based combination of SAR data with temporal aggregation of optical data (maxNDVI) using ROM. Previous studies for monsoon crop monitoring using combination of SAR and optical data were limited to water-intensive rice crops and/or small geographic regions where it was possible to obtain at least one cloud free optical image [34,53,57]. The results presented in this work are important, as this will provide the first high-resolution (10 m) monsoon cropland map generation, and can also be transferred to other agro-ecoregions. This method shows an improvement over existing methods that are primarily focused on non-monsoon/winter cropland mapping at 30 m or coarse resolution [20,24,27,85].

Several sources of error might have affected the results from the S1+S2 combined methodology presented in this study. One of the reasons may be due to the lack of cloud free S2 pixels during the crop growing season. It is possible that in some regions the classification results were solely generated from SAR data due to the non-availability of a single cloud-free S2 image, and could result in inconsistencies in accuracy. There might be errors while training the model and/or due to mixed pixels which may reduce the classification accuracy. The study area is complex with varying farming practices which may result in misclassification of land cover classes in SAR data. Also, the variations in SAR backscattering due to geometric errors (layover, shadow) over hilly terrain affects the accuracy [31]. Using temporal SAR data, along with stratified random sampling and running multiple iterations of the RF classifier reduces the biases, however doesn't completely eliminate it. Using the automatic Otsu thresholding method to extract vegetation cover and segregate vegetation from low NDVI values representing soil, water, and other non-vegetated regions in optical data reduces the overall uncertainty as well.

The ROM generated from S1 SAR data, addresses the issue of miss-classification of spectrally similar plantation and forested vegetation with monsoon crops as visually interpreted (Figure 6). During the monsoon season, optical datasets are only available towards the end of season when crops have already been harvested or are in their peak growth stages [26]. During the peak-growth stage, the spectral signatures of these crops are similar to plantation or other non-crop vegetation, thus making it difficult to segregate the monsoon crops from natural vegetation [31–33]. The usage of temporal SAR-based phenology to generate five land cover classes to produce ROM, masks out the vegetation from monsoon cropland, and improves the classification accuracy (Figures 6 and 8). ROM helps in segregating monsoon cropland from plantation and natural vegetation (forest/grassland) and can be utilized for large regions, as it is not affected by clouds. The ROM produced here is dynamic and can be regularly updated based on the available SAR images. The ROM may also have applications in LULC change monitoring and segregation of non-dynamic LULC features from dynamic croplands.

Overall, our method of integrating SAR composite with seasonal NDVImask for monsoon cropland mapping overcomes four main challenges of mapping smallholder agriculture across large spatio-temporal scales: (i) the method works well in different agro-ecological regions as it takes into consideration of the crop planting time and duration, (ii) it can be used in regions with high cloud cover, such as most tropical countries, (iii) it reduces the sub-pixel heterogeneity in mapping monsoon cropland as the resolution of the output cropland map (10 m) better matches the small farm sizes in most developing countries, and (iv) it helps in distinguishing between monsoon cropland areas from plantation/natural vegetation which has similar signatures during the peak crop growing season. The high-resolution monsoon cropland map produced in this work has the potential to assist government agencies, landscape managers, and researchers in monitoring monsoon crops, which in turn would help us to better understand the factors influencing the production of these crops. Until now, it takes more than a year to make these crop estimates available for decision makers and researchers. This study also has the potential to support global agriculture monitoring missions of Sen2Agri and Geo Global Agriculture Monitoring (GEOGLAM). The objective of GEOGLAM is to provide timely, easily accessible scientifically validated remotely sensed data and derived products for crop-condition monitoring and production assessment. Also, one of the requirements of Sen2Agri mission is to produce national scale dynamic cropland masks other than producing cloud free composites, crop type maps and to indicate the status of current vegetation at 10 m resolution [27,86]. Previous research conducted for Sen2Agri mission to generate dynamic cropland was limited in scope in tropical regions as they relied only on optical datasets [86,87]. This work supports the GEOGLAM and Sen2Agri mission as it produces high-resolution monsoon cropland map over large region comprised of different crop-growing regions. The methodology developed here is also suitable for generating national level dynamic cropland masks.

### 4.2. ROM Uncertainty

In this study, S1-based classification was performed using the RF classifier and training and testing data set-1 for generating ROM. The accuracy of ROM and in turn NDVImask depends on how accurately the non-dynamic land use/cover classes are classified. Based on the classification accuracy (Table 2), it was observed that PA was lowest for the urban class. There were many instances where urban area on the ground was misclassified as other classes including vegetation, likely due to the presence of tree canopy cover in urban centers. The accuracy of ROM will vary depending on whether these omitted 'urban' points are being classified as 'vegetation' or other classes. Our results indicate that UA was the lowest for vegetation class (Table 2) which shows that points from other classes were committed to the vegetation class. The overall accuracy of the classification will also affect the performance of ROM. The F-scores for urban (0.64) and vegetation (0.71) show low discriminative capability compared to the water (0.96) and monsoon crop (0.84) classes (Table 2). Thus, this may also have affected the accuracy of ROM. Visual inspection of the output maps revealed that the classification accuracy of S1 data to obtain ROM was high for plantation compared to forested regions (Figure 6c–f). This is due to the fact that in our study region, forested regions are mainly found in hilly and mountainous regions, which are affected by geometric errors such as layover or shadow and thus affect the classification accuracy. Also, the forested regions in this part of India is either open forest or scrubland which has open spaces or bare soil in between the canopies, affecting the accuracy of ROM [88].

To improve the classification accuracy of S1 for ROM generation, second order texture measures, which involves using Grey-Level Co-occurrence Matrices (GLCM), can be included for improving the classification accuracy of SAR data, especially for discriminating forest and plantation regions [53]. In addition, with advancement in technology and availability of large amount of satellite data, more powerful deep learning methods such as long short-term memory (LSTM), which efficiently handles time series data, may be utilized for improving the overall classification accuracy and in producing ROM in particular [89,90].

## 5. Conclusions

This study presents a synergistic approach of combining SAR with optical data for monsoon crops over different agro-ecological regions in India utilizing the GEE platform. The overall accuracy of 93%, for the binary cropland/non-cropland map, suggests that the combined approach introduced in this research is reliable for monsoon cropland mapping, especially in regions dominated by rainfed-dryland crops. The combined approach provides classification accuracy of 90% or more in different agro-ecological regions dominated by diverse crops. The ROM proposed here has overcome the challenge of differentiating natural vegetation from monsoon cropland mapped during the peak growth stages in monsoon season. Thus, it has applications for segregating cropland from vegetation cover, and may assist in generating a non-crop mask in regions affected by cloud cover. This study can provide important information for decision makers and researchers as monitoring these crops is a challenging task due to small farm size and frequent cloud cover during the crop-growing season.

**Author Contributions:** Conceptualization, A.Q. and P.M.; methodology, A.Q.; writing—original draft preparation, A.Q.; writing—review and editing, P.M.; funding acquisition, P.M. All authors have read and agreed to the published version of the manuscript.

**Funding:** This research was funded by the University of Delaware Research Fund awarded to P.M.

**Acknowledgments:** The authors would like to acknowledge four anonymous reviewers for their constructive feedback on the earlier version of this manuscript. The authors are thankful to Delphis Levia (University of Delaware) and Rodrigo Vargas (University of Delaware) for their constructive feedback and intellectual support during this research work, and to the Department of Geography and Spatial Sciences (University of Delaware) for supporting Open Access Publishing by funding the article processing charge. The authors are grateful to the officials of District Agriculture Department of Uttar Pradesh, Madhya Pradesh, Andhra Pradesh, the scientists of Acharya N G Ranga Agricultural University, Guntur and Touseef Ahmed (Acharya Nagarjuna University College of Engineering and Technology) for their assistance and co-operation during the field surveys. The authors appreciate constant support of Meghavi Prashnani (University of Maryland) and Matthew Walter (University of Delaware) during this research.

**Conflicts of Interest:** The authors declare no conflict of interest.

## References

1. Madhusudhan, L. Agriculture Role on Indian Economy. *Bus. Econ. J.* **2015**, *6*, 1.
2. Gadgil, S.; Gadgil, S. The Indian monsoon, GDP and agriculture. *Econ. Political Wkly.* **2006**, *41*, 4887–4895.
3. Gadgil, S.; Rupa Kumar, K. The Asian monsoon—Agriculture and economy. In *The Asian Monsoon*; Springer: Berlin/Heidelberg, Germany, 2006; pp. 651–683.
4. Krishna Kumar, K.; Rupa Kumar, K.; Ashrit, R.G.; Deshpande, N.R.; Hansen, J.W. Climate impacts on Indian agriculture. *Int. J. Climatol.* **2004**, *24*, 1375–1393. [CrossRef]

5. Fritz, S.; See, L.; Bayas, J.C.L.; Waldner, F.; Jacques, D.; Becker-Reshef, I.; Whitcraft, A.; Baruth, B.; Bonifacio, R.; Crutchfield, J.; et al. A comparison of global agricultural monitoring systems and current gaps. *Agric. Syst.* **2019**, *168*, 258–272. [CrossRef]
6. Ramaswami, B.; Ravi, S.; Chopra, S.D. Risk management in agriculture. In *Discussion Papers*; Indian Statistical Institute: Delhi unit, India, 2003.
7. Venkateswarlu, B.; Prasad, J. Carrying capacity of Indian agriculture: Issues related to rainfed agriculture. *Curr. Sci.* **2012**, *102*, 882–888.
8. Suresh, A.; Raju, S.S.; Chauhan, S.; Chaudhary, K.R. Rainfed agriculture in India: An analysis of performance and implications. *Indian J. Agric. Sci.* **2014**, *84*, 1415–1422.
9. Srinivasarao, C.; Venkateswarlu, B.; Lal, R.; Singh, A.K.; Kundu, S. Sustainable management of soils of dryland ecosystems of india for enhancing agronomic productivity and sequestering carbon. In *Advances in Agronomy*; Academic Press Inc.: Cambridge, MA, USA, 2013; Volume 121, pp. 253–329.
10. Department of Agriculture Cooperation and Farmers Welfare, Government of India. Annual Report. Available online: http://agricoop.nic.in/annual-report (accessed on 30 August 2019).
11. Division, A.I. District Wise Land Use Statistics. Available online: http://aps.dac.gov.in/APY/Index.htm (accessed on 30 August 2019).
12. Craig, M.; Atkinson, D. A Literature Review of Crop Area Estimation; UN-FAO Report. 2013. Available online: http://www.fao.org/fileadmin/templates/ess/documents/meetings_and_workshops/GS_SAC_2013/Improving_methods_for_crops_estimates/Crop_Area_Estimation_Lit_review.pdf (accessed on 20 October 2019).
13. Laamrani, A.; Pardo Lara, R.; Berg, A.A.; Branson, D.; Joosse, P. Using a Mobile Device "App" and Proximal Remote Sensing Technologies to Assess Soil Cover Fractions on Agricultural Fields. *Sensors* **2018**, *18*, 708. [CrossRef]
14. Strengthening Agricultural Support Services for Small Farmers, Report of the APO Seminar on Strengthening Agricultural Support Services for Small Farmers Held in Japan. 2001. Available online: https://www.apo-tokyo.org/publications/wp-content/uploads/sites/5/pjrep-sem-28-01.pdf (accessed on 30 August 2019).
15. Dadhwal, V.K.; Singh, R.P.; Dutta, S.; Parihar, J.S. Remote sensing based crop inventory: A review of Indian experience. *Trop. Ecol.* **2002**, *43*, 107–122.
16. Parihar, J.S.; Oza, M.P. FASAL: An integrated approach for crop assessment and production forecasting. In *Agriculture and Hydrology Applications of Remote Sensing, Proceedings of the SPIE Asia-Pacific Remote Sensing, Goa, India, 13–17 November 2006*; Kuligowski, R.J., Parihar, J.S., Saito, G., Eds.; SPIE: Bellingham, WA, USA, 2006; Volume 6411, p. 641101.
17. Deschamps, B.; McNairn, H.; Shang, J.; Jiao, X. Towards operational radar-only crop type classification: Comparison of a traditional decision tree with a random forest classifier. *Can. J. Remote Sens.* **2012**, *38*, 60–68. [CrossRef]
18. Shanahan, J. Use of Remote-Sensing Imagery to Estimate Corn Grain Yield Agronomy—Faculty Publications Use of Remote-Sensing Imagery to Estimate Corn Grain Yield. *Agron. J.* **2001**, *93*, 583–589. [CrossRef]
19. Jones, J.W.; Antle, J.M.; Basso, B.; Boote, K.J.; Conant, R.T.; Foster, I.; Godfray, H.C.J.; Herrero, M.; Howitt, R.E.; Janssen, S.; et al. Toward a new generation of agricultural system data, models, and knowledge products: State of agricultural systems science. *Agric. Syst.* **2017**, *155*, 269–288. [CrossRef] [PubMed]
20. Pittman, K.; Hansen, M.C.; Becker-Reshef, I.; Potapov, P.V.; Justice, C.O. Estimating global cropland extent with multi-year MODIS data. *Remote Sens.* **2010**, *2*, 1844–1863. [CrossRef]
21. Granados-Ramírez, R.; Reyna-Trujillo, T.; Gómez-Rodríguez, G.; Soria-Ruiz, J. Analysis of NOAA-AVHRR-NDVI images for crops monitoring. *Int. J. Remote Sens.* **2004**, *25*, 1615–1627. [CrossRef]
22. Jiang, D.; Wang, N.B.; Yang, X.H.; Wang, J.H. Study on the interaction between NDVI profile and the growing status of crops. *Chin. Geogr. Sci.* **2003**, *13*, 62–65. [CrossRef]
23. Delrue, J.; Bydekerke, L.; Eerens, H.; Gilliams, S.; Piccard, I.; Swinnen, E. Crop mapping in countries with small-scale farming: A case study for West Shewa, Ethiopia. *Int. J. Remote Sens.* **2013**, *34*, 2566–2582. [CrossRef]
24. Jain, M.; Mondal, P.; DeFries, R.S.; Small, C.; Galford, G.L. Mapping cropping intensity of smallholder farms: A comparison of methods using multiple sensors. *Remote Sens. Environ.* **2013**, *134*, 210–223. [CrossRef]

25. Debats, S.R.; Luo, D.; Estes, L.D.; Fuchs, T.J.; Caylor, K.K. A generalized computer vision approach to mapping crop fields in heterogeneous agricultural landscapes. *Remote Sens. Environ.* **2016**, *179*, 210–221. [CrossRef]
26. Whitcraft, A.K.; Vermote, E.F.; Becker-Reshef, I.; Justice, C.O. Cloud cover throughout the agricultural growing season: Impacts on passive optical earth observations. *Remote Sens. Environ.* **2015**, *156*, 438–447. [CrossRef]
27. Becker-Reshef, I.; Justice, C.; Sullivan, M.; Vermote, E.; Tucker, C.; Anyamba, A.; Small, J.; Pak, E.; Masuoka, E.; Schmaltz, J.; et al. Monitoring Global Croplands with Coarse Resolution Earth Observations: The Global Agriculture Monitoring (GLAM) Project. *Remote Sens.* **2010**, *2*, 1589–1609. [CrossRef]
28. Quarmby, N.A.; Milnes, M.; Hindle, T.L.; Silleos, N. The use of multi-temporal NDVI measurements from AVHRR data for crop yield estimation and prediction. *Int. J. Remote Sens.* **1993**, *14*, 199–210. [CrossRef]
29. Shang, R.; Liu, R.; Xu, M.; Liu, Y.; Dash, J.; Ge, Q. Determining the start of the growing season from MODIS data in the Indian Monsoon Region: Identifying available data in the rainy season and modeling the varied vegetation growth trajectories. *Remote Sens.* **2018**, *11*, 939. [CrossRef]
30. Chakraborty, A.; Seshasai, M.V.R.; Dadhwal, V.K. Geo-spatial analysis of the temporal trends of kharif crop phenology metrics over India and its relationships with rainfall parameters. *Environ. Monit. Assess.* **2014**, *186*, 4531–4542. [CrossRef] [PubMed]
31. Singha, M.; Dong, J.; Zhang, G.; Xiao, X. High resolution paddy rice maps in cloud-prone Bangladesh and Northeast India using Sentinel-1 data. *Sci. Data* **2019**, *6*, 26. [CrossRef] [PubMed]
32. Singha, M.; Wu, B.; Zhang, M. An object-based paddy rice classification using multi-spectral data and crop phenology in Assam, Northeast India. *Remote Sens.* **2016**, *8*, 479. [CrossRef]
33. Mercier, A.; Betbeder, J.; Rumiano, F.; Baudry, J.; Gond, V.; Blanc, L.; Bourgoin, C.; Cornu, G.; Ciudad, C.; Marchamalo, M.; et al. Evaluation of Sentinel-1 and 2 Time Series for Land Cover Classification of Forest–Agriculture Mosaics in Temperate and Tropical Landscapes. *Remote Sens.* **2019**, *11*, 979. [CrossRef]
34. Haldar, D.; Patnaik, C.; Chakraborty, M. Jute Crop Discrimination and Biophysical Parameter Monitoring Using Multi-Parametric SAR Data in West Bengal, India. *Open Access Lib. J.* **2014**, *1*, 1. [CrossRef]
35. Maity, S.; Patnaik, C.; Chakraborty, M.; Panigrahy, S. Analysis of temporal backscattering of cotton crops using a semiempirical model. *IEEE Trans. Geosci. Remote Sens.* **2004**, *42*, 577–587. [CrossRef]
36. Lone, J.M.; Sivasankar, T.; Sarma, K.K.; Qadir, A.; Raju, P.L.N. Influence of Slope Aspect on above Ground Biomass Estimation using ALOS-2 Data. *Int. J. Sci. Res.* **2017**, *6*, 1422–1428.
37. Alonso-González, A.; Hajnsek, I. Radar Remote Sensing of Land Surface Parameters. In *Observation and Measurement of Ecohydrological Processes. Ecohydrology*; Li, X., Vereecken, H., Eds.; Springer: Berlin, Germany, 2018; pp. 1–38.
38. Woodhouse, I.H. *Introduction to Microwave Remote Sensing*; Taylor & Francis: Didcot, UK, 2006; ISBN 9780415271233.
39. Orynbaikyzy, A.; Gessner, U.; Conrad, C. Crop type classification using a combination of optical and radar remote sensing data: A review. *Int. J. Remote Sens.* **2019**, *40*, 6553–6595. [CrossRef]
40. Van Tricht, K.; Gobin, A.; Gilliams, S.; Piccard, I. Synergistic use of radar sentinel-1 and optical sentinel-2 imagery for crop mapping: A case study for Belgium. *Remote Sens.* **2018**, *10*, 1642. [CrossRef]
41. Clevers, J.G.P.W.; Van Leeuwen, H.J.C. Combined use of optical and microwave remote sensing data for crop growth monitoring. *Remote Sens. Environ.* **1996**, *56*, 42–51. [CrossRef]
42. Susan Moran, M.; Alonso, L.; Moreno, J.F.; Cendrero Mateo, M.P.; Fernando De La Cruz, D.; Montoro, A. A RADARSAT-2 quad-polarized time series for monitoring crop and soil conditions in Barrax, Spain. *IEEE Trans. Geosci. Remote Sens.* **2012**, *50*, 1057–1070. [CrossRef]
43. Larrañaga, A.; Álvarez-Mozos, J. On the Added Value of Quad-Pol Data in a Multi-Temporal Crop Classification Framework Based on RADARSAT-2 Imagery. *Remote Sens.* **2016**, *8*, 335. [CrossRef]
44. Useya, J.; Chen, S. Exploring the Potential of Mapping Cropping Patterns on Smallholder Scale Croplands Using Sentinel-1 SAR Data. *Chin. Geogr. Sci.* **2019**, *29*, 626–639. [CrossRef]
45. Skakun, S.; Kussul, N.; Shelestov, A.Y.; Lavreniuk, M.; Kussul, O. Efficiency Assessment of Multitemporal C-Band Radarsat-2 Intensity and Landsat-8 Surface Reflectance Satellite Imagery for Crop Classification in Ukraine. *IEEE J. Sel. Top. Appl. Earth Obs. Remote Sens.* **2016**, *9*, 3712–3719. [CrossRef]

46. Kussul, N.; Mykola, L.; Shelestov, A.; Skakun, S. Crop inventory at regional scale in Ukraine: Developing in season and end of season crop maps with multi-temporal optical and SAR satellite imagery. *Eur. J. Remote Sens.* **2018**, *51*, 627–636. [CrossRef]
47. Skriver, H.; Mattia, F.; Satalino, G.; Balenzano, A.; Pauwels, V.R.N.; Verhoest, N.E.C.; Davidson, M. Crop Classification Using Short-Revisit Multitemporal SAR Data. *IEEE J. Sel. Top. Appl. Earth Obs. Remote Sens.* **2011**, *4*, 423–431. [CrossRef]
48. Sun, Z.; Wang, D.; Zhou, Q. Dryland crop recognition based on multi-temporal polarization SAR data. In Proceedings of the 2019 8th International Conference on Agro-Geoinformatics (Agro-Geoinformatics), Istanbul, Turkey, 16–19 July 2019; Institute of Electrical and Electronics Engineers Inc.: Piscataway, NJ, USA, 2019.
49. Wang, D.; Su, Y.; Zhou, Q.; Chen, Z. Advances in research on crop identification using SAR. In Proceedings of the 2015 4th International Conference on Agro-Geoinformatics (Agro-Geoinformatics), Istanbul, Turkey, 20–24 July 2015; Institute of Electrical and Electronics Engineers Inc.: Piscataway, NJ, USA, 2015; pp. 312–317.
50. Sirro, L.; Häme, T.; Rauste, Y.; Kilpi, J.; Hämäläinen, J.; Gunia, K.; de Jong, B.; Pellat, F.P. Potential of different optical and SAR data in forest and land cover classification to support REDD+ MRV. *Remote Sens.* **2018**, *10*, 942. [CrossRef]
51. Tavares, P.A.; Beltrão, N.E.S.; Guimarães, U.S.; Teodoro, A.C. Integration of sentinel-1 and sentinel-2 for classification and LULC mapping in the urban area of Belém, eastern Brazilian Amazon. *Sensors* **2019**, *19*, 1140. [CrossRef]
52. McNairn, H.; Champagne, C.; Shang, J.; Holmstrom, D.; Reichert, G. Integration of optical and Synthetic Aperture Radar (SAR) imagery for delivering operational annual crop inventories. *ISPRS J. Photogramm. Remote Sens.* **2009**, *64*, 434–449. [CrossRef]
53. Ranjan, A.K.; Parida, B.R. Paddy acreage mapping and yield prediction using sentinel-based optical and SAR data in Sahibganj district, Jharkhand (India). *Spat. Inf. Res.* **2019**, *27*, 399–410. [CrossRef]
54. Monitoring Cotton (Gossypium sps.) Crop Condition through Synergy of Optical and Radar Remote Sensing | Publons. Available online: https://publons.com/publon/2200411/ (accessed on 26 December 2019).
55. Gu, L.; He, F.; Yang, S. Crop classification based on deep learning in northeast China using sar and optical imagery. In Proceedings of the 2019 SAR in Big Data Era (BIGSARDATA), Beijing, China, 5–6 August 2019; Institute of Electrical and Electronics Engineers Inc.: Piscataway, NJ, USA, 2019.
56. Verma, A.; Kumar, A.; Lal, K. Kharif crop characterization using combination of SAR and MSI Optical Sentinel Satellite datasets. *J. Earth Syst. Sci.* **2019**, *128*, 230. [CrossRef]
57. Kumari, M.; Murthy, C.S.; Pandey, V.; Bairagi, G.D. Soybean Cropland mapping using Multi-Temporal Sentinel-1 data. *ISPRS Int. Arch. Photogramm. Remote Sens. Spat. Inf. Sci.* **2019**, *XLII-3/W6*, 109–114. [CrossRef]
58. Gajbhiye, K.S.; Mandal, C. *Agro-Ecological Zones, Their Soil Resource and Cropping Systems*; National Bureau of Soil Survey and Land Use Planning: Nagpur, India, 2000; Available online: http://www.indiawaterportal.org/sites/indiawaterportal.org/files/01jan00sfm1.pdf (accessed on 11 July 2019).
59. Beck, H.E.; Zimmermann, N.E.; McVicar, T.R.; Vergopolan, N.; Berg, A.; Wood, E.F. Present and future köppen-geiger climate classification maps at 1-km resolution. *Sci. Data* **2018**, *5*, 180214. [CrossRef] [PubMed]
60. NFSM: National Food Security Mission. Available online: https://www.nfsm.gov.in/ (accessed on 31 October 2019).
61. Land Use Statistics Information System. Available online: https://aps.dac.gov.in/APY/Index.htm (accessed on 30 June 2019).
62. Google Developers Image Collection Reductions. Available online: https://developers.google.com/earth-engine/reducers_image_collection (accessed on 19 July 2019).
63. Liu, D.; Xia, F. Assessing object-based classification: Advantages and limitations. *Remote Sens. Lett.* **2010**, *1*, 187–194. [CrossRef]
64. Memarian, H.; Balasundram, S.K.; Khosla, R. Comparison between pixel- and object-based image classification of a tropical landscape using Système Pour l'Observation de la Terre-5 imagery. *J. Appl. Remote Sens.* **2013**, *7*, 073512. [CrossRef]
65. Tian, F.; Wu, B.; Zeng, H.; Zhang, X.; Xu, J. Efficient Identification of Corn Cultivation Area with Multitemporal Synthetic Aperture Radar and Optical Images in the Google Earth Engine Cloud Platform. *Remote Sens.* **2019**, *11*, 629. [CrossRef]

66. Tutorials-Sentinel-1 Toolbox-Sentinel Online. Available online: https://sentinel.esa.int/web/sentinel/toolboxes/sentinel-1/tutorials (accessed on 30 June 2019).
67. Gatti, A.; Bertolini, A. Sentinel-2 Products Specification Document. 2015. Available online: https://sentinel.esa.int/documents/247904/349490/S2_MSI_Product_Specification.pdf (accessed on 30 June 2019).
68. Carrasco, L.; O'Neil, A.W.; Daniel Morton, R.; Rowland, C.S. Evaluating combinations of temporally aggregated Sentinel-1, Sentinel-2 and Landsat 8 for land cover mapping with Google Earth Engine. *Remote Sens.* **2019**, *11*, 288. [CrossRef]
69. Scanlon, T.M.; Albertson, J.D.; Caylor, K.K.; Williams, C.A. Determining land surface fractional cover from NDVI and rainfall time series for a savanna ecosystem. *Remote Sens. Environ.* **2002**, *82*, 376–388. [CrossRef]
70. De Alban, J.D.T.; Connette, G.M.; Oswald, P.; Webb, E.L. Combined Landsat and L-band SAR data improves land cover classification and change detection in dynamic tropical landscapes. *Remote Sens.* **2018**, *10*, 306. [CrossRef]
71. Breiman, L. *Machine Learning*; Statistics Department, University of Berkeley: Berkeley, CA, USA, 2001; Volume 45, pp. 5–32.
72. Liaw Merck, A.; Liaw, A.; Wiener, M. Classification and Regression by RandomForest. *R News* **2002**, *2*, 18–22. Available online: http://cogns.northwestern.edu/cbmg/LiawAndWiener2002.pdf (accessed on 10 October 2019).
73. Rodriguez-Galiano, V.F.; Chica-Olmo, M.; Abarca-Hernandez, F.; Atkinson, P.M.; Jeganathan, C. Random Forest classification of Mediterranean land cover using multi-seasonal imagery and multi-seasonal texture. *Remote Sens. Environ.* **2012**, *121*, 93–107. [CrossRef]
74. Mondal, P.; McDermid, S.S.; Qadir, A. A reporting framework for Sustainable Development Goal 15: Multi-scale monitoring of forest degradation using MODIS, Landsat and Sentinel data. *Remote Sens. Environ.* **2020**, *237*, 111592. [CrossRef]
75. Kamusoko, C.; Gamba, J.; Murakami, H. Mapping Woodland Cover in the Miombo Ecosystem: A Comparison of Machine Learning Classifiers. *Land* **2014**, *3*, 524–540. [CrossRef]
76. Toosi, N.B.; Soffianian, A.R.; Fakheran, S.; Pourmanafi, S.; Ginzler, C.; Waser, L.T. Comparing different classification algorithms for monitoring mangrove cover changes in southern Iran. *Glob. Ecol. Conserv.* **2019**, *19*, e00662. [CrossRef]
77. Pal, M. Random forest classifier for remote sensing classification. *Int. J. Remote Sens.* **2005**, *26*, 217–222. [CrossRef]
78. Akar, Ö.; Güngör, O. Classification of multispectral images using Random Forest algorithm. *J. Geodesy Geoinf.* **2012**, *1*, 105–112. [CrossRef]
79. Tucker, C.J. Red and photographic infrared linear combinations for monitoring vegetation. *Remote Sens. Environ.* **1979**, *8*, 127–150. [CrossRef]
80. Benedetti, R.; Rossini, P. On the use of NDVI profiles as a tool for agricultural statistics: The case study of wheat yield estimate and forecast in Emilia Romagna. *Remote Sens. Environ.* **1993**, *45*, 311–326. [CrossRef]
81. Pericak, A.A.; Thomas, C.J.; Kroodsma, D.A.; Wasson, M.F.; Ross, M.R.V.; Clinton, N.E.; Campagna, D.J.; Franklin, Y.; Bernhardt, E.S.; Amos, J.F. Mapping the yearly extent of surface coal mining in central appalachia using landsat and google earth engine. *PLoS ONE* **2018**, *13*, e0197758. [CrossRef]
82. Liu, J.; Li, W.; Tian, Y. *Automatic Thresholding of Gray-Level Pictures Using Two-Dimension Otsu Method*; Institute of Electrical and Electronics Engineers (IEEE): Piscataway, NJ, USA, 2002; pp. 325–327.
83. Congalton, R.G. A review of assessing the accuracy of classifications of remotely sensed data. *Remote Sens. Environ.* **1991**, *37*, 35–46. [CrossRef]
84. Powers, D.M. Evaluation: From Precision, Recall and F-Measure to ROC, Informedness, Markedness and Correlation. *J. Mach. Learn. Technol.* **2011**, *1*, 37–63.
85. Gumma, M.K.; Thenkabail, P.S.; Teluguntla, P.G.; Oliphant, A.; Xiong, J.; Giri, C.; Pyla, V.; Dixit, S.; Whitbread, A.M. Agricultural cropland extent and areas of South Asia derived using Landsat satellite 30-m time-series big-data using random forest machine learning algorithms on the Google Earth Engine cloud. *GISci. Remote Sens.* **2019**, 1–21. [CrossRef]
86. Defourny, P.; Bontemps, S.; Bellemans, N.; Cara, C.; Dedieu, G.; Guzzonato, E.; Hagolle, O.; Inglada, J.; Nicola, L.; Rabaute, T.; et al. Near real-time agriculture monitoring at national scale at parcel resolution: Performance assessment of the Sen2-Agri automated system in various cropping systems around the world. *Remote Sens. Environ.* **2019**, *221*, 551–568. [CrossRef]

87. Inglada, J.; Arias, M.; Tardy, B.; Hagolle, O.; Valero, S.; Morin, D.; Dedieu, G.; Sepulcre, G.; Bontemps, S.; Defourny, P.; et al. Assessment of an Operational System for Crop Type Map Production Using High Temporal and Spatial Resolution Satellite Optical Imagery. *Remote Sens.* **2015**, *7*, 12356–12379. [CrossRef]
88. Roy, P.S.; Sharma, K.P.; Jain, A. Stratification of density in dry deciduous forest using satellite remote sensing digital data—An approach based on spectral indices. *J. Biosci.* **1996**, *21*, 723–734. [CrossRef]
89. Massey, R.; Sankey, T.T.; Yadav, K.; Congalton, R.G.; Tilton, J.C. Integrating cloud-based workflows in continental-scale cropland extent classification. *Remote Sens. Environ.* **2018**, *219*, 162–179. [CrossRef]
90. Rußwurm, M.; Korner, M. Temporal Vegetation Modelling Using Long Short-Term Memory Networks for Crop Identification from Medium-Resolution Multi-spectral Satellite Images. In Proceedings of the IEEE Conference on Computer Vision and Pattern Recognition Workshops, Honolulu, HI, USA, 21–26 July 2017; Institute of Electrical and Electronics Engineers (IEEE): Piscataway, NJ, USA, 2017; pp. 1496–1504.

remote sensing

MDPI

Article

# Monitoring Agricultural Fields Using Sentinel-1 and Temperature Data in Peru: Case Study of Asparagus (*Asparagus officinalis* L.)

**Cristian Silva-Perez [1,*], Armando Marino [1] and Iain Cameron [2]**

[1] Biological and Environmental Sciences, University of Stirling, Scotland FK9 4LJ, UK; armando.marino@stir.ac.uk

[2] Environment Systems LTD, Aberystwyth, Wales TA6 6LL, UK; iain.cameron@envsys.co.uk

* Correspondence: c.j.silva.perez@stir.ac.uk

Received: 14 May 2020; Accepted: 18 June 2020; Published: 21 June 2020

**Abstract:** This paper presents the analysis and a methodology for monitoring asparagus crops from remote sensing observations in a tropical region, where the local climatological conditions allow farmers to grow two production cycles per year. We used the freely available dual-polarisation GRD data provided by the Sentinel-1 satellite, temperature from a ground station and ground truth from January to August of 2019 to perform the analysis. We showed how particularly the VH polarisation can be used for monitoring the canopy formation, density and the growth rate, revealing connections with temperature. We also present a multi-output machine learning regression algorithm trained on a rich spatio-temporal dataset in which each output estimates the number of asparagus stems that are present in each of the pre-defined crop phenological stages. We tested several scenarios that evaluated the importance of each input data source and feature, with results that showed that the methodology was able to retrieve the number of asparagus stems in each crop stage when using information about starting date and temperature as predictors with coefficients of determination ($R^2$) between 0.84 and 0.86 and root mean squared error (RMSE) between 2.9 and 2.7. For the multitemporal SAR scenario, results showed a maximum $R^2$ of 0.87 when using up to 5 images as input and an RMSE that maintains approximately the same values as the number of images increased. This suggests that for the conditions evaluated in this paper, the use of multitemporal SAR data only improved mildly the retrieval when the season start date and accumulated temperature are used to complement the backscatter.

**Keywords:** tropical agricultural monitoring; canopy development analysis; phenology retrieval; Sentinel-1; multitemporal SAR; multi-task machine learning

## 1. Introduction

Due to the recent and future growth of freely available satellite remote sensing data, there is an opportunity to implement near real time agricultural monitoring systems to increase yield and crop management efficiency. This is based on informed decision making with information derived fully or partially from satellite sensors.

Such a system is particularly important in tropical regions which highly contribute to the global food production but for many crops with considerably lower yields per hectare compared with temperate regions [1]. It is also essential given the necessity in the tropics to preserve natural ecosystems by increasing yield in existing crop areas rather than transforming tropical forests to low yield croplands [1]. A distinctive operational characteristic in tropical and subtropical regions for several crop types is the uninterrupted production cycles, with cultivation of more than one cycle per year. Each of these production cycles or campaigns may be under slightly different meteorological conditions due to a "soft seasonality", e.g., mild winters, thus modifying to a certain extent the crop growth rate and structure (as will be shown later in this paper). This leads to the different responses captured by the satellites at each campaign. This paper considers the above-mentioned conditions for agricultural monitoring, particularly analysing the case of asparagus crops in Peru. *Asparagus officinalis* L. is a key crop for the country's agricultural exports, being the largest exporter in the world, the second largest producer after China [2] and an important source of job [3].

In this context, crop phenology has been used as a tool to measure crop status at any given time during the cultivation period and to measure the development rate relative to previous campaigns or relative to neighbour plots [4]. Accordingly, monitoring phenological evolution and accurately knowing crop status and development rate, the farmers can strategically plan the management.

Given the importance of monitoring phenology during the campaign without interruptions, synthetic aperture radar (SAR) emerges as a potential technology for this task. The capabilities to acquire images at day and night and under nearly all-weather conditions of SAR satellites offer significant opportunities for systematic monitoring regardless of cloud coverage [5]. On the other hand, to consider the impact of different climatological conditions on each campaign of a year, temperature records can be analysed to support the crop development monitoring.

### 1.1. Related Work

An initial step for crop monitoring from SAR data is to understand the time series evolution of the features derived from sensor observations. To this end, ground truth surveys are used as a reference and for validation, correlating measurements in the field with the SAR response.

In the case of Quad-polarimetric data, multi-temporal polarimetric SAR (PolSAR) analysis is used to characterise a crop signature in terms of evolution of scattering mechanisms along the season identifying key moments [6–8].

Recently, more attention is being given to dual polarimetric systems given the free access to these data (Sentinel-1). Research to understand the interaction of Sentinel-1 signal response to crop evolution has been presented for several crop types. The authors found sensitivity of the VH, VV backscatter and the ratio between the two polarisations with different crop biophysical parameters [9–12].

For the case of crop phenology retrieval, once the temporal evolution of the SAR indices has been analysed, an initial approach is to use the polarimetric features as inputs to a statistical or a machine learning model. These models are trained with SAR and ground truth from past campaigns [13,14].

Other authors have proposed the use of distance measures to compare the covariance matrix of a given SAR resolution cell inside a parcel with a set of previously characterised covariance matrices that are associated with a phenological stage [15]. The aim is to find the most similar predefined covariance matrix and assign the pixel under analysis, the crop stage with the most similar covariance matrix.

However, these approaches consider the phenology retrieval as a classification application, aiming at classifying the current parcel state as one of the previously defined pool of possible states (e.g., emergence, vegetative stage, maturation, etc.). This generates inconveniences selecting the appropriate boundaries for each stage in a process that may be subjective, often selecting ones (biasing) that the algorithm can actually identify. On the other hand, if standard phenological scales are used, such as the BBCH scale [16], the algorithm may not be able to disentangle every stage since the SAR response may not be sensitive to all these agronomic processes. Likewise, previous approaches ignore the fact that it is possible to have simultaneously more than a single stage in a parcel due to different plant growth rates and the fact that in the real evolution adjacent stages overlap (e.g., a parcel with some plants in flowering and some in fructification simultaneously). These approaches predict current phenology based on a single SAR image without considering the multi-temporal information. This leads for instance to stages at the beginning of the season being miss-classified by the model with final crop stages as they may have similar SAR response, as occurred in [13,17].

To overcome this, a hidden Markov model technique was proposed in [18,19] so that a prediction of the current stage is dependant on the previous stage, following a Markov property. However, the problem of subjectively selecting the boundaries for the crop stages still remains. Other authors have proposed a different approach whereby they consider the crop evolution as a time dependant dynamical process that follows a trajectory governed by the crop underlying dynamics [20–22]. The aim of these methodologies is to retrieve the crop state in that trajectory when a new SAR image is analysed. However, these last methods have only been proven using polarimetric SAR data, which provides a much richer amount of information to characterise a target compared to the freely available data from Sentinel-1. Studies associated with the potential of space borne radar remote sensing concerning asparagus fields have been presented in [23–27], although all of them focus on the crop type classification problem rather than in the analysis of individual crop stages as we present in this paper. In [23–25], C-band ENVISAR ASAR satellite data in VH and VV polarisations are used to identify land use of two agricultural regions, with asparagus being one of several crop types to identify. However, neither description of the crop stages nor the backscatter response over time is presented. In [26], time series of HH and VV polarisations of X-band data from TerraSAR-X satellite are reported. This study aims to evaluate the potential for classification of agricultural areas by analysing the crop signature of several crop types. Among them, 12 parcels of asparagus are studied and both the HH and the VV polarisations using X-band data are presented. The backscatter seems to increase during the period of vegetation growth, with a widespread distribution among the 12 parcels. This increase happens during the summer similar to several other crop types which was identified as an inconvenient to classify asparagus. This same effect was reported in [23,24], where asparagus response is particularly similar to sugar beet. In [27], among other 13 crop types, the multitemporal response of asparagus was evaluated to use the crop signature for agricultural fields classification purposes using Sentinel-1 data. The authors report an increase of the VH backscatter during the periods of vegetation growth. Interestingly, the authors also report a more constant backscatter response during the whole cultivation period using the VV polarisation. This is aligned to what is presented in Section 2.6.1 of this paper and in contrast to [26] although it has to be highlighted that in [26], X-band data was used. The authors, also report low accuracies for classifying asparagus due to similarities to summer crop types as reported in [23,24].

Please note that none of the reported work in asparagus focuses on monitoring growth development or stages classification. Also note that since the climatological conditions in Peru are different to [23–27], direct comparisons are not straight forward. For instance, in [27], authors report that the agricultural season length is more than a year, which is not the case for our test site. Similarly, the crop signature is inherently different since the senescence periods for crops located in temperate regions are not present in our test site.

### *1.2. Objectives of the Study*

The main objectives of this paper are the following:

1. To analyse the SAR response to the asparagus crop evolution.
2. To present examples of how the seasonal climatological conditions influence the crop development in the test site (tropical conditions).
3. To present the implementation of a data-driven methodology that captures the recurrent patterns in the SAR response and the temperature to provide an approximation of the crop development at every new SAR acquisition. It consists of a Multi-output machine learning regression algorithm in which each output estimates the number of asparagus stems that are present in each of the predefined phenological stages at a given date.

## 2. Materials and Methods

### *2.1. Asparagus Crop Development and Production Cycles*

In this section, we briefly introduce the asparagus crop and the main phases of development. For a more detailed description of the asparagus growth and physiology the reader is directed to [28,29].

Asparagus is a vegetable perennial crop which once is in a productive phase, re-emerges after harvest without the need to re-plant it. The cultivation process begins by transplanting to the fields the seedlings grown in a nursery. The roots system below ground, also known as the crown, and the fern above ground begin to grow and after approximately 2 years of development and establishment, the ferns are cut, the asparagus spears emerge and the crop is lightly harvested for the first time [30,31].

After the first harvest, at the emergence crop stage, the next stems that emerge from the buds of the crown develop into a fern as shown in Figure 1. The asparagus stems grow vertically and will start producing the horizontal branches in a crop stage known as ramification. From this point, the cladophylls (leaf-like structures in the branches) will develop during the aperture stage. The aggregation of several consecutive individual asparagus stems that emerged from the root system, with their respective side branches and cladophylls compose the fern.

Subsequently, the fern thickens and covers the sandy soil intercepting light and beginning the production of carbohydrates which are sent down via translocation to replenish the roots system and be stored [29]. The fern development is followed by the short appearance of small yellow flowers and a maturation period, which corresponds to the longest crop stage. In total, each production cycle takes between four to five months before the crop is ready to be harvested.

During harvest, the spears that emerge from the buds are harvested after reaching approximately 20 cm until a minimum level of carbohydrates depletion in the root system. At this point, the new asparagus stems are left to develop and grow again to begin a new production cycle, in a life cycle that can last up to 15 years [30,31].

The crop stages in Figure 1 are the same stages that are recorded during the ground surveys and that are estimated in Section 2.7.

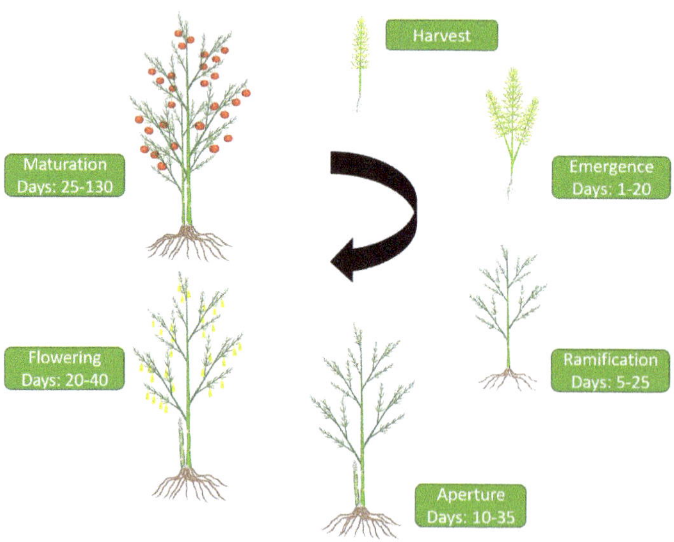


**Figure 1.** Asparagus crop growth and production cycles. Every season after harvest, new asparagus stems emerge to begin a new production cycle

*2.2. Test Site*

The asparagus fields (Figure 2) are located in the north of Peru in a dry coastal zone with sandy soil, divided in plots of an average of two hectares. We considered approximately 442 parcels in total, where timing and management practices such as starting and harvest dates as well as application of nutrients and pesticides among other activities, are performed simultaneously in groups of around four to six neighbouring parcels.

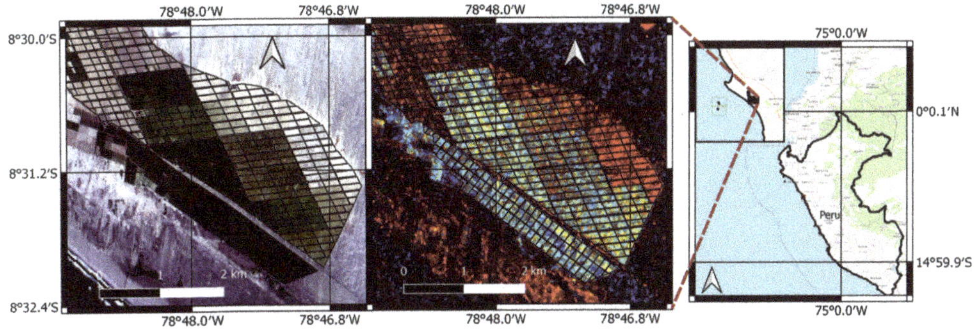


**Figure 2.** Test site. (**Left**): Sentinel 2 RGB image acquired the 29/09/2018. (**Center**): Sentinel-1 RGB image acquired the 30/09/2018. (**Right**): Location of test site in Peru

### 2.3. Climatological Conditions

The temperature and solar radiation present the maximum variability in the local test site, following the seasons of the southern hemisphere with a maximum peak of temperature around mid-February (summer) reaching up to 26 degrees Celsius and lowest values in August (winter) with temperatures of around 15 degrees (Figure 3).

Since the winter never reaches extremely low temperatures, the asparagus crop does not reach a dormant stage which permits growers to have two productive cycles per year. However, since the conditions along the year are not exactly the same, there may be a difference in the crop evolution of the same plot in two consecutive productive cycles, in response to these changing conditions. This is an important factor that will be analysed in Section 2.6.2.

The rainfall levels are extremely low given the desert conditions where the parcels are located with an average of less than 1 mm per month.

### 2.4. Ground Truth

A survey campaign to collect asparagus phenological information was carried over a period of 8 months, from January to August of 2019, for 442 asparagus parcels. For each parcel evaluated, field surveyors randomly selected two transects to assess, with each transect being a metre in length. Within each transect the surveyors identified the phenological stage of each stem and counted the total number of stems in each stage present. Results for the two transects were averaged and recorded. In this way, the proportion of stems per parcel in any of the phenological stages identified in Figure 1 can be recorded, and a proxy for the evolution of crop stage over time can be established. The average temporal evolution of asparagus stems in each crop stage is presented in Figure 4 for all the production cycles covered in the ground truth. Please note that after the asparagus stems have emerged from the ground, the total number of stems remains approximately the same throughout the season while the number of stems present at each stage changes as the crop develops.

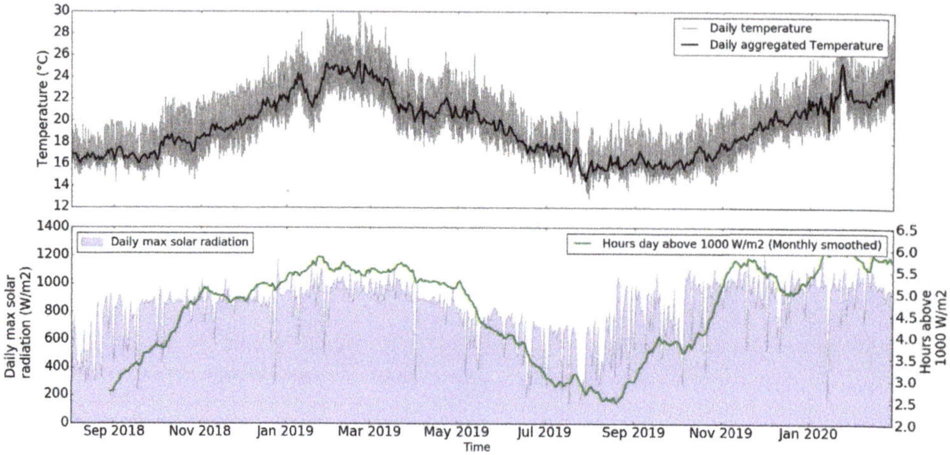


**Figure 3.** Average temperature and solar radiation conditions in the test site.

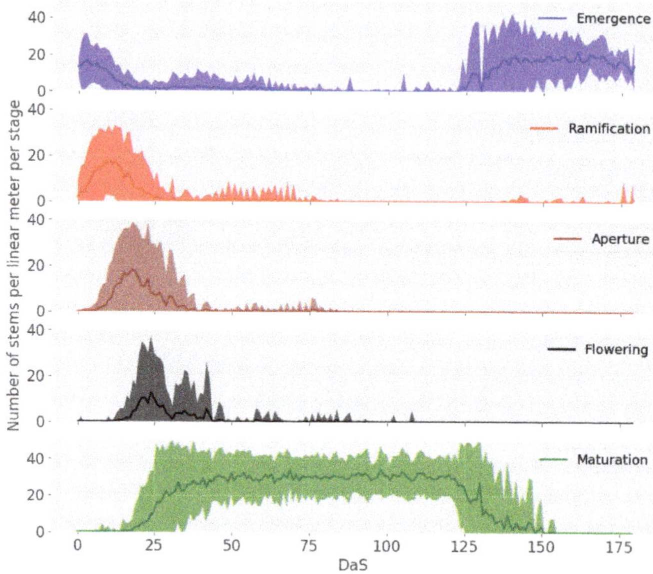


**Figure 4.** Number of asparagus stems in each of the crop stages. The solid lines represent the averages of all the ground measurements collected in 442 plots during the first 8 months of 2019. The shaded regions represent plus and minus two standard deviations.

An average of 36 surveys were performed per parcel during the eight months of the ground data collection, i.e, approximately one survey per week per parcel. However, given the complexity of the operation not all surveys were carried on systematically, e.g., exactly every week, but rather with irregular sampling. Also, note that given the climatological conditions as explained in the previous section, it is possible to grow up to two production cycles per year independently of the starting month and consequently the resulting surveys contain information to characterise multiple times an entire asparagus campaign or season. This enable us to create a rich spatio-temporal dataset to characterise the crop development.

### 2.5. Sar Datasets

The Sentinel-1 dataset used for the analysis was built using the Level-1 Ground Range Detected (GRD) and the Interferometric Wide swath (IW) acquisition mode, with VV and VH polarisation channels. The data was collected using the Google Earth Engine (GEE) platform [32] in which the data pre-processing steps include applying orbit file, GRD border noise removal, thermal noise removal, Radiometric calibration and terrain correction. After obtaining the data from the GEE, a 3 × 3 boxcar averaging window was used as speckle filter.

Table 1 shows the three acquisition geometries available for the test site with the corresponding average incidence angles and acquisition times, including the orbit 142 in descending pass direction and the orbits 18 and 91 in ascending pass. Using the three available orbits the revisit frequency corresponds to between 3 and 5 days while it is 12 days using a single incidence angle.

**Table 1.** Sentinel-1 acquisition geometries available in the test site

| Pass Direction | Relative Orbit | Inc. Angle | Acquisition Time |
|---|---|---|---|
| Descending | 142 | 35 | 10:54 |
| Ascending | 18 | 31 | 23:34 |
| Ascending | 91 | 45 | 23:42 |

In order to analyse the long-term behaviour of the backscatter signal (e.g., seasonality), a time series of nearly two years was built from Sentinel-1 data over a typical parcel of asparagus where the ground truth is known and which includes four consecutive agricultural seasons. For the methodology to monitor asparagus development presented in Section 2.7 data from January to August of 2019 was used for the 442 plots when ground truth is available. Figure 5 shows the average temporal evolution of the backscatter for all the production cycles covered in the ground truth. The crop characteristics that cause the SAR observations in Figure 5 will be further explained in Section 2.6.1. Please note that the VH polarisation presents more significant changes through time and with less statistical variance than the VV polarisation as described in Section 2.6.1. Also note that after the day 125 there is a significant increase in the variance of the response. This is due to different cultivation period lengths as will be explained in Section 2.6.2. Figure 7 shows the VH backscatter profile with the corresponding ground truth of a typical asparagus parcel over time, covering 4 consecutive production cycles.

### 2.6. Methodology for Estimating Asparagus Stems Per Stage

The methodology developed in the present analysis consists of three main parts: (1) Understanding of Sentinel-1 signal interaction and sensitivity with the asparagus temporal evolution, (2) Analysis of the impact that the local climatological conditions, particularly the temperature, have over the canopy development and (3) the multi-output machine learning regression model training and use to estimate the number of asparagus stems in the phenological stages of Figure 1.

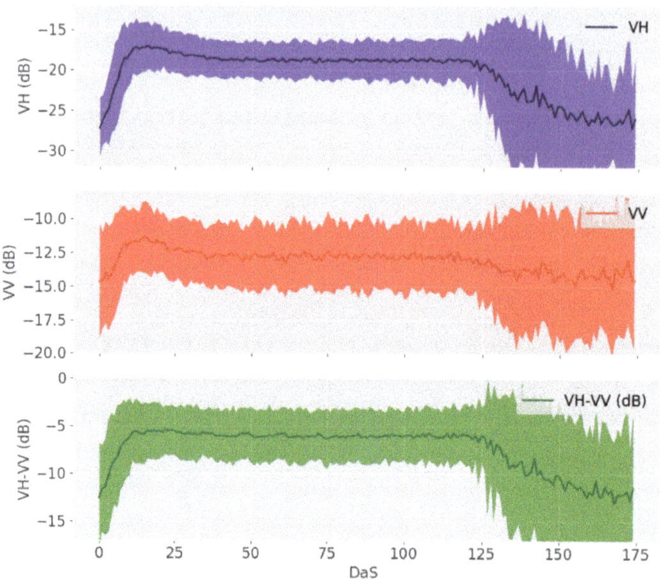


**Figure 5.** Sentinel-1 temporal backscatter evolution. This chart shows the mean backscatter plus/minus two standard deviations of 442 parcels backscatter time series.

2.6.1. Sar Sensitivity to Crop Evolution

Due to the mild winters with minimum temperatures of around 16 degrees Celsius, the asparagus crop does not reach a dormant stage naturally and there is not a harvest time forced by the climatological conditions. This allows growers to plan the starting and ending dates of the season (occasionally for individual parcels) so that is possible to have more than one production cycles per year (normally two) and if required, adjust to the market needs and contractual planning. Given this, at a single SAR acquisition there are plots at almost all possible crop stages.

The photographs presented in Figure 6 were taken the same date as a SAR acquisition (28/05/2019). Based on this, it is possible to locate in the time series of each parcel where the picture was taken, what the SAR response to the crop is and compare it with the crop status recorded in the footage. Please note that all the images were taken the same date to parcels at different growth stages, a possibility enabled by the local climatological conditions. Since the parcels are at different crop stages, the backscatter is also different as shown in the VH polarisation time series of Figure 6, in which the lowest backscatter of the season is present during the harvest periods given that the fern has been mechanically removed and the new emerging spears are being harvested (Figure 6, images 1 and 2). At this point, the SAR signal interacts only with the soil presumably with a predominant surface scattering mechanism.

The SAR response to these conditions has been shown to depend on the dielectric and geometric properties of the soil, e.g., surface moisture and roughness [33,34]. In the test site, the moisture is mainly dependant on the irrigation since the rainfall levels are extremely low given the desert conditions (average of less than 1 mm per month). On the other hand, the roughness is determined by the ridges and furrows created by the rows where the plants were sowed. In this case, the height of these rows, the plot age (the younger the crop, the more sand present), together with the row orientation and the incidence angle define the soil-SAR signal interaction. This effect is particularly evident with the VV polarisation (although not shown here).

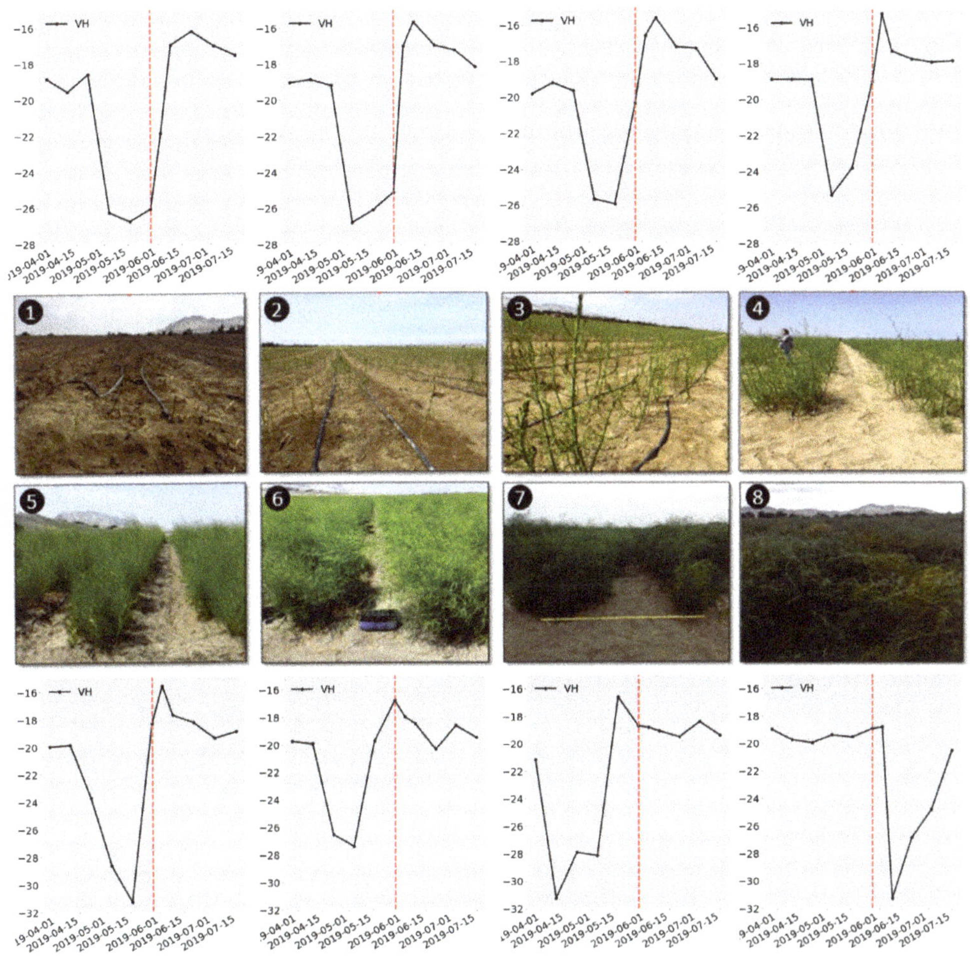


**Figure 6.** Images taken in the fields the 28/05/2019 (red vertical line) when a Sentinel-1 image was also acquired. The time series correspond to the VH backscatter. Each pair of image and time series correspond to a different parcel. The parcels are at different growth stages taking advantage of the local climate.

Both polarisations, but particularly the VH polarisation increase significantly as the asparagus stems start emerging and vertically elongating up to two meters height. This increase may be a consequence of the double bounce created with the SAR signal reaching the soil and bouncing off the vertical spears back to the satellite (Figure 6, image 3), although a more detailed polarimetric analysis would be required to confirm it. Please note that as show in Figure 4, the VV backscatter has less increment in time than the VH backscatter possibly due to the VH being more sensitive to the quasi-horizontal branches that grow sideways from the main vertical asparagus stems. A similar result was reported in [27] where authors present an almost constant VV response during the periods of vegetation growth.

However, when the crop reaches approximately the peak of the aperture stage (see Figure 7), the backscatter also reaches the peak in the entire cultivation period. At this point, the fern already has developed branches and the leaf-like structures in the stems are developing. From this moment, the contact of the SAR signal with the soil decreases thus also reducing the backscatter measured.

Subsequently, at the flowering stage (image 5 of Figure 6) the fern is fully developed and denser covering the soil and presumably creating a volume scattering response. The latter is less intense than the previous double bounce at the aperture stage, causing a decrease in the overall backscatter. Once the crop has reached the mature stage (images 6 to 8 of Figure 6), no significant changes happen in the biomass of the canopy hence the SAR signal remains at approximately the same level until the end of the season. An additional aspect to highlight is that as presented in Figure 4 and mentioned in Section 2.5, during the maturation stage the VV channel presents more statistical variation presumably since the VV backscatter still has an important ground component present as opposed to the VH backscatter which after the fern develops and stabilises, seems to have a strong contribution from the canopy and less from the ground. As a consequence, the VV captures features related to the soil, such as changes in moisture or roughness.

### 2.6.2. Impact of Temperature on the Crop and the Sar Response

In this section, we analyse the impact of the seasonality and the variable climatological conditions on the crop evolution. This analysis is relevant to build a phenology retrieval algorithm since these factors change the crop behaviour and/or the SAR response in time thus modifying the inputs for an algorithm and affecting the accuracy results. Please note that in this analysis of effect of temperature, we have used the VH channel given that it shows a more dynamic and cleaner signal than the VV polarisation.

Based on empirical observations, growers have noticed that the crop evolution during a "winter" and a "summer" campaign in the same year are different, in terms of canopy volume and development rate, possibly due to the different climatological conditions. Similarly, previous research showed that temperature influences asparagus plant growth rates and may cause growth depression [35,36]. On the other hand, mechanistic models of asparagus shoots height have been developed as a function of the temperature [36].

To investigate this effect in our test site using remotely sensed observations, time series of meteorological information, ground truth and SAR backscatter were analysed. Figure 7 shows the SAR and ground truth data for a typical parcel during four consecutive campaigns.

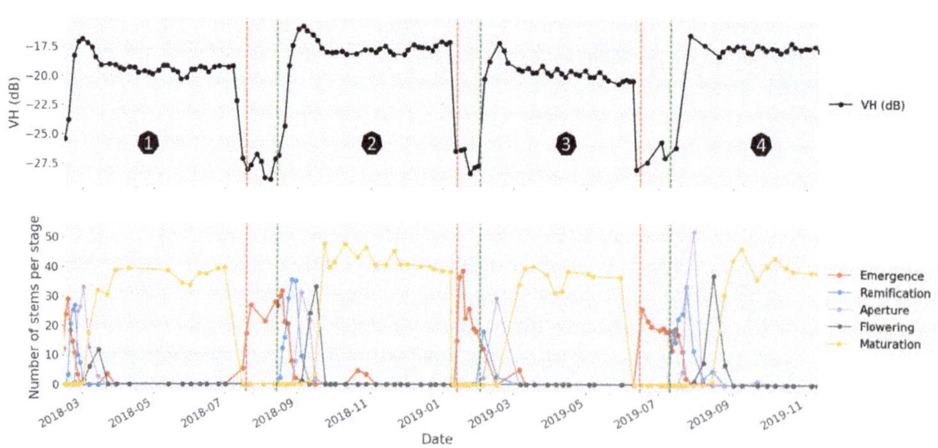


**Figure 7.** VH polarisation and ground truth observations for a typical parcel during four consecutive campaigns. Both the backscatter and the ground truth show a seasonal behaviour. The green and red vertical lines represent the start and end of the cultivation period respectively. The campaigns 1 and 3 correspond to summer season while the campaigns 2 and 4 represent winter campaigns.

The bottom plot of Figure 7 shows that the number of asparagus stems in maturation recorded during surveys are lower in the first semesters of 2018 and 2019 compared to the corresponding second semesters of the same years. Looking closely at the backscatter level for the same periods in the plot of the same figure, it is possible to see that the same pattern is followed in the time series of the VH polarisation once the crop has reached the maturation stage. This provides initial evidence of sensitivity of the SAR signal to the changes in the canopy volume (measured as number of stems in maturation, which would represent the total number of stems in the parcel for this period).

On the other hand, the left side of Figure 8 shows the VH backscatter response for the same parcel, during the winter campaign (second semester of 2018) and the summer campaign (first semester of 2019) as a function of the number of days after the cultivation started (DaS). It is possible to see the difference in the growth rate at the beginning of the cultivation, where in the summer campaign (red line) the crop reaches the peak of the VH time series faster relative to this same point in the cold season (blue line).

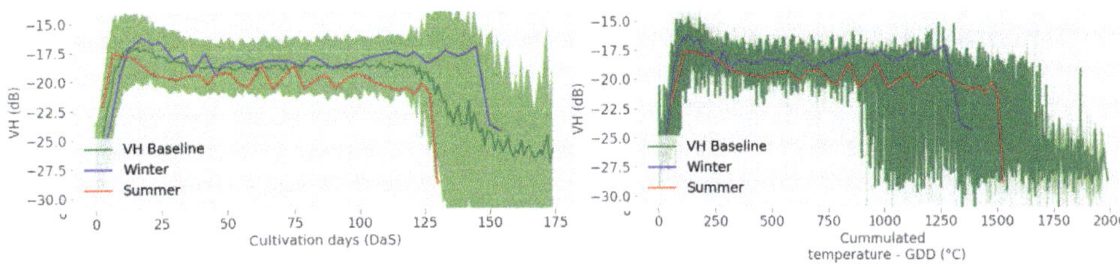


**Figure 8.** SAR backscatter of two consecutive campaigns aligned as function of cultivation days (DaS) and accumulated temperature (GDD). The blue line corresponds to the campaign one that grew during the colder season ("winter") and red line corresponds to one in the warmer season ("summer"). The plots at the bottom show the cultivation period length measured as a function of time (**left**) and temperature (**right**).

In order to confirm that the temperature impacts the crop growth rate, a test was done using the temperature as independent variable instead of the number of days after the campaign start. A measure of daily accumulated heat has been previously used in the literature for this purpose. It considers an averaged measure of the daily maximum and minimum temperature to determine how much heat the crop receive in a day (Growing Degree Day) and how much it accumulates day after day during a period of time [37]. For the present study, 10 degrees Celsius was considered to be base temperature [30]. By accumulating the Growing degree days (GDD) and using it as independent variable, the plot on the right side of Figure 8 shows that the VH backscatter observed for the winter and summer campaigns are approximately aligned. This suggests that the temperature drives the rate of canopy formation as suggested in other studies [35,36], and that it is observable with the VH polarisation measurements. It also provides insights about the potential usefulness of the temperature as input feature for a remote sensing algorithm to retrieve crop stage information.

The boxplot of Figure 9 presents the median accumulated temperature (GDD) in all the campaigns registered in the ground truth, from the campaign start to harvest. It can be seen that depending on the month when the cultivation started, there is a seasonal trend in the accumulated temperature. This information is key to for example estimate the harvest date given the campaign starting month, based on the required accumulated temperature.

To summarise, in principle there are three visible effects of the temperature on the crop. The first one, corresponds to the canopy volume developed, being less biomass during hotter temperatures. The SAR backscatter signal is sensitive to this by measuring lower backscatter intensity during the maturation period (Figure 7).

The second effect is associated with the growth rate, since as shown in the left side plot of Figure 8, it causes the stages at the beginning of the season to develop faster in a warmer campaign (red line). This effect is also visible from the backscatter response.

The third effect is related to the season length depending on the accumulated temperature during the cultivation period (Figure 9). This accumulated temperature in turn depends on the month of the year when the campaign started.

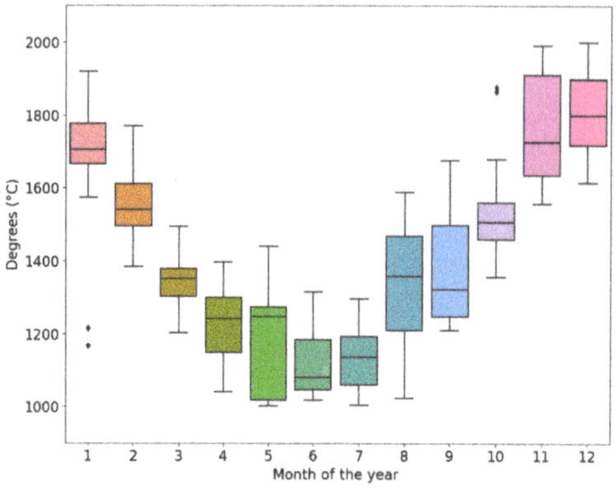


**Figure 9.** Campaign length measured in degrees Celsius (accumulated temperature) as a function of the production cycle starting month. As an example, if a campaign starts in January it normally accumulates around 600 degrees more than a campaign that starts in July. A total 442 campaigns were considered to generate this plot.

## 2.7. Estimation of Number of Asparagus Stems in Each Crop Stage

This section presents the methodology used for monitoring asparagus development as described by the number of asparagus stems in each of the stages of Figure 1. We use a data driven model to estimate ground measurements measurements from SAR observations. Please note that a deterministic physical model inversion is an ill posed problem since the number of unknowns are greater than the number of independent SAR measurements [38,39]. However, we exploit the patterns and correlations found in the SAR observations and temperature together with the ground truth to build empirical models.

We consider the retrieving the number of asparagus stems as a regression problem since it is possible to avoid selecting crop stage boundaries and allow soft transition between adjacent phenological stages. Please note that the ground truth used for training corresponding to phenological information is given by multiple and correlated variables (Figure 4). These variables do not evolve in time independently but rather they have temporal co-variation since they are produced by the same underlying process, i.e., the crop growth.

To exploit this structure in the output data, we use a multi-output regression algorithm that considers this interdependence of the individual outputs before making predictions. In this context, multitask learning (MTL) has been used in several applications precisely with this objective [40,41].

It is expected that not only the accuracy of a single multi-task learner increases compared to individual single-task learners (i.e., fitting an individual model for each output), but also since the model captures the structure of the data, it is able to generalize or interpolate better when the model is presented with unseen data [40,41].

In the remote sensing community multitask learning has been previously implemented using different machine learning algorithms [42,43]. Specifically for the case of SAR, in [42] the authors show how a multitask learner is able to make more accurately predictions of soil moisture and plant water content than individual learners.

### 2.7.1. Model Development

We chose initially a multi-task Random Forest Regressor [44] from the available MTL algorithms due to its power for capturing both the non-linear relationships and the correlation between multiple outputs.

A Random Forest Regressor [45] is known as an ensemble algorithm in which several individual decision regression trees (n-estimator trees) are built from individual bootstrapped datasets (datasets in which the samples are randomly selected from the original training dataset) [46]. Each regression tree uses a random subset of the input feature variables (m-features) from the original number of input features, and from this subset an optimal feature with an associated threshold is selected for each node [47]. In the single output case of a regression decision tree, both the optimal feature and the node threshold (i.e., the threshold that decides whether to go to the left or right child node) are found by minimizing a split function (also known as node cost) based on a euclidean distance error measure [47]. In the case of a multiple-output regression decision tree, an additional term is added to the node cost to account for the correlations in the output data. Specifically, in the framework proposed by [44], a Mahalanobis distance [48] is added to the split function to consider the multiple dimensions in the output data in the minimization cost function to select the corresponding thresholds. Please note that other split functions have been developed [49] but the in this paper the framework proposed in [44] is used. A final estimation is obtained from the multi-task Random Forest Regressor by averaging the estimations provided by the leaf nodes in each individual tree in the forest.

In this paper, the objective is to estimate the number asparagus stems in each of five possible stages. Each of these five estimations corresponds to an output predicted by the multi-output random forest regression. On the other hand, the algorithm uses historical SAR, temperature and ground truth data to learn the corresponding mapping functions. For this purpose, the scenarios in Table 2 have been considered in this paper.

**Table 2.** Scenarios considered for asparagus growth estimation. Please note that each of the scenarios in B and C categories is tested using one image as well as sequence of multiple images.

| Category | Scenario | Input | Description |
|---|---|---|---|
| A | A1 | DaS | Only number of days after cultivation started |
| A | A2 | DaS, DoY | Days after cultivation started, Day of year when cultivation started |
| A | A3 | DaS, DoY, AGDD | Days after cultivation started, Day of year when cultivation started, accumulated temperature |
| B | B1 | VH | Only VH polarisation |
| B | B2 | VH, VV, VH/VV | VH, VV polarisations and the VH/VV ratio |
| C | C1 | VH, VV, VH/VV, DaS | VH, VV, Ratio, Days after cultivation started |
| C | C2 | VH, VV, VH/VV, DaS, DoY | VH, VV, Ratio, Days after cultivation started, Day of year when cultivation started |
| C | C3 | VH, VV, VH/VV, DaS, DoY, AGDD | All previous features |

### 2.7.2. Inputs

An initial category identified as category A, does not use remote sensing as input for the multi-task regression but only uses ground data. In the scenario A1 the number of days after the cultivation started (DaS) is used to estimate the number of asparagus stems present in each phenological stage at every image (Table 2), similar to how farmers traditionally execute their planning in the test site and in general for farms with low adoption of technology. Considering that multiple production cycles per year occur and grow under different climatological conditions, as shown in Figures 8 and 9, the day of the year (DoY) when the cultivation season starts impacts the canopy development. We tested the value of using this information as input for the algorithm, given that for instance, 20 days of cultivation in summer may differ from 20 days of cultivation in winter. This corresponds to the scenario (A2).

A more robust scenario, the scenario A3, uses additionally the accumulated temperature during the cultivation period (from cultivation start to the SAR acquisition date) or accumulated growing degree-days (AGDD), since as it was shown in Section 2.6.2, the temperature drives the growth rate and canopy volume. Several other methodologies have used AGDD to account for the impact of climatic conditions in the crop growth [22,50,51]. Please note that using these input data sources the model learns the mapping function to give a theoretical estimation of the number of asparagus stems in each growth stage. This estimation may be accurate only if no external abnormal conditions affect the crop, such as extreme weather events including droughts, hail, etc., plant diseases, pests or changes in the management practices. Similarly, it does not provide any spatial information of crop status but a single prediction for the entire field. Although this information may be valuable for planning, it is not sufficient for operational crop monitoring.

A remote sensing-based approach that uses SAR images was considered to be the second category (B), where the Sentinel-1 backscatter including VH and VV polarisation channels with their corresponding ratio was used. This was tested using a single image and a sequence of images, from two to five. In this case, the near-real time data acquired by the satellite provides the capabilities for operational monitoring. In this scenario the VH polarisation was tested individually as well as together with the VH and VV ratio (scenarios B1 and B2). This scenario only uses SAR data as input for the multi-output regression.

A third category (C), includes each of the previous data sources SAR, DaS, DoY, and AGDD as input for the algorithm hoping to integrate their individual advantages (scenarios C1 to C3). Please note that the scenario C1 is included with the aim of quantifying the usefulness of the DaS feature when using multi-temporal SAR data and the scenario C2 quantifies the usefulness of the DoY feature.

#### 2.7.3. Outputs

In all the scenarios, the aim is to produce estimates of the number of asparagus stems present in each of the following phenological stages: Emergence, ramification, aperture, flowering and maturation at any given SAR acquisition (same stages shown in Figure 1). Because in this case we require to simulatenoulsy predict the multiple outputs, we chose the model described in Section 2.7.1. On the other hand, as described in Section 2.4, these values would correspond to measurements taken in a randomly selected linear meter within the parcel. Here we assume that the ground data collected in this way is representative of the entire parcel.

#### 2.7.4. Training and Testing Data

The ground truth described in Section 2.4 collected between January and August 2019, Sentinel-1 data and temperature measurements from the same periods are used to create the datasets. The dataset $D$ takes the form $D = (X, y)$ where $X$ is a matrix of dimension $mxn$, $m$ is the number of ground truth measurements available, $n$ is defined by the features being used according to each scenario of Table 2 and $y$ is the matrix of ground truth data with dimension $mx5$, where the number 5 corresponds to the number of asparagus stems in each stage of Figure 1 recorded in the ground truth surveys.

In order to separate training and testing data, we randomly select plots based on the ID to complete approximately 70% of them in a training dataset and 30% in a testing dataset so that we guarantee the unseen data required for the testing phase. In total, from the 442 plots (of 2 Ha on average) with available ground truth, approximately 309 plots randomly selected are used for training and the remaining 133 plots for testing.

Given that several ground truth measurement dates do not coincide with the Sentinel-1. acquisition dates, a three order spline interpolation was used to interpolate daily the ground truth so that an associated ground truth measurement can be obtained for every SAR observation. In total, there exist 4023 training data-points and 1739 testing data-points.

### 2.7.5. Model Hyper Parameters

Tuning of the optimal model hyper-parameters was done using 5-fold cross-validation with grid search. Table 3 presents the selected hyperparameters for the scenario C3.

**Table 3.** Selected model hyperparameters

| Hyperparameter | Selected |
|---|---|
| Bootstrap | True |
| The number of trees in the forest | 800 (a) |
| Split funciton | Mahalanobis Distance |
| max-depth | 30 (a) |

(a): Tunned hyper-parameter.

### 2.7.6. Accuracy Metrics

The coefficient of determination $R^2$ computed with Equation (1) was used to measure the model performance, both for the individual outputs and for the model as a whole, by averaging the scores of the five outputs.

$$R^2 = 1 - \frac{\Sigma_{i=1}^{n}\left(y_i - \widehat{y}_i\right)^2}{\Sigma_{i=1}^{n}\left(y_i - \widehat{y}\right)^2} \tag{1}$$

where $y_i$ corresponds to the $i - th$ ground truth test sample, $\widehat{y}_i$ a prediction made with the model for this sample after training, and $\widehat{y}$ the mean value of the n-ground truth test samples.

Similarly, the root mean squared error RMSE calculated with Equation (2), was computed between predicted and testing values.

$$RMSE = \sqrt{\frac{1}{n}\Sigma_{i=1}^{n}\left(y_i - \widehat{y}_i\right)^2} \tag{2}$$

where $y_i$ corresponds to the $i - th$ ground truth test sample, $\widehat{y}_i$ a prediction made with the model for this sample after training.

## 3. Results

### 3.1. *Single-Sar Image Results*

The obtained results when using a single SAR image as input for the model are reported in Tables 4 and 5.

**Table 4.** Summary of coefficients of determination $R^2$ for the predicted number of stems in each crop stage when using a single SAR image.

| Stage | A1 | A2 | A3 | B1 | B2 | C1 | C2 | C3 |
|---|---|---|---|---|---|---|---|---|
| Emergence | 0.72 | 0.83 | 0.88 | 0.54 | 0.68 | 0.78 | 0.83 | 0.84 |
| Aperture | 0.65 | 0.9 | 0.92 | 0.04 | 0.27 | 0.71 | 0.87 | 0.9 |
| Ramification | 0.74 | 0.9 | 0.9 | −0.22 | 0.11 | 0.81 | 0.9 | 0.9 |
| Flowering | 0.41 | 0.79 | 0.82 | −0.36 | −0.06 | 0.49 | 0.7 | 0.76 |
| Maturation | 0.79 | 0.91 | 0.94 | 0.36 | 0.52 | 0.82 | 0.89 | 0.9 |
| Overall | 0.66 | 0.87 | 0.89 | 0.07 | 0.3 | 0.72 | 0.84 | 0.86 |

**Table 5.** Summary of root mean square error RMSE for the predicted number of stems in each crop stage when using a single SAR image.

| Stage | A1 | A2 | A3 | B1 | B2 | C1 | C2 | C3 |
|---|---|---|---|---|---|---|---|---|
| Emergence | 4.16 | 3.23 | 2.72 | 5.36 | 4.42 | 3.69 | 3.25 | 3.13 |
| Aperture | 3.18 | 1.67 | 1.54 | 5.26 | 4.61 | 2.87 | 1.96 | 1.66 |
| Ramification | 2.2 | 1.34 | 1.34 | 4.78 | 4.09 | 1.87 | 1.36 | 1.35 |
| Flowering | 2.91 | 1.74 | 1.61 | 4.43 | 3.92 | 2.71 | 2.08 | 1.85 |
| Maturation | 6.48 | 4.34 | 3.48 | 11.39 | 9.88 | 6.08 | 4.63 | 4.36 |
| Overall | 4.07 | 2.72 | 2.3 | 6.77 | 5.84 | 3.73 | 2.9 | 2.72 |

When using one SAR image (see Table 4), particularly the scenarios A2, A3, C2 and C3 achieve satisfactory predictive capabilities with overall coefficients of determination $R^2$ between 0.84 and 0.89. This is confirmed with the RMSE's that are also the lowest for these scenarios.

Regarding individual outputs of the multi-task regression for these same scenarios, the maturation phase has the best performance achieving an $R^2$ of more than 0.9 in almost all of them and flowering the lowest accuracy between 0.70 and 0.8. Note from Figure 4 that comparing flowering to any other stage the temporal shape described in the ground truth by this measurement is more irregular and reaches on average fewer stems than the other stages. This could be due to an agronomic reason that requires further analysis or due to a systematic error affecting flowering when surveying the fields. This is also possibly the reason causing the predictions of asparagus stems in flowering less accurate than in the other stages.

On the other hand, the results of scenario C1 are substantially higher than the scenarios in category B. By providing to the regressor the number of days after the season started (DaS) as in scenario C1, the algorithm improves the retrieval with respect to category B potentially since it would be possible to disentangle similar backscatters at different dates.

An additional increase in $R^2$ (from 0.72 to 0.84) and reduction of RMSE (from 3.73 to 2.9) is achieved in the scenario C2 only by specifying the day of the year when the agricultural season starts. This feature indirectly provides information about the seasonality present in the test site and shown in Figures 7 and 8. The $R^2$ and RMSE are further improved and decreased respectively in the scenario C3 after the addition of the AGDD feature, although not significantly. This low increase may be explained by considering that providing DoY (as in scenario C2) we already provide information about seasonality and given that as mentioned in Section 2.6.2, the impact of higher and lower accumulated temperature in the canopy is perceived by the VH backscatter as shown in Figures 7 and 8. Consequently, the algorithm indirectly receives information about the accumulated temperature through the use of VH and DoY. This is an important result given that it implies that not using the AGDD feature (as in scenario C2), the temperature data from a ground station is not needed, without sacrificing substantially the model performance.

Figure 10 summarises the model performance for the scenario C3. For this same scenario, Figure 11 shows the test and predicted data-points as a function of the days after the season started (using the cultivation days associated with the test data-points as x-axis). It can be seen that in general the predicted values (in red) follow the timing and the expected number of stems of the testing data-points (in blue). It is also possible to see however, that they are not exactly the same, indicating that although the model is accurately making predictions, it is not over-fitting to produce identical values as the testing points nor predicting extreme values that may correspond to outliers.

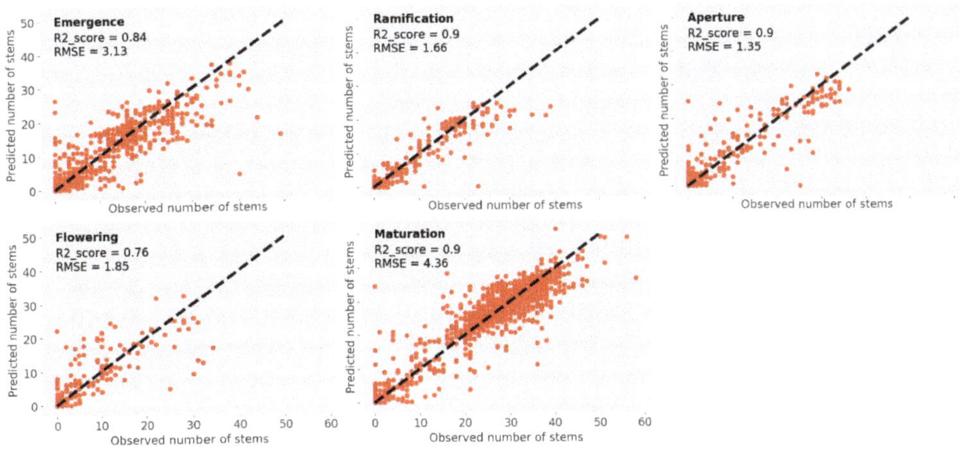


**Figure 10.** Observed vs. Predicted number of asparagus stems per stage, with the corresponding overall coefficient of determination and root mean squared error, using features of scenario C3 to train the model .

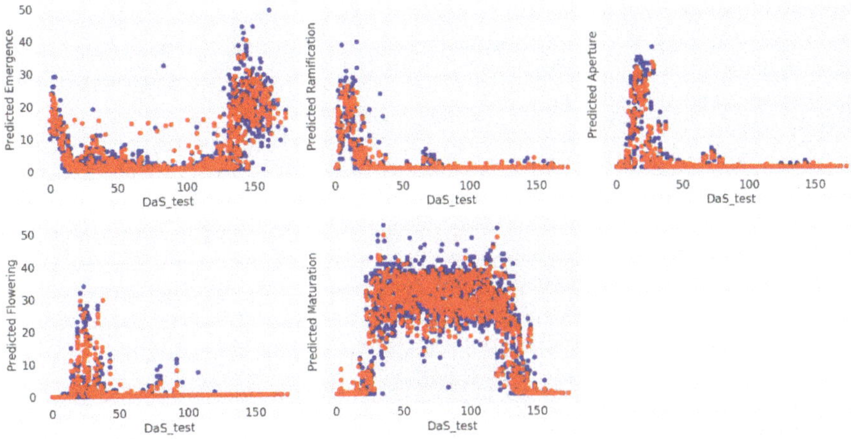


**Figure 11.** Predicted (red) vs. test (blue) number of asparagus stems per stage, using the cultivation days associated with the testing data-points as x-axis.

## 3.2. Multi-Temporal Sar Results

In order to quantify the performance of all the scenarios considered when using multiple Sentinel-1 images, we tested increasing the number of images to create the training time series, e.g., from only using the latest SAR image to using the 5 latest available images. In this case, the dataset takes the form $D = ([X_t, X_{t-1}, X_{t-2}, X_{t-3}, X_{t-4}], y_t)$ where $t$ represents the index of the date when the ground truth was collected or in a prediction setting, when the prediction is desired. Figure 12 show the corresponding overall, $R^2$ scores and RMSE for each scenario of Table 2 when using from one to five Sentinel-1 images to estimate the number of asparagus stems in each phenological stage. Please note that since the scenarios of the category A do not use Sentinel-1 data, we only consider the categories B and C for this part of the analysis.

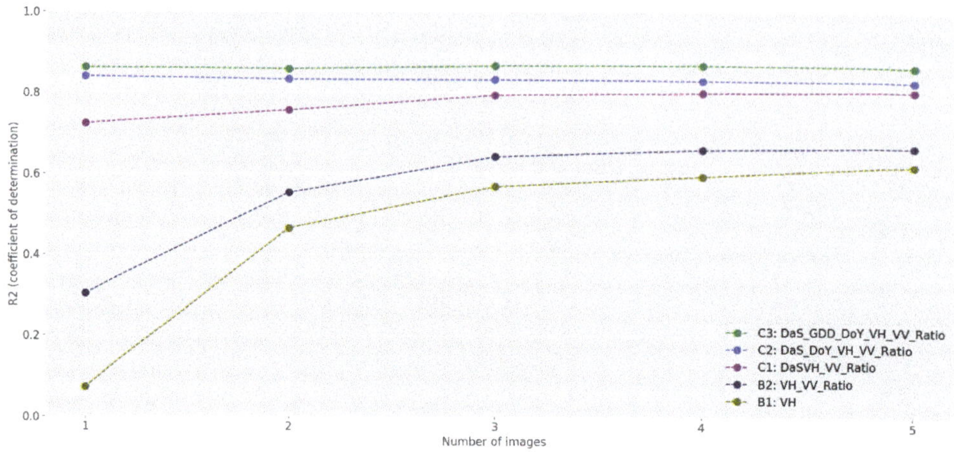


(a) Coefficient of determination $((R^2))$

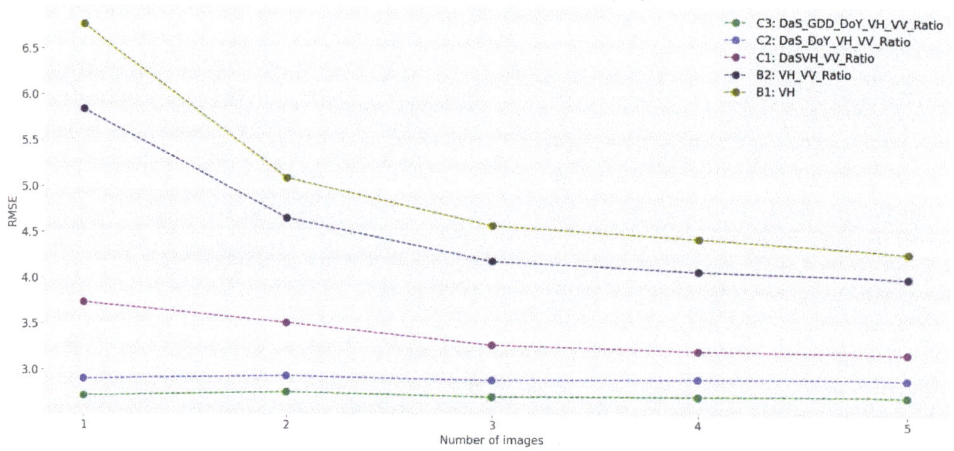


(b) Root mean square error (RMSE)

**Figure 12.** Multi-task regression performance metrics as a function of the number of images used to train the model for each of the scenarios of Table 2.

In this case, the scenarios that use the VH, VV and ratio as features (B1 and B2), increase their performance when more images are used, with the scenario B2 achieving an $R^2$ of 0.66. This is not surprising since by using the sequence of the backscatter evolution provides additional information for the algorithm to disentangle similar backscatter present at different time of the season. An additional significant result obtained here is that no substantial improvement is achieved when more than 4 images are used.

However, all the other scenarios considered in this paper do not increase the performance when increasing the number of images used as it would be expected, but rather maintain the same accuracy achieved when using a single SAR image, temperature and the start of the season information as input features. This may be due to the fact that the information provided by DaS, DoY and AGDD is sufficient for helping to disentangle the backscatter of a single image and thus no further images are required for this purpose.

### 3.3. Growth Stage Estimation Maps

Figure 13 shows for all the parcels in the test site, the estimation of asparagus stems present in each phenological stage, obtained using the trained multi-task random forest of the scenario C3, for the Sentinel-1 image acquired the 2018/10/12. This is the same acquisition date as in Figure 2 and the intermediate subplot of Figure 14, which in turn shows an RGB composite of the same information using the predicted asparagus stems in emergence in the blue channel, the predicted asparagus stems in maturation in the green channel and the sum of the predicted asparagus stems in ramification, aperture and flowering in the red channel ( given their short duration).

Figure 14 shows additionally the RGB composites of 4 other Sentinel-1 acquisitions in order to visualise the change in time of the predicted crop stages due to crop development. This composite reveals the crop stage of each parcel in an intuitive and fast way while the number of asparagus stems predictions map of Figure 13 shows more detailed information for every individual crop stage.

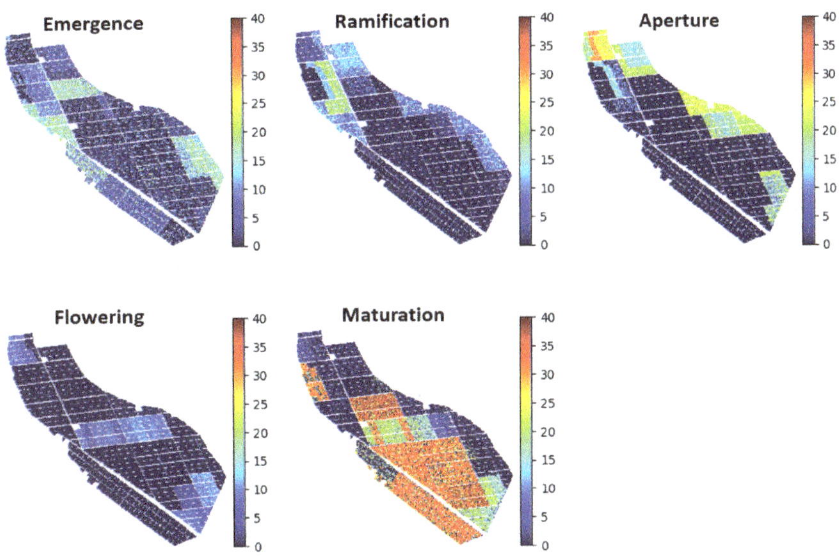


**Figure 13.** Number of asparagus stems estimated for each of the crop stages for the 2018/10/12 Sentinel-1 image (Same as Figure 2 and intermediate plot of Figure 14).

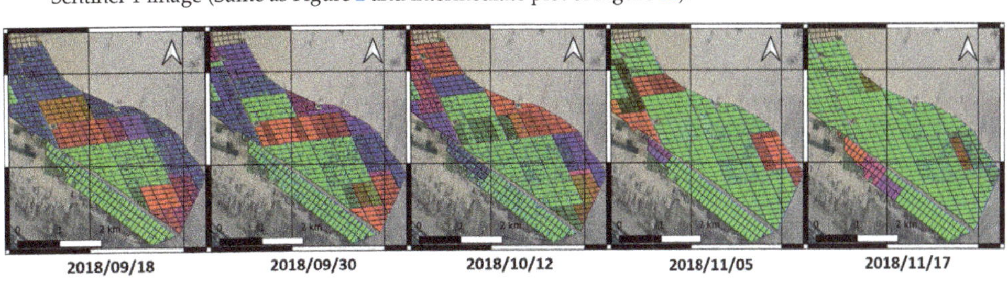


**Figure 14.** RGB composites of the estimated crop stage. Red: (Ramification+Aperture+flowering), Green: Maturation, Blue: Emergence.

## 4. Discussion

We have provided the analysis of the SAR backscatter response to asparagus growth development and canopy formation as shown in Figure 4. Similarly, Figure 7 presents the seasonality effect both in the sensor response and the ground truth due to consecutive production cycles that grow under different meteorological conditions.

Figure 8 shows how the VH polarisation is used for crop monitoring in order to visualise the canopy growth rate, revealing that it is faster in summer than in winter, but with less canopy volume (biomass). It also shows that although the production cycles in winter are longer, those cycles accumulate less temperature measured in GDD compared to the summer campaigns. Based on this information, the season length varies depending on the cycle starting month as shown in Figure 9. The backscatter response is sensitive to all these events as shown in Section 2.6.2.

With respect to the algorithm to retrieve the crop stage algorithm, several scenarios were considered in the analysis to understand the relevance of each data source and input feature and to determine the best way to combine the available information.

The scenarios of category A, which do not include remote sensing data, show that using accumulated growing degree days improves the predicting capabilities of an algorithm given that temperature is an important factor driving the crop evolution. In fact, the scenario A3 provides the highest $R^2$-scores as well as the lowest RMSE's of the scenarios tested (0.89 and 2.3 respectively). This is aligned with the well known techniques to estimate the timing of phenological events using thermal calendars [52,53]. However, this estimation may be accurate only if no external anomalous conditions affect the crop, such as extreme weather events, diseases or changes in the management practices. Similarly, it does not provide spatial information of crop status but a single prediction for the entire field. In consequence, although this information may be valuable for planning, it is not sufficient for operational crop monitoring.

With regard to the scenarios of category B (SAR only), although using multiple images as input for a model the predictive capabilities improve significantly, it achieves poorer results than the categories A and C. The scenarios in category C which use all available input sources, achieve similar accuracy retrieving the number of asparagus stems in each phenological stage than the scenarios of category A, according to the results of Tables 4 and 5, but additionally providing spatial resolution and the ability to determine growth anomalies. Focusing particularly in scenarios C2 and C3, the difference in their performance is not significant thus the scenario C2 which do not use temperature may be preferred.

To summarise the scenarios considered and based on the results obtained, we grouped four users cases that may benefit differently for each scenario depending on their specific needs and data availability as shown in Table 6. Depending on the farm size a scenario evaluated in this paper can be adopted, considering additionally the availability of temperature data and whether a larger scale monitoring system is required.

With regards to the current literature available, although previous studies have considered the possibility of identifying asparagus from other crops using radar remote sensing data [23–27], they did not focus on analysing the backscatter response relative the crop evolution due to the transition between phenological stages as was presented in Sections 2.6.1 and 2.6.2. Similarly, no previous study evaluated the possibility to retrieve the crop stage. In this context, the present study contributes to the literature with a more detailed analysis of asparagus providing an interpretation of backscatter evolution that may offer tools for better crop classification (since so far low accuracy have been reported [23,24,27]), through a better understanding of the temporal crop signature.

The VV backscatter throughout the agricultural season does not have significant changes relative to the VH polarisation channel. This has been previously reported in [27] and it differs from what was presented in [26] using TerraSAR-X. This suggest that X-band is able to capture events in the crop development not visible in C-band as happens for instance in rice fields [6].

It is important to highlight that the current methodology is limited in part by the availability of the season starting date information as input, being used for calculating the AGDD as well as input for the multitemporal regressor (DaS). A potential solution to this is the estimation of starting dates from remote sensing as it has been previously investigated with satisfactory results in [54–59]. A further limitation in the current analysis is the lower accuracy of retrieving the number of asparagus stems in flowering, with respect to other stages such as emergence and maturation (Tables 4 and 5), which as mentioned in Section 3.1 might be caused by the unclear ground truth samples used to train the model. Statistical tests to better understand the characteristics of the training data and variations between measurements in each stage may be used to make decisions about better strategies to use the same data.

Future research will focus on the automatic detection of starting date from SAR and the use of quad-polarimetric data to better understand the scattering mechanisms throughout the season. Additionally, the use of better ways to deal with the sequential nature of data generated from agricultural fields and multitemporal remote sensing data will be considered including dynamical modelling [20,22] and more advanced machine learning models .

**Table 6.** User cases and scenarios considered.

| User Case | Season Start Date | Temperature Data | Scenario |
|---|---|---|---|
| Small farm interested in estimating occurrence of key dates for planning | known from ground truth | known from ground station | A * |
| Medium or large size farms | known from ground truth | known from ground station | C2 and C3 ** |
| Large scale monitoring (regional or national level) without temperature | Automatically detected | - | B2 and C2 *** |
| Large scale monitoring (regional or national level) and satellite measurements of land surface temperature | Automatically detected | known from satellite measurements of land surface temperature | C3 **** |

* It is assumed that the spatial resolution is not critical since anomalies can be identified by inexpensive field surveys. ** Spatial resolution is important since growth anomalies cannot be identified by simple visual inspection or the field surveys are expensive. *** The scenario C2 is the same as B2 by adding information about season start date. The lost of accuracy for not using temperature is not signicant based on the findings of Section 3. **** The accuracy results may decrease since automatically detecting start date and using low spatial resolution LST measurements introduce uncertainty.

## 5. Conclusions

In this paper, we provided an interpretation of the SAR backscatter response to asparagus crop growth and analysed the impact that temperature has on the canopy volume, its development rate, and the cultivation length. It was shown how the VH backscatter is sensitive to all these effects. We also presented a multi-output machine learning regression algorithm trained to retrieve the number of asparagus stems present in each of five possible phenological stages. We tested several operational scenarios finding that using the VH, VV, VH/VV, and information about season start date (scenario C2), the model is able to retrieve the number of asparagus stems with an overall $R^2$ of 0.84 and RMSE of 2.9. Adding the accumulated temperature (AGDD) as in the scenario C3, improved slightly the accuracy resulting in overall $R^2$ of 0.86 and RMSE of 2.7. However, given that this increase is not substantial, the scenario C2 might be preferred since the temperature feature is not required and therefore this additional data source could be removed without losing significant model performance.

Additionally, as shown in Figure 12, for the conditions evaluated in this paper, the use of multitemporal SAR data is not critical when using information about the season start date and temperature as crop stage predictors to complement the backscatter, since these features provide similar information for the algorithm to disentangle events in the temporal dimension.

**Author Contributions:** Conceptualization, C.S.-P.; Methodology, C.S.-P., A.M. and I.C.; Software, C.S.-P.; Validation, C.S.-P., A.M. and I.C.; Formal Analysis, C.S.-P.; Investigation, C.S.; Resources, A.M. and I.C.; Data Curation, C.S.-P.; Writing—Original Draft Preparation, C.S.-P., Writing—Review & Editing, A.M. and I.C.; Visualization, C.S.-P.; Supervision, A.M. and I.C.; Project Administration, I.C.; Funding Acquisition, I.C. All authors have read and agreed to the published version of the manuscript.

**Funding:** This research was supported by the Project EO4cultivar, led by Environment Systems Ltd. and co-funded by the UK Space Agency (https://www.envsys.co.uk/projects/eo4cultivar/).

**Conflicts of Interest:** The authors declare no conflict of interest.

## References

1. Pedregosa, F.; Varoquaux, G.; Gramfort, A.; AU West, P.C.; Gibbs, H.K.; Monfreda, C.; Wagner, J.; Barford, C. C.and Carpenter, S.R.F.J.A. Trading carbon for food: Global comparison of carbon stocks vs. crop yields on agricultural land. *Proc. Natl. Acad. Sci. USA* **2010**, *107*, 19645–19648.
2. Food and Agriculture Organization of the United Nations. *FAOSTAT Crops*; FAO: Rome, Italy, 2019.
3. Terán-Velazco, C.A. Impactos sociales del espárrago en el Perú. 2017. Available online: http://repositorio.ulima.edu.pe/bitstream/handle/ulima/6003/Teran_Esparragos_Peru.pdf?sequence=1&isAllowed=y (accessed on 15 May 2020).
4. Hodges, T. *Predicting Crop Phenology*; CRC Press: Boca Raton, FL, USA, 1990.
5. Jong-Sen Lee, E.P. *Polarimetric Radar Imaging: From Basics to Applications*; CRC Press: Boca Raton, FL, USA, 2017.
6. Lopez-Sanchez, J.M.; Vicente-Guijalba, F.; Ballester-Berman, J.D.; Cloude, S.R. Polarimetric Response of Rice Fields at C-Band: Analysis and Phenology Retrieval. *IEEE Trans. Geos. Rem. Sen.* **2014**, *52*, 2977–2993. [CrossRef]
7. Alonso-Gonzalez, A.; Joerg, H.; Papathanassiou, K.; Hajnsek, I. Change Analysis and Interpretation in Polarimetric Time Series over Agricultural Fields at C-band. In Proceedings of the EUSAR 2016: 11th European Conference on Synthetic Aperture Radar, Hamburg, Germany, 6–9 June 2016; pp. 1–6.
8. Wiseman, G.; McNairn, H.; Homayouni, S.; Shang, J. RADARSAT-2 Polarimetric SAR Response to Crop Biomass for Agricultural Production Monitoring. *IEEE J. Sel. Top. Appl. Earth Obs. Remote Sens.* **2014**, *7*, 4461–4471. [CrossRef]
9. Veloso, A.; Mermoz, S.; Bouvet, A.; Le Toan, T.; Planells, M.; Dejoux, J.F.; Ceschia, E. Understanding the temporal behavior of crops using Sentinel-1 and Sentinel-2-like data for agricultural applications. *Remote Sens. Environ.* **2017**, *199*, 415–426. [CrossRef]
10. Steele-Dunne, S.C.; Khabbazan, S.; Vermunt, P.C.; Arntz, L.R.; Marinetti, C.; Iannini, L.; Westerdijk, K.; van der Sande, C. Monitoring Key Agricultural CROPS in the Netherlands using Sentinel-1. In Proceedings of the IGARSS 2018—2018 IEEE International Geoscience and Remote Sensing Symposium, Valencia, Spain, 22–27 July 2018; pp. 6639–6642. [CrossRef]
11. Khabbazan, S.; Vermunt, P.; Steele-Dunne, S.; Ratering Arntz, L.; Marinetti, C.; van der Valk, D.; Iannini, L.; Molijn, R.; Westerdijk, K.; van der Sande, C. Crop Monitoring Using Sentinel-1 Data: A Case Study from The Netherlands. *Remote Sens.* **2019**, *11*, 1887. [CrossRef]
12. Harfenmeister, K.; Spengler, D.; Weltzien, C. Analyzing Temporal and Spatial Characteristics of Crop Parameters Using Sentinel-1 Backscatter Data. *Remote Sens.* **2019**, *11*, 1569. [CrossRef]
13. Küçük, Ç.; Taşkın, G.; Erten, E. Paddy-Rice Phenology Classification Based on Machine-Learning Methods Using Multitemporal Co-Polar X-Band SAR Images. *IEEE J. Sel. Top. Appl. Earth Obs. Remote Sens.* **2016**, *9*, 2509–2519. [CrossRef]
14. Wang, H.; Magagi, R.; Goïta, K.; Trudel, M.; McNairn, H.; Powers, J. Crop phenology retrieval via polarimetric SAR decomposition and Random Forest algorithm. *Remote Sens. Environ.* **2019**, *231*, 111234. [CrossRef]
15. Mascolo, L.; Lopez-Sanchez, J.M.; Vicente-Guijalba, F.; Nunziata, F.; Migliaccio, M.; Mazzarella, G. A Complete Procedure for Crop Phenology Estimation With PolSAR Data Based on the Complex Wishart Classifier. *IEEE Trans. Geosci. Remote Sens.* **2016**, *54*, 6505–6515. [CrossRef]
16. Meier, U. *Growth Stages of Mono-And Dicotyledonous Plants*; Blackwell Wissenschafts-Verlag: Berlin, Germany, 1997.
17. Lopez-Sanchez, J.M.; Ballester-Berman, J.D.; Hajnsek, I. First Results of Rice Monitoring Practices in Spain by Means of Time Series of TerraSAR-X Dual-Pol Images. *IEEE J. Sel. Top. Appl. Earth Obs. Remote Sens.* **2011**, *4*, 412–422. [CrossRef]

18. Cota, N.; Kasetkasem, T.; Rakwatin, P.; Chanwimaluang, T.; Kumazawa, I. Rice phenology estimation based on statistical models for time-series SAR data. In Proceedings of the 2015 12th International Conference on Electrical Engineering/Electronics, Computer, Telecommunications and Information Technology (ECTI-CON), Hua Hin, Thailand, 24–27 June 2015; pp. 1–6. [CrossRef]
19. Siachalou, S.; Mallinis, G.; Tsakiri-Strati, M. A hidden Markov models approach for crop classification: Linking crop phenology to time series of multi-sensor remote sensing data. *Remote Sens.* **2015**, *7*, 3633–3650. [CrossRef]
20. Vicente-Guijalba, F.; Martinez-Marin, T.; Lopez-Sanchez, J.M. Dynamical Approach for Real-Time Monitoring of Agricultural Crops. *IEEE Trans. Geosci. Remote Sens.* **2015**, *53*, 3278–3293. [CrossRef]
21. De Bernardis, C.; Vicente-Guijalba, F.; Martinez-Marin, T.; Lopez-Sanchez, J.M. Contribution to Real-Time Estimation of Crop Phenological States in a Dynamical Framework Based on NDVI Time Series: Data Fusion With SAR and Temperature. *IEEE J. Sel. Top. Appl. Earth Obs. Remote Sens.* **2016**, *9*, 3512–3523. [CrossRef]
22. McNairn, H.; Jiao, X.; Pacheco, A.; Sinha, A.; Tan, W.; Li, Y. Estimating canola phenology using synthetic aperture radar. *Remote Sens. Environ.* **2018**, *219*, 196–205. [CrossRef]
23. Tavakkoli, M.; Lohmann, P.; Soergel, U. Mapping of agricultural activities using Multi-Temporal ASAR ENVISAT DATA. In Proceedings of the ISPRS Hannover Workshop 2007: High-Resolution Earth Imaging for Geospatial Information', Hannover, Germany, 29 May—1 June 2007, ISPRS Archives—XXXVI-1/W51.
24. Tavakkoli, M.; Lohmann, P. Multi-temporal classification of ASAR images in agricultural areas. In Proceedings of the ISPRS Commission VII Symposium: 'Remote Sensing: From Pixels to Processes', Enschede, The Netherlands, 8–11 June 2006.
25. Sabour, S.T.; Lohmann, P.; Soergel, U. Monitoring agricultural activities using multi-temporal ASAR ENVISAT data. *Int. Arch. Photogramm. Remote Sens. Spat. Inf. Sci.* **2008**, *37*, 735–742.
26. Bargiel, D.; Herrmann, S.; Lohmann, P.; Sörgel, U. Land use classification with high-resolution satellite radar for estimating the impacts of land use change on the quality of ecosystem services. *Int. Arch. Photogramm. Remote Sens. Spat. Inf. Sci.* **2010**, *38*, 68–73.
27. Arias, M.; Campo-Bescós, M.Á.; Álvarez-Mozos, J. Crop Classification Based on Temporal Signatures of Sentinel-1 Observations over Navarre Province, Spain. *Remote Sens.* **2020**, *12*, 278. [CrossRef]
28. Wien, H.C.; Stützel, H. *The Physiology of Vegetable Crops*; CABI: Oxfordshire, UK, 2020.
29. Wilson, D.; Sinton, S.; Butler, R.; Drost, D.; Paschold, P.J.; van Kruistum, G.; Poll, J.; Garcin, C.; Pertierra, R.; Vidal, I.; et al. Carbohydrates and yield physiology of asparagus–A global overview. In *XI International Asparagus Symposium 776, Horst, Netherlands*; International Society for Horticultural Science (ISHS): Leuven, Belgium, 2005; pp. 413–428.
30. Wilson, D.; Cloughley, C.; Jamieson, P.; Sinton, S. A model of asparagus growth physiology. In *X International Asparagus Symposium 589, Niigata, Japan*; International Society for Horticultural Science (ISHS): Leuven, Belgium, 2001; pp. 297–301.
31. Casas, A. El Cultivo del Espárrago en la Costa Peruana. Available online: http://www.lamolina.edu.pe/agronomia/dhorticultura/html/apuntesdeclase/Casas/El%20Cultivo%20del%20esp%C3%A1rrago%20en%20la%20Costa%20Peruana.pdf (accessed on 15 May 2020).
32. Gorelick, N.; Hancher, M.; Dixon, M.; Ilyushchenko, S.; Thau, D.; Moore, R. Google Earth Engine: Planetary-scale geospatial analysis for everyone. *Remote Sens. Environ.* **2017**, *202*, 18–27. [CrossRef]
33. Oh, Y.; Sarabandi, K.; Ulaby, F.T. An empirical model and an inversion technique for radar scattering from bare soil surfaces. *IEEE Trans. Geosci. Remote Sens.* **1992**, *30*, 370–381. [CrossRef]
34. Davidson, M.W.; Le Toan, T.; Mattia, F.; Satalino, G.; Manninen, T.; Borgeaud, M. On the characterization of agricultural soil roughness for radar remote sensing studies. *IEEE Trans. Geosci. Remote Sens.* **2000**, *38*, 630–640. [CrossRef]
35. Yen, Y.f.; Nichols, M.; Woolley, D. Growth of asparagus spears and ferns at high temperatures. In *VIII International Asparagus Symposium 415, Palmerston North, New Zealand*; International Society for Horticultural Science (ISHS): Leuven, Belgium, 1993; pp. 163–174.
36. Wilson, D.; Cloughley, C.; Sinton, S. Model of the influence of temperature on the elongation rate of asparagus spears. In *IX International Asparagus Symposium 479, Pasco, Washington, USA*; International Society for Horticultural Science (ISHS): Leuven, Belgium, 1997; pp. 297–304.
37. McMaster, G.S.; Wilhelm, W. Growing degree-days: One equation, two interpretations. *Agric. For. Meteorol.* **1997**, *87*, 291–300. [CrossRef]

38. Liang, S. *Advances in Land Remote Sensing: System, Modeling, Inversion and Application*; Springer Science & Business Media: Berlin/Heidelberg, Germany, 2008.
39. Oliver, C.; Quegan, S. *Understanding Synthetic Aperture Radar Images*; SciTech Publishing: Raleigh, NC, USA, 2004.
40. Caruana, R. Multitask Learning. *Mach. Learn.* **1997**, *28*, 41–75. [CrossRef]
41. Ruder, S. An Overview of Multi-Task Learning in Deep Neural Networks. *arXiv* **2017**, arXiv:1706.05098.
42. Camps-Valls, G.; Martino, L.; Svendsen, D.H.; Campos-Taberner, M.; Muñoz-Marí, J.; Laparra, V.; Luengo, D.; García-Haro, F.J. Physics-aware Gaussian processes in remote sensing. *Appl. Soft Comput.* **2018**, *68*, 69–82. [CrossRef]
43. Leiva-Murillo, J.M.; Gomez-Chova, L.; Camps-Valls, G. Multitask Remote Sensing Data Classification. *IEEE Trans. Geosci. Remote Sens.* **2013**, *51*, 151–161. [CrossRef]
44. Segal, M.; Xiao, Y. Multivariate random forests. *Wiley Interdiscip. Rev. Data Min. Knowl. Discov.* **2011**, *1*, 80–87. [CrossRef]
45. Breiman, L. Random forests. *Mach. Learn.* **2001**, *45*, 5–32. [CrossRef]
46. Johnson, R.W. An introduction to the bootstrap. *Teach. Stat.* **2001**, *23*, 49–54. [CrossRef]
47. Breiman, L.; Friedman, J.; Stone, C.J.; Olshen, R.A. *Classification and Regression Trees*; CRC Press: Boca Raton, FL, USA, 1984.
48. Mahalanobis, P.C. *On the Generalized Distance in Statistics*; National Institute of Science of India: Brahmapur, India, 1936.
49. Kim, S.J.; Lee, K.B. Constructing decision trees with multiple response variables. *Int. J. Manag. Decis. Mak.* **2003**, *4*, 337–353. [CrossRef]
50. d'Andrimont, R.; Taymans, M.; Lemoine, G.; Ceglar, A.; Yordanov, M.; van der Velde, M. Detecting flowering phenology in oil seed rape parcels with Sentinel-1 and-2 time series. *Remote Sens. Environ.* **2020**, *239*, 111660. [CrossRef]
51. Skakun, S.; Vermote, E.; Franch, B.; Roger, J.C.; Kussul, N.; Ju, J.; Masek, J. Winter Wheat Yield Assessment from Landsat 8 and Sentinel-2 Data: Incorporating Surface Reflectance, Through Phenological Fitting, into Regression Yield Models. *Remote Sens.* **2019**, *11*, 1768. [CrossRef]
52. Ahmad, L.; Kanth, R.H.; Parvaze, S.; Mahdi, S.S. Growing degree days to forecast crop stages. In *Experimental Agrometeorology: A Practical Manual*; Springer: Berlin, Germany, 2017; pp. 95–98.
53. University, M.S. Using Growing Degree Days to Predict Plant Stages. Available online: http://landresources.montana.edu/soilfertility/documents/PDF/pub/GDDPlantStagesMT200103AG.pdf (accessed on 15 May 2020).
54. Phan, H.; Le Toan, T.; Bouvet, A.; Nguyen, L.D.; Pham Duy, T.; Zribi, M. Mapping of rice varieties and sowing date using x-band SAR data. *Sensors* **2018**, *18*, 316. [CrossRef] [PubMed]
55. Mascolo, L.; Forino, G.; Nunziata, F.; Pugliano, G.; Migliaccio, M. A New Methodology for Rice Area Monitoring With COSMO-SkyMed HH–VV PingPong Mode SAR Data. *IEEE J. Sel. Top. Appl. Earth Obs. Remote Sens.* **2019**, *12*, 1076–1084. [CrossRef]
56. De Bernardis, C.G.; Vicente-Guijalba, F.; Martinez-Marin, T.; Lopez-Sanchez, J.M. Estimation of key dates and stages in rice crops using dual-polarization SAR time series and a particle filtering approach. *IEEE J. Sel. Top. Appl. Earth Obs. Remote Sens.* **2014**, *8*, 1008–1018. [CrossRef]
57. Tan, B.; Morisette, J.T.; Wolfe, R.E.; Gao, F.; Ederer, G.A.; Nightingale, J.; Pedelty, J.A. Vegetation phenology metrics derived from temporally smoothed and gap-filled MODIS data. In Proceedings of the IGARSS 2008—2008 IEEE International Geoscience and Remote Sensing Symposium, Boston, MA, USA, 7–11 July 2008; IEEE: Piscataway, NJ, USA, 2008; Volume 3, p. 593.
58. Boschetti, M.; Stroppiana, D.; Brivio, P.; Bocchi, S. Multi-year monitoring of rice crop phenology through time series analysis of MODIS images. *Int. J. Remote Sens.* **2009**, *30*, 4643–4662. [CrossRef]
59. Ozer, H. Sowing date and nitrogen rate effects on growth, yield and yield components of two summer rapeseed cultivars. *Eur. J. Agron.* **2003**, *19*, 453–463. [CrossRef]

remote sensing

MDPI

Article

# Deriving Wheat Crop Productivity Indicators Using Sentinel-1 Time Series

**Nikolaos-Christos Vavlas [1,2,*], Toby W. Waine [2], Jeroen Meersmans [2,3], Paul J. Burgess [2], Giacomo Fontanelli [1,4] and Goetz M. Richter [1]**

1 Sustainable Agriculture Sciences, Rothamsted Research, Harpenden AL5 2JQ, UK; g.fontanelli@ifac.cnr.it (G.F.); goetz.richter@rothamsted.ac.uk (G.M.R.)

2 School of Water, Energy and Environment, Cranfield University, Cranfield, Bedfordshire MK43 0AL, UK; t.w.waine@cranfield.ac.uk (T.W.W.); jeroen.meersmans@uliege.be (J.M.); p.burgess@cranfield.ac.uk (P.J.B.)

3 TERRA Teaching and Research Centre, Gembloux Agro-Bio Tech, University of Liège, 5030 Gembloux, Belgium

4 Institute of Applied Physics (IFAC) - National Research Council (CNR), Via Madonna del Piano, 10, 50019 Sesto Fiorentino, Italy

* Correspondence: nikolaos.vavlas@rothamsted.ac.uk

Received: 25 June 2020; Accepted: 22 July 2020; Published: 24 July 2020

**Abstract:** High-frequency Earth observation (EO) data have been shown to be effective in identifying crops and monitoring their development. The purpose of this paper is to derive quantitative indicators of crop productivity using synthetic aperture radar (SAR). This study shows that the field-specific SAR time series can be used to characterise growth and maturation periods and to estimate the performance of cereals. Winter wheat fields on the Rothamsted Research farm in Harpenden (UK) were selected for the analysis during three cropping seasons (2017 to 2019). Average SAR backscatter from Sentinel-1 satellites was extracted for each field and temporal analysis was applied to the backscatter cross-polarisation ratio (VH/VV). The calculation of the different curve parameters during the growing period involves (i) fitting of two logistic curves to the dynamics of the SAR time series, which describe timing and intensity of growth and maturation, respectively; (ii) plotting the associated first and second derivative in order to assist the determination of key stages in the crop development; and (iii) exploring the correlation matrix for the derived indicators and their predictive power for yield. The results show that the day of the year of the maximum VH/VV value was negatively correlated with yield ($r = -0.56$), and the duration of "full" vegetation was positively correlated with yield ($r = 0.61$). Significant seasonal variation in the timing of peak vegetation ($p = 0.042$), the midpoint of growth ($p = 0.037$), the duration of the growing season ($p = 0.039$) and yield ($p = 0.016$) were observed and were consistent with observations of crop phenology. Further research is required to obtain a more detailed picture of the uncertainty of the presented novel methodology, as well as its validity across a wider range of agroecosystems.

**Keywords:** Sentinel-1; crop development; remote sensing; productivity indicators; wheat; SAR; growth dynamics

---

## 1. Introduction

Time series analysis of satellite remote sensing (RS) images allows detailed crop monitoring, an enhanced understanding of the impact of agricultural practices, and can provide early warnings of low yields [1,2]. There is a wide range of RS data and associate products [3,4], and the most appropriate form will depend on the challenge, such as crop classification, vegetation modelling, and understanding water dynamics such as drought stress and irrigation [5]. Data can also support farming decisions due to the ability to gather and display environmental variables across large areas, providing spatial information [5,6]. The increasing availability and frequency of satellite data also support the application of RS in the context of crop modelling [6–8] to improve regional production estimates. However, in most cases the use of RS for crop management remains limited as the relationship between the data and crop development and growth varies with the environment [9–11], and with time and location [12].

Synthetic aperture radar (SAR) data have become freely available from the European Space Agency (ESA) with the Sentinel-1 (S1) constellation under the Copernicus program [13]. The data comprise high-temporal resolution images, which can be used for RS applications in agriculture [2,5,14], although its potential has not been fully established [5,15,16]. The main advantages of active radar sensors over optical sensors (Sentinel-2 satellite) are the ability to penetrate the clouds and independence from sun illumination. For these reasons, SAR can provide high-density time series (every 5–6 days). In addition, SAR can be used to monitor the biophysical properties of agricultural fields [17], as it is sensitive to changes in the canopy structure and biomass [14,18]. As radar is also sensitive to the water content in the observed surface [5,19,20], it can be used to quantify the moisture and structural change in fields [7,21–23].

Another important application of SAR data in agriculture is crop identification [24–26]. In many cases, the SAR time series were combined with optical images to increase the accuracy of crop identification [25,27–29]. In some instances, SAR time series were used to derive metrics (e.g., mean and variance) for identifying irrigated fields [30] or models to simulate the backscatter interaction with vegetation and soil conditions [31–36] like the water cloud model [37]. Some studies used backscatter to describe vegetation dynamics, taking into consideration L-, C-, and X-microwave bands [23,38,39], and there is the potential to use SAR to identify crop development stages in the field [24,27]. Such information could be used in combination with crop modelling using different data assimilation techniques [1,8,40]. The use of time series and logistic based methods to simulate development and growth stages of vegetation can improve the accuracy of the crop development simulations [41–43].

In Europe, S1 C-band (5.405 GHz) SAR instruments support operation in dual polarisation (VV+VH), implemented through one vertical transmit chain (V) and two parallel receive chains for H and V polarisation (horizontal and vertical, respectively) over the land [44]. Thanks to the constellation, made by two satellites orbiting in near-polar, sun-synchronous orbits, it is possible to benefit from six-days return time of satellites on the same orbit, leading to a very frequent description of vegetation growth. The VH/VV ratio, which is derived from the two polarisations, shows great potential to describe the dynamics of crop development because it is relatively insensitive to changes in soil moisture [2,20,28,45]. However, so far, there are few quantitative analyses of the dynamic changes of the SAR cross-polarisation ratio (VH/VV) in the literature.

This study is focused on the comparison of VH/VV ratio time series among wheat fields on an experimental farm (Rothamsted Research, UK) across three different years to understand and identify changes in backscatter values that can be related to crop growth and development. The assumption is that the SAR VH/VV ratio can be used to derive indicators that are related to vegetation dynamics, which then be used to improve crop management. The objective of the paper is to identify new SAR-derived indicators of wheat crop development and productivity that can provide insights for crop yield formation at field scale.

## 2. Materials and Methods

### 2.1. Study Area

In total, 18 winter wheat fields were selected for the analysis of S1 backscatter interaction with vegetation during the 2017 to 2019 agronomic seasons (nine fields were selected in 2017, five fields in 2018, and four fields in 2019 (Figure 1, Table 1). The fields belong to the Rothamsted Research experimental farm in Harpenden, UK (51°48′37.3″ N, 0°22′36.0″ W), located 35 km north of London (Figure 1). The site comprises slightly acid loamy and clayey soils with restricted drainage [46], and the main soil series at Rothamsted is the Batcombe series [47]. The annual rainfall at the site ranged from

662 mm in 2017 to 704 mm in 2018, and 616 mm in 2019. June and July 2018 were particularly dry [48] and the potential soil moisture deficit (PSMD) reached a maximum value of almost 325 mm. The mean annual air temperature was 10.6 °C, ranging from 2.2 °C in February to 19.9 °C in July (Figure 2).

**Figure 1.** Google Earth image of Rothamsted Research experimental farm (coordinate system EPSG:27700), with wheat field boundaries from 2017, 2018, and 2019 indicated using red lines (see also Table 1).

**Table 1.** Wheat field names and number of 10 × 10 m pixels per field on Rothamsted farm.

| Field Name | No. S1 Pixels in Field | Perimeter (m) | Area (ha) | Ground Data Collected |
|---|---|---|---|---|
| Great_Knott_1 | 333 | 805 | 3.37 | 2017 |
| Great_Knott_2A | 146 | 555 | 1.46 | 2017 |
| Great_Knott_3A | 112 | 457 | 1.12 | 2017 |
| Little_Knott_1 | 122 | 482 | 1.22 | 2017 |
| Osier_1_2_3 | 479 | 1046 | 4.75 | 2017 |
| Sawyers_3 | 228 | 645 | 2.27 | 2017, 2019 |
| Whitehorse_2B | 191 | 649 | 1.90 | 2017, 2019 |
| Bones_Close | 437 | 810 | 4.39 | 2018 |
| Sawyers_2 | 116 | 469 | 1.16 | 2018 |
| Sawyers_4 | 131 | 465 | 1.30 | 2017, 2018 |
| Stackyard | 276 | 739 | 2.77 | 2017, 2018 |
| West_Barnfield_1_2 | 365 | 910 | 3.62 | 2018, 2019 |
| Drapers | 399 | 839 | 3.93 | 2019 |

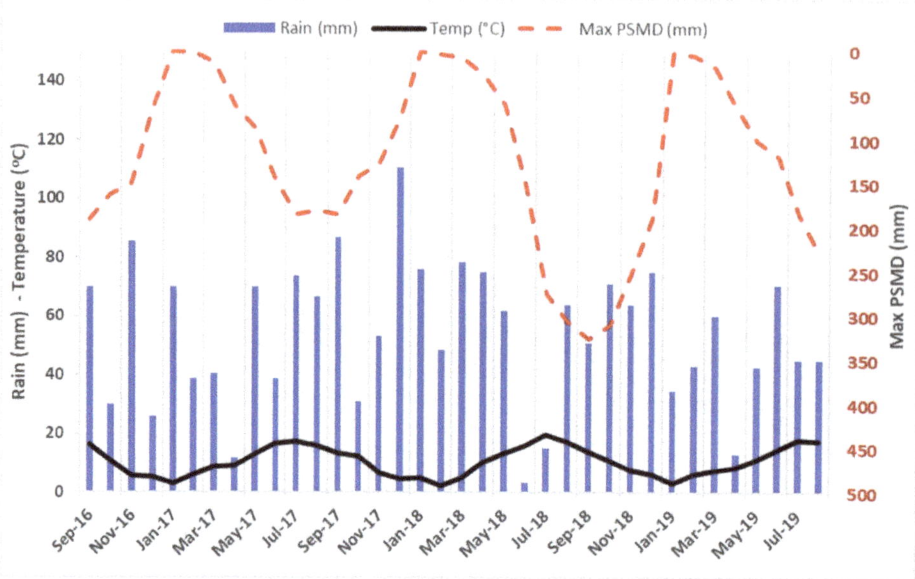


**Figure 2.** Total monthly rainfall (mm) and mean air temperatures (°C) on the primary axis (left) and maximum potential soil moisture deficit (PSMD; mm) on the secondary axis (right) at Rothamsted farm from September 2016 to August 2019.

### 2.2. Site Ground Measurements of Wheat Development and Yield

In situ observations of phenology were made weekly during the growing season to monitor the wheat development and growth stages. In each field, measurements of grain yield were recorded by the harvest machinery. These data were used as reference points in the methodology to describe the response of the VH/VV ratio to the dynamics of the winter wheat development and growth, and explore the different parameters of the segments with respect to their contribution to yield.

### 2.3. SAR Data to Obtain VH/VV Time Series in Each Wheat Field

The Sentinel-1 SAR data acquired for the Rothamsted Research farm, over the three-year study period were derived from orbit 132 (ascending) with a temporal resolution of six days. One of the reasons for selecting a single orbit is the orientation of the view of the satellite, which plays a significant role in the direction of the emitted beam from the SAR antenna [49,50]. Using the same orbit also had the advantage of a similar incidence angle (range of 5°) for adjacent fields on even terrain [49]. The ascending orbit 132, instead of descending 81, was also used to avoid early morning measurements when dew may become a confounding factor [14,45]. Temporal profiles of the S1 VH/VV backscatter ratio were plotted on each field to define the curves' shapes and amplitude, which were related to ground observations of wheat at key growth stages, assisting in the definition of periods defined by the calculated parameters. Figure 3 displays the flow diagram with all the needed steps of the analysis.

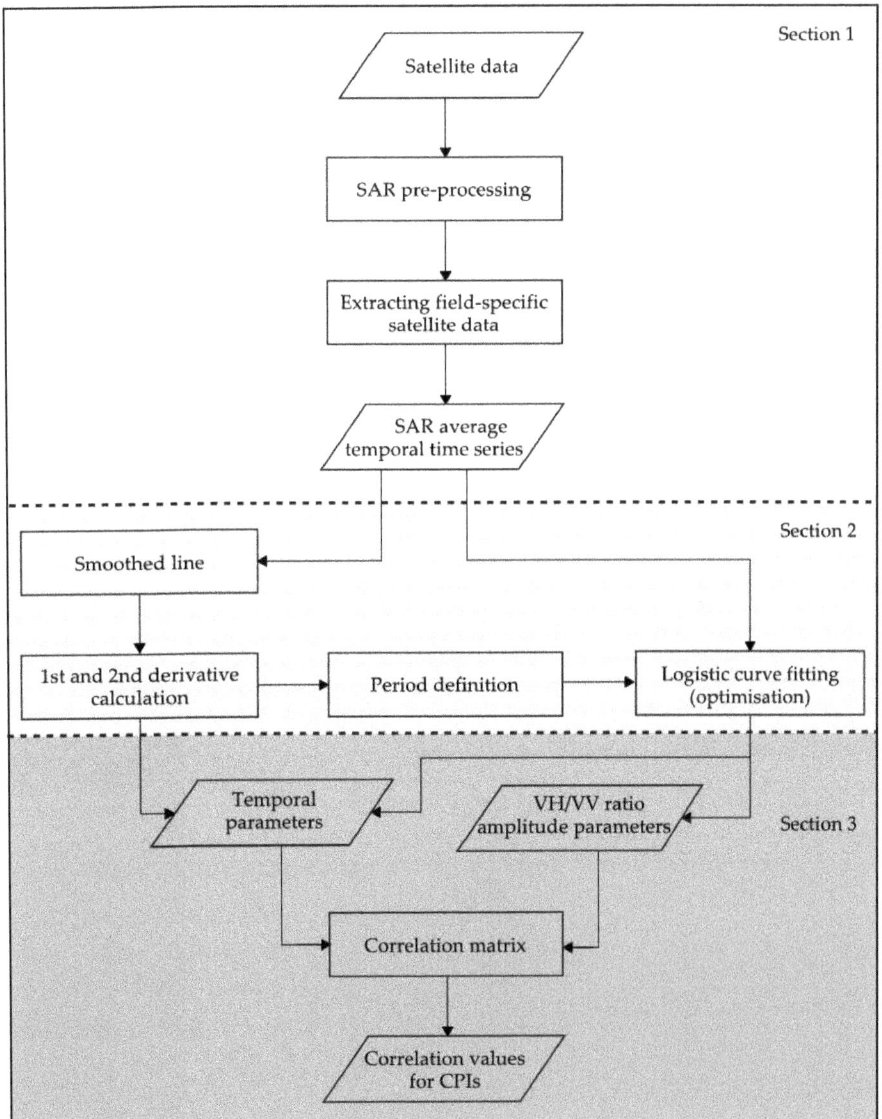


**Figure 3.** Flow diagram explaining the steps for the temporal analysis of synthetic aperture radar (SAR) time series, including the data and process.

#### 2.3.1. SAR Preprocessing for Field Specific VH/VV Time Series

The SAR data were derived from the Level-1 Ground Range Detected Interferometric Wide Swath (GRD-IW) product from Sentinel-1 satellite. The Sentinel-1 toolbox was used to preprocess the data, including border and thermal noise removal, slice assembly, radiometric calibration, terrain flattening, speckle filtering using refined Lee filter with a $3 \times 3$ window, and terrain correction to produce gamma naught ($\gamma°$) data in VH and VV polarisations. Lastly, each scene was clipped and the $\gamma°$ VH/VV ratio was calculated. Based on the scale selected for the analysis, a mean value for the whole field (Table 1) was used to minimise the speckle effect, and the area of each field was buffered to minimise the influence of surrounding fields. All of the calculations were conducted using untransformed data in a linear form. In the presentation of the results, the initial graphs were developed to allow the reading of both linear and decibel (dB) values, and the final results were presented on decibel axes. In addition, temporal filtering (in the form of smoothing, Savitzky–Golay filter) was used to reduce localised weather patterns, such as heavy rainfall, ice or snow, that can temporarily lower the SAR backscatter. This step avoided the use of additional data (e.g., rainfall used to clean the data) for automation.

2.3.2. Temporal Analysis and Wheat Development Definition

The curve of the VH/VV ratio time series was used to produce indicators that could describe wheat crop growth and development across the season. Smoothing approaches for time series of SAR [51,52] and optical data [42,53] have been used in various studies to reduce noise or fill the gaps of the datasets. Here, the Savitzky–Golay filter was applied over the whole period with a moving window of two months and using a second-order polynomial function [54]. Then the associated first and second derivatives were calculated to define additional wheat growth stages, linked to the SAR temporal characteristics. Key periods of these three curves were identified, as shown as vertical lines in Figure 4, and matched with crop development observed in the field.

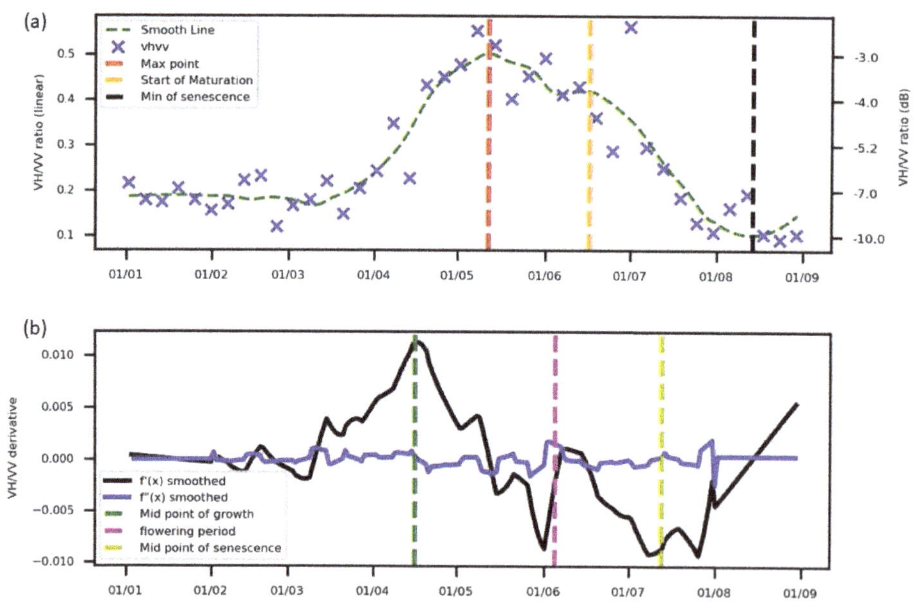


**Figure 4.** (**a**) Mean VH/VV ratio from Sentinel-1 orbit 132 of wheat field (for the Swayers_4 field in 2018) together with a smoothed curve of the mean (with the date of the maximum point indicated) and (**b**) first (f′x) and second (f″x) derivative (bottom) of the smoothed field-averaged line. The date of the midpoint of the growth period, and the date of the start, midpoint, and minimum value of the maturation period are indicated as vertical dotted lines. The secondary y axis in the first graph displays the VH/VV ratio on a decibel scale for reference.

The next stage was to further parameterise and automate the extraction of these dates and test if they were linked to wheat development. The smoothed VH/VV curve was divided into two periods: (i) growth and (ii) maturation and senescence to address an observed change in maximum values in the middle of the season. The splitting of the analysis into two periods was also used by Che et al. (2014) [41], who analysed the temporal evolution of the leaf area index (LAI) of vegetation during development and senescence in Shandong Province, China. The breakpoint of two curves occurs during the flowering period (anthesis) when the wheat has reached its full height. The calculation of the breakpoint specifies the start for the VH/VV ratio stabilisation period close to the end of spring.

### 2.3.3. Logistic Curve Fitting

In order to quantify the characteristics of the temporal curve of VH/VV ratio for winter wheat, a logistic model was used during both the growth and maturation phases. Two separate curves, rather than a single double logistic curve, were chosen because the starting value of the VH/VV ratio of the maturation phase (which coincided with flowering) tended to be lower than the finishing point of the growth stage (which coincided with booting) (Figure 5). In addition, the baseline at the beginning of the season can differ from the end. The two sigmoid logistic curves describing the VH/VV ratio in relation to time $t$ for each stage was characterised by four parameters (Equation (1)):

$$f(t) = base + \frac{Max}{1 + e^{-b(t-t_0)}} \tag{1}$$

where *base* is the minimum VH/VV ratio value, *Max* is the maximum VH/VV ratio value, $b$ is the steepness of the logistic curve, and $t_o$ is the x-value of the sigmoid inflexion point. The calculated values are sensitive to a clear definition of the period represented by each sigmoid curve (Figure 5). The growth period curve was defined from the beginning of the season (Parameter 1 in Figure 5) up to the time of maximum value of the smoothed curve (Parameter 5). The maturation period curve was defined as the period after flowering to the minimum value in the maturation period of the temporally smoothed curve.

**Table 2.** Definitions of the VH/VV ratio curve parameters and the anticipated associated crop development stage (Biologische Bundesanstalt, Bundessortenamt und CHemische Industrie - BBCH or Zadoks scale).

| No. | Symbol | Parameter Name | Definition | Derived From | Associated Crop Development Stage |
|---|---|---|---|---|---|
| 1 | G_base | Baseline value for the growth stage | VH/VV ratio at the beginning of the season | Logistic curve | Tillering (GS20) |
| 2 | G_steep | Steepness of logistic curve for growth period | Rate of coefficient in Equation (1) (b G) | Logistic curve | Stem elongation |
| 3 | G_midP | Time of midpoint of growth period ($t_0$, G) | DOY when the midpoint of the logistic curve occurs in the growth period | Logistic curve | Stem elongation |
| 4 | G_max | Max value for growth stage | Maximum VH/VV ratio value for the full season | Logistic curve | End of stem elongation (GS39) and booting (GS49) |
| 5 | TZmax | Time of maximum point | DOY of maximum smoothed value of VH/VV | Smoothed curve | Time of booting, flag leaf unrolled |
| 6 | S_max | Value at the start of grain filling | Period of backscatter stabilisation | Logistic curve | Post anthesis: start of grain filling (GS71) |
| 7 | S_steep | Steepness of logistic curve for maturation period | Rate of coefficient in Equation (1) (b S) | Logistic curve | Maturation rate |
| 8 | S_midP | Time of midpoint of maturation ($t_0$, S) | DOY when the midpoint of the logistic curve occurs in the maturation period | Logistic curve | Ripening (GS 85–89) |
| 9 | S_base | Baseline value at the end of the season | Background value of the VH/VV ratio | Logistic curve | After harvest<br>Period with soil exposed |
| 10 | Duration | Duration of "full" vegetation to maturation | Time difference between midpoints (3, 8) | Combination | Period of most of the photosynthate accumulation and translocation |
| 11 | D_max | Structure change (Inflorescence) | VH/VV ratio value differences between booting and grain filling periods (4–6) | Combination | Backscatter change during the period when the ear emerges |
| 12 | D_base | Tillering backscatter | VH/VV ratio value differences between tillering and bare soil (1–9) | Combination | Tillering with reduced impact of soil |

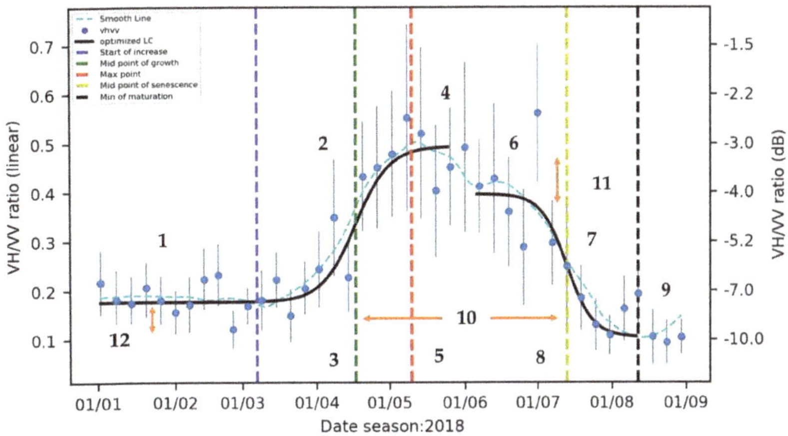


**Figure 5.** Mean SAR VH/VV ratio for a wheat field (Swayers_4 in 2018) with 228 pixels (1.3 ha), between January and September 2018, showing the standard deviation for each date (n = 41), and displaying 12 parameters as calculated using a smoothed line and two logistic curves (see Table 2). All calculations were completed using a linear scale and the secondary y-axis displays the decibel scale for reference.

Combining parameters extracted from the two logistic curves allowed us to derive three additional vegetation development-related parameters (10, 11, and 12 in Figure 5). The period between the two midpoints of the logistic curves (Parameter 10 in Figure 5) was defined as the duration of "full" vegetation, which starts with stem elongation and ends with the reduction of the green canopy and translocation of carbohydrates into the grain during maturation. The difference in baselines is related to the backscatter from the biomass of the crop during tillering. The summary of the extracted parameters and their definitions as well as the associated crop development stage is given in Table 2.

### 2.3.4. Parameter Optimisation

The calculation of the logistic curve fitting parameters in Table 2 was based on a weighted least squares (WLS) estimator, which considers the uncertainty of each point by minimising the sum of the squared difference divided by the respective standard deviation (σ) in each point (Equation (2)). The incorporation of the standard deviation in the estimator was used to minimise the outlier effects. The VH/VV ratio points were selected as the observation values in each defined period.

$$min \sum_{i} \left( \frac{obs_i - y_i}{\sigma_i} \right)^2 \quad (2)$$

### 2.3.5. Automatic Curve Extraction and Correlation Analysis

By scripting the SAR processing using Python, it was possible to automatically derive a smoothed VH/VV ratio curve and two logistic curves. Then the 12 parameters were derived for each field site across three years. A Pearson correlation coefficient (r) was calculated for each pair in a matrix. This approach allows the illustration of the interactions between the parameters and the direct effect on final crop yield. A significant test of the r was performed and highlighted in the plots.

## 3. Results

### *3.1. Annual Analysis of VH/VV Ratio Curve Parameters, 2017 to 2019*

Two logistic curves for the nine fields selected in 2017 (Figure 6), the five fields selected in 2018 (Figure 7), and the four fields selected in 2019 (Figure 8), were plotted. The smoothed time series were used to identify the start and end of (i) the growth (start of the year to red line) and (ii) the maturation periods (purple to min senescence value). Then the midpoints were calculated by fitting the logistic curves in these periods, creating the time parameters shown in Figures 6–8.

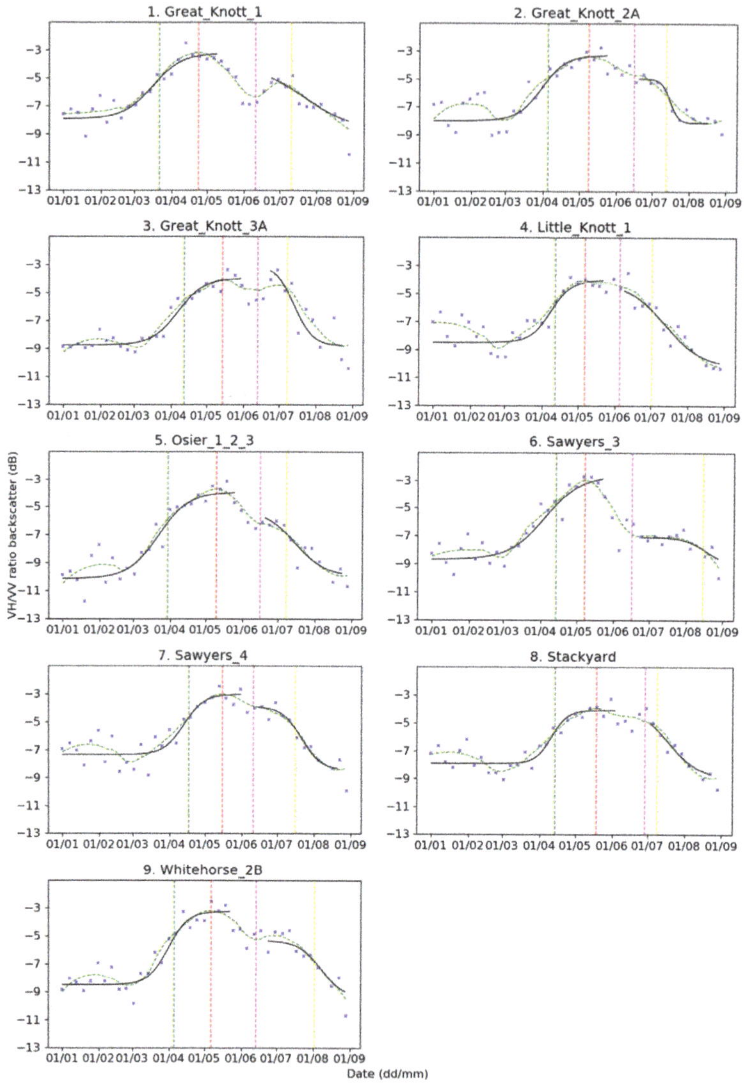


**Figure 6.** VH/VV ratio time series (blue dots) with the smoothed curve (dotted green line) and two logistic curves fitted (black line) for nine fields (subplots 1–9) during 2017, orbit 132. The vertical dotted lines indicate the dates of G_midP (green), TZmax (red), inflorescence (purple), and S_midP (yellow), as determined from the VH/VV backscatter time series analysis.

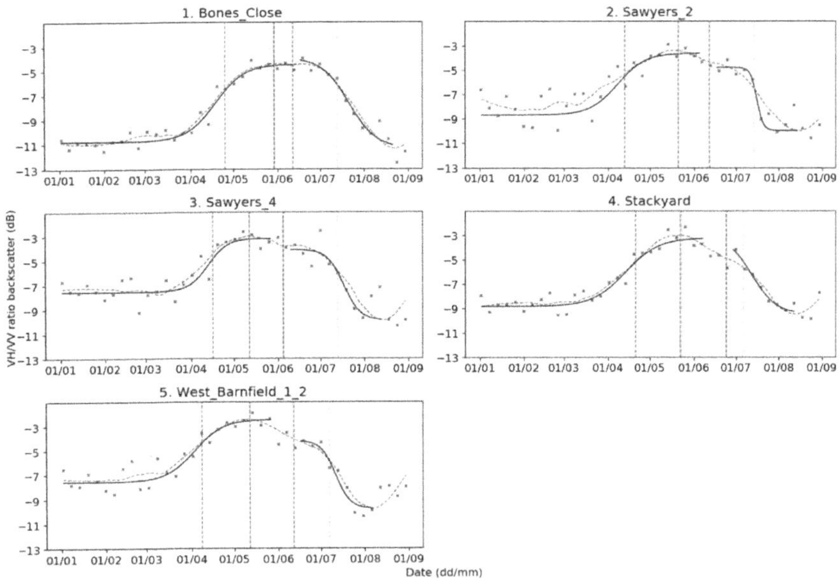


**Figure 7.** VH/VV ratio time series (blue dots) with the smoothed curve (green dashed line) and two logistic curves fitted (black line) for five fields (subplots 1–5) during 2018, orbit 132. The vertical dotted lines indicate the dates of G_midP (green), TZmax (red), inflorescence (purple), and S_midP (yellow), as determined from the VH/VV backscatter time series analysis.

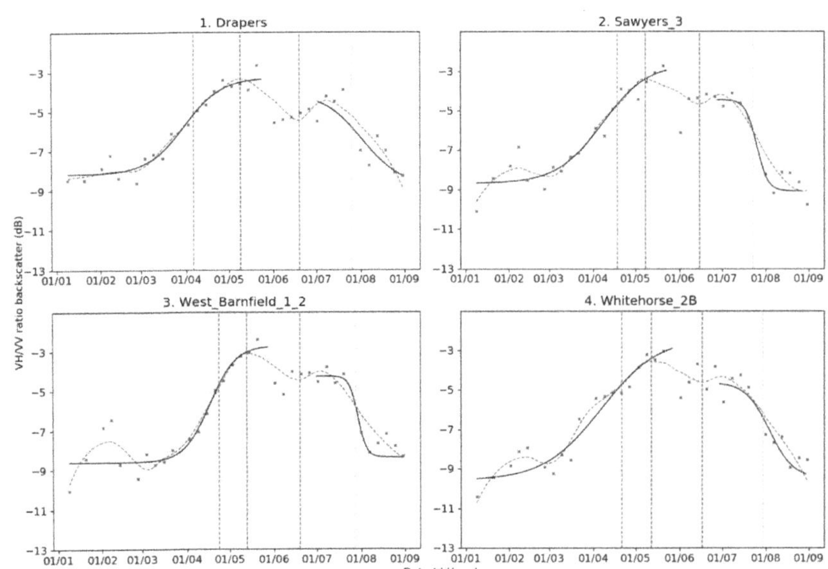


**Figure 8.** VH/VV ratio time series (blue dots) with the smoothed curve (green dashed line) and two logistic curves fitted (black line) for four fields (subplots 1–4) in 2019, orbit 132. The vertical dotted lines indicate the dates of G_midP (green), TZmax (red), inflorescence (purple), and S_midP (yellow), as determined from the VH/VV backscatter time series analysis.

The timing of the mean maximum VH/VV ratio ($TZ_{max}$) was relatively consistent varying from day-of-year (DOY) 129 in 2017, to DOY 140 in 2018, and DOY 131 in 2019 (Table 3). The mean values (Table 4) of the three parameters of the logistic curve for the growth period were also relatively consistent from year to year: G_base (range 0.13–0.15), G_steep (range: 0.08–0.11), and G_midP (range: DOY 99 to 109). There was greater variation in G_max, which ranged from a mean of 0.31 in 2017 to 0.40 in 2019. The mean values of two of the parameters of the logistic curve for the maturation period were also relatively similar across the three years: S_base (range: 0.10–0.13), and S_midP (range: DOY 199 to DOY 208). By contrast, the mean values of S_max (range: 0.23–0.30) and S_steep (range: –0.12 to –0.21) showed greater interannual variation (Table 4).

The root mean square error (RMSE) of the growth (G) and maturation (S) logistic curves are shown in the last two columns of Table 3 for the seasons 2017, 2018, and 2019. Statistical comparison between the parameters of different years showed that TZmax ($p = 0.042$), the G_midP ($p = 0.037$), S_base ($p = 0.030$) are significantly different at 95% confidence using ANOVA or the Kruskal–Wallis H-test if there was a non-normal distribution. There was also a significant interannual variation in the Duration ($p = 0.039$) and Yield ($p = 0.016$).

### 3.2. Relationship Between SAR-Derived Parameters and Yield

The interactions between each of the parameters and yield were further displayed using a correlation matrix (Figure 9). There were various significant correlations between parameters. As would be expected, the inflorescence indicator (D_max = G_max – S_max) was correlated with the two parameters used in the calculation (G_max and S_max), with r values of 0.76 and −0.75, respectively. At the same time, significant correlations were found between D_max and G_steep (r = −0.54), S_midP (r = 0.68), and Duration (r = 0.49). By contrast, D_base displayed no significant correlations with the other parameters, with the exception of G_base (r = 0.65) that was used for its calculation. The definition of D_base is G_base minus S_base, where S_base represents the VH/VV ratio value on bare soil (that should remain relatively consistent) and the value of G_base is determined by the combination of soil and low vegetation during the tillering period.

Starting from the first row in Figure 9, the baseline of growth (G_base) shows a significant positive correlation (r = 0.58) with the baseline of maturation (S_base). This is because G_base depends on the shift of the VH/VV ratio during the initial bare/sparse/small vegetation period and the soil background (S_base). In the second row, the slope of the midpoint of growth (G_steep) has a negative correlation with the VH/VV ratio at G_max value (r = −0.68). In the third row, the midpoint of growth (G_midP) is highly correlated with the timing of the max (TZmax) smoothed VH/VV ratio value (r = 0.69), highlighting the interaction of stem elongation and booting stage. G_midP was also correlated with the calculated Duration (r = −0.68) and the maturation midpoint (S_midP; r = 0.78). This is because the Duration value is defined by the difference between these two time indicators. The value of TZmax, the timing of the maximum value of VH/VV, was positively correlated with S_max (r = 0.50), but negatively correlated with low Duration values (r = −0.64).

During the maturation period, S_max was negatively correlated with the midpoint S_midP (r = −0.72) and Duration (r = −0.80). The only correlation shown by the steepness of the curve in this period was with the yield. The value of S_midP was positively correlated with Duration (r = 0.76) and D_max (r = 0.68). The value of S_base was not significantly correlated with any of the indicators except G-base.

**Table 3.** Values of derived parameters (linear scale), yield data, and the root mean standard error of the growth ($RMSE_G$) and maturation ($RMSE_S$) stages, for each of the winter wheat fields in 2017, 2018, and 2019. Note: day-of-year, DOY.

| Field | Season | $RMSE_G$ | $RMSE_S$ | G_base | G_steep | G_max | S_max | S_steep | S_base | G_midP | S_midP | TZmax | Duration | D_max | D_base | Yield |
|---|---|---|---|---|---|---|---|---|---|---|---|---|---|---|---|---|
| | | | | | | | | | | DOY | DOY | DOY | days | | | t/ha |
| Great_Knott_1 | 2017 | 0.051 | 0.004 | 0.16 | 0.077 | 0.36 | 0.25 | −0.05 | 0.14 | 85 | 194 | 114 | 109 | 0.11 | 0.020 | 8.67 |
| Great_Knott_2A | 2017 | 0.092 | 0.022 | 0.16 | 0.115 | 0.28 | 0.17 | −0.29 | 0.15 | 93 | 196 | 129 | 103 | 0.12 | 0.010 | 6.43 |
| Great_Knott_3A | 2017 | 0.020 | 0.047 | 0.13 | 0.11 | 0.24 | 0.37 | −0.13 | 0.13 | 99 | 190 | 135 | 91 | −0.13 | 0.000 | 9.26 |
| Little_Knott_1 | 2017 | 0.090 | 0.033 | 0.14 | 0.108 | 0.31 | 0.28 | −0.07 | 0.10 | 106 | 184 | 127 | 78 | 0.02 | 0.040 | 6.17 |
| Osier_1_2_3 | 2017 | 0.054 | 0.004 | 0.10 | 0.072 | 0.34 | 0.20 | −0.08 | 0.10 | 93 | 190 | 130 | 97 | 0.14 | 0.000 | 6.63 |
| Sawyers_3 | 2017 | 0.070 | 0.012 | 0.14 | 0.066 | 0.45 | 0.07 | −0.11 | 0.12 | 108 | 228 | 128 | 120 | 0.38 | 0.020 | 6.36 |
| Sawyers_4 | 2017 | 0.047 | 0.004 | 0.18 | 0.115 | 0.33 | 0.26 | −0.12 | 0.14 | 108 | 199 | 136 | 90 | 0.07 | 0.040 | 6.85 |
| Stackyard | 2017 | 0.037 | 0.002 | 0.16 | 0.159 | 0.22 | 0.30 | −0.08 | 0.13 | 103 | 191 | 139 | 88 | −0.08 | 0.030 | 5.04 |
| Whitehorse_2B | 2017 | 0.061 | 0.035 | 0.14 | 0.143 | 0.30 | 0.18 | −0.11 | 0.11 | 93 | 215 | 127 | 122 | 0.12 | 0.030 | 9.59 |
| Bones_Close | 2018 | 0.021 | 0.017 | 0.08 | 0.098 | 0.28 | 0.33 | −0.11 | 0.07 | 117 | 194 | 150 | 78 | −0.05 | 0.010 | 4.09 |
| Sawyers_2 | 2018 | 0.118 | 0.035 | 0.13 | 0.102 | 0.31 | 0.23 | −0.45 | 0.10 | 104 | 196 | 141 | 91 | 0.08 | 0.030 | 6.58 |
| Sawyers_4 | 2018 | 0.051 | 0.069 | 0.18 | 0.12 | 0.40 | 0.29 | −0.18 | 0.10 | 111 | 195 | 132 | 84 | 0.11 | 0.080 | 5.49 |
| Stackyard | 2018 | 0.044 | 0.009 | 0.13 | 0.071 | 0.49 | 0.39 | −0.12 | 0.11 | 121 | 187 | 143 | 66 | 0.10 | 0.020 | 4.81 |
| W_Barnfield_1_2 | 2018 | 0.057 | 0.012 | 0.17 | 0.094 | 0.43 | 0.28 | −0.20 | 0.11 | 101 | 189 | 132 | 88 | 0.15 | 0.060 | 7.90 |
| Drapers | 2019 | 0.018 | 0.047 | 0.15 | 0.079 | 0.32 | 0.26 | −0.08 | 0.13 | 97 | 207 | 129 | 110 | 0.06 | 0.018 | 9.65 |
| Sawyers_3 | 2019 | 0.007 | 0.026 | 0.14 | 0.065 | 0.41 | 0.24 | −0.26 | 0.12 | 108 | 205 | 128 | 97 | 0.17 | 0.012 | 7.99 |
| W_Barnfield_1_2 | 2019 | 0.030 | 0.036 | 0.14 | 0.114 | 0.40 | 0.23 | −0.34 | 0.15 | 113 | 209 | 133 | 96 | 0.17 | −0.008 | 8.28 |
| Whitehorse_2B | 2019 | 0.021 | 0.029 | 0.11 | 0.053 | 0.47 | 0.23 | −0.12 | 0.11 | 112 | 211 | 132 | 99 | 0.23 | −0.002 | 10.17 |

**Table 4.** Mean and standard deviation values of 12 derived parameters (linear scale) and the wheat yield for each season of the analysis.

| Season | | G_base | G_steep | G_max | S_max | S_steep | S_base | G_midP | S_midP | TZmax | Duration | D_max | D_base | Yield |
|---|---|---|---|---|---|---|---|---|---|---|---|---|---|---|
| | | | | | | | | DOY | DOY | DOY | days | | | t/ha |
| 2017 | Mean | 0.15 | 0.11 | 0.31 | 0.23 | −0.12 | 0.12 | 99 | 199 | 129 | 100 | 0.08 | 0.02 | 7.22 |
| | Std dev | 0.02 | 0.03 | 0.07 | 0.09 | 0.07 | 0.02 | 8 | 14 | 7 | 15 | 0.14 | 0.02 | 1.56 |
| 2018 | Mean | 0.14 | 0.10 | 0.38 | 0.30 | −0.21 | 0.10 | 111 | 192 | 140 | 81 | 0.08 | 0.04 | 5.77 |
| | Std dev | 0.04 | 0.02 | 0.09 | 0.06 | 0.14 | 0.02 | 8 | 4 | 8 | 10 | 0.08 | 0.03 | 1.50 |
| 2019 | Mean | 0.13 | 0.08 | 0.40 | 0.24 | −0.20 | 0.13 | 107 | 208 | 131 | 101 | 0.16 | 0.01 | 9.02 |
| | Std dev | 0.02 | 0.03 | 0.06 | 0.01 | 0.12 | 0.01 | 7 | 3 | 2 | 7 | 0.07 | 0.01 | 1.00 |

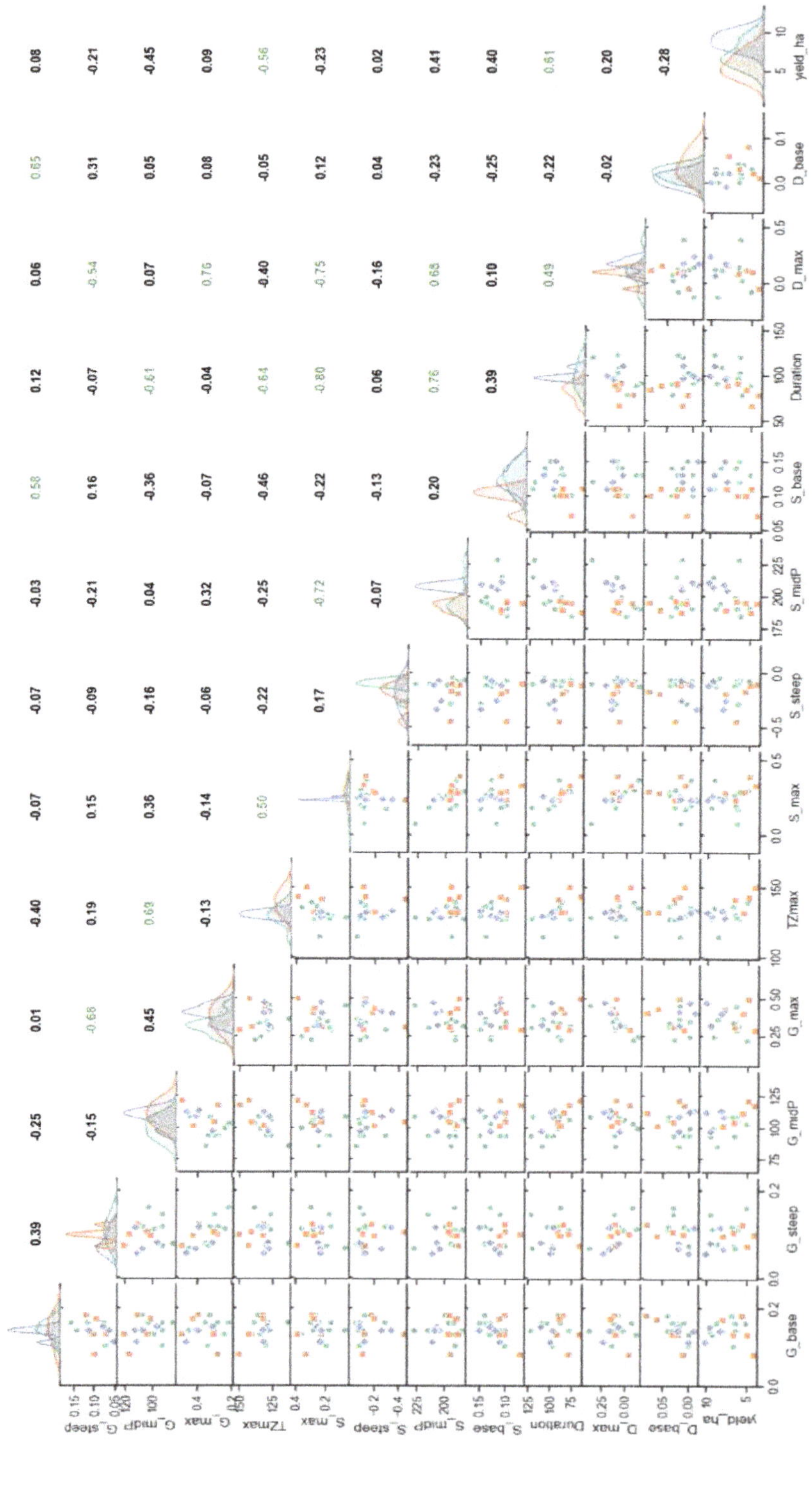


**Figure 9.** Correlation matrix with 2017 (green circles), 2018 (red squares), and 2019 (blue diamonds), using the 18 fields. R values are colour-coded green when there was a 95% level of significance for a two-tailed test.

As illustrated by the matrix (Figure 9), multiple correlations exist among the individual parameters and between individual parameters and yield (last column). The results suggest that the two parameters, G_midP and TZmax (which are correlated), had the strongest association with grain yield, i.e., r = −0.45 and r = −0.56, respectively, after Duration (r = 0.61). The negative correlation suggests that earlier growth, i.e., lower values of G_midP and TZmax were associated with higher end-of-season grain yields. Figure 10 displays the two most significant parameters that are sensitive to yield.

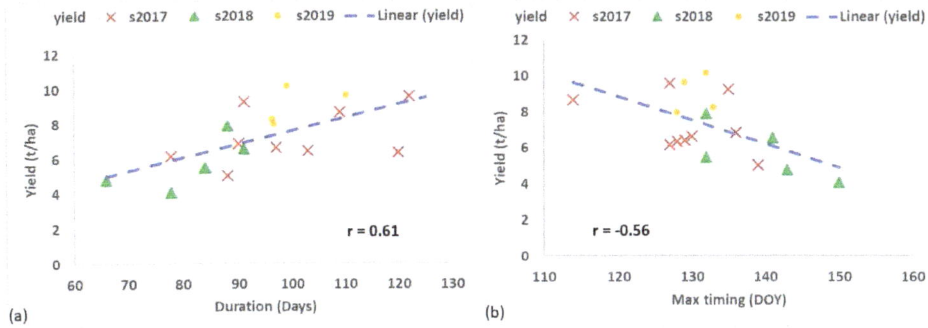


**Figure 10.** Scatter plot showing the relationship between Yield and (**a**) Duration and (**b**) the timing of the maximum value of VH/VV (Tmax). The r value and best fit linear line for each relationship, across the three seasons, is shown.

## 4. Discussion

Using the example of winter wheat over three growing seasons, it was possible to derive field-specific annual coefficients of a generalised two-component empirical model of crop productivity from the SAR VH/VV ratio. Due to the ubiquitous and high-frequency availability of S1 data, this approach could offer new opportunities for monitoring and predicting productivity in areas affected by frequent cloud cover. The characterisation of VH/VV response over a season used three stages: (i) the creation of a smooth VH/VV time series for each field; (ii) the definition of the timing of the maximum value using temporal characteristics of the time series, growth, and maturation period using first and second derivatives; and (iii) the creation of two logistic curves using the observation in the predefined periods of growth and maturation [41]. The results are discussed as follows: first, in terms of the robustness of our mathematical method over three seasons to consider multiple factors of variation; second, the gain of biophysical understanding using this method; and finally, the agronomic potential of our results.

### 4.1. Technical and Statistical Process, the Goodness of Fit, Robustness, and Uncertainty

Since 2016, SAR data from Sentinel-1 have become twice as frequent as data used in previous studies [2,20], which improves the robustness of the possible analysis. However, to reduce the variability of SAR data we chose to use imagery from only the ascending satellite passing as images from the descending orbit were often affected by morning dew. Similar observations have been made in other studies [14,45], indicating greater certainty when using the ascending rather the descending phase of the satellite for predicting crop state variables in this area. Even so, the results demonstrate substantial variation of SAR backscatter, which can be due to differences in topography affecting the incidence angle or changes in canopy architecture [49,50].

As pointed out in an earlier review [5], SAR variability is also connected to diurnal variation in plant–soil–water relations, which can affect the erectness of the crop canopy and the dielectric properties of the system. To reduce the effect of such fluctuations in the SAR time series, our approach used a step of smoothing the data, which resulted in a set of relatively consistent dynamic curves for winter cereals [20], and limited the extreme values of the VH/VV (Figure 6). The subsequent logistic fitting approach enabled the calculation of a specific backscatter value and time-specific parameters from the VH/VV ratio data (Figure 5). Accounting for the standard deviation during the fitting of the logistic curves made it possible to give lower weight to data with high uncertainty and improved the precision of the derived curves. The average RMSE values of the logistic curve fitting in growth (G) and senescence/maturation (S) periods were 0.05 and 0.07, respectively.

The creation of the fitted curves allowed the derivation of 12 parameters that can be related to the development, growth, and yield of winter wheat. Two temporal parameters from the analysis (TZmax and Duration) were significantly correlated with yield (Figure 9). Further analysis could focus on the opportunity of using a combination of parameters or a combination of VV and VH/VV values. The approach was effective in each of three years that were characterised by different environmental conditions and yields. In 2018, there was minimal rainfall in June and July, and the grain yields obtained in 2018 (4.09–7.01 t $ha^{-1}$) were substantially lower than in 2017 (5.04–9.59 t $ha^{-1}$) and 2019 (7.99–10.17 t $ha^{-1}$).

### 4.2. Environmental and Biophysical Understanding

The sensitivity of SAR backscatter to surface changes in the structure and the soil and crop water content means that SAR can be an effective tool to monitor crop development [5,24] starting from early growth, to the increase in the water volume of the crop through stem elongation, and its decrease during maturity. In this paper, we assigned two separate logistic curves during the cropping season to examine whether the resulting parameters were correlated with observations in the field. A specific feature of creating two curves is the derivation of the D_max value, which would not be captured by assuming a single curve. Our interpretation is that the change in the VH/VV value at flowering is a real effect, which equates to a change in the structure of the crop (Figure 6) during the inflorescence period, as mentioned by other authors [2,14,18]. Even so, the value of D_max was variable, ranging from –0.13 to 0.38 for an individual field and year (Table 3). Additional research, including perhaps data focused on structure, is required to further understand this variability.

The correlation analysis (Figure 9) was able to identify two time-based parameters, one derived from the growth period (G_midP) and one derived from the smoothed VH/VV curve (TZmax), which were negatively associated with the final grain yield ($r = -0.45$ and $-0.56$, respectively). This means that a delayed occurrence of the midpoint of the growth curve and the timing of the maximum VH/VV ratio was associated with lower grain yields. The late or slow growth of the crop canopy is likely to result in a reduced capacity of the crop to intercept solar radiation and use photosynthesis to produce assimilates required for final yield. For example, in 2018, the late occurrence of G_midP and TZmax was associated with an unusually dry period in June and July 2018 (Figure 2), which reduced canopy growth and photosynthesis and resulted in substantially lower grain yields than in the other two years.

The analysis also showed that the duration of the period of "full" vegetation (Duration) was positively related to grain yield ($r = 0.61$). This can be explained by the yield benefitting from a longer period of photosynthesis and a longer grain-filling period when assimilates are transferred from green tissue to the grain [55,56]. These results suggest that using the VH/VV ratio to identify the timing of specific crop periods, which are related to the structural change of the wheat canopy [2,14], may be a more useful determinant of yield than the actual values of the VH/VV ratio, per se.

### 4.3. Opportunities for Agronomic Management and Modelling

The analysis raises two questions for crop management. First, what can be learnt about crop dynamics from monitoring satellite-based SAR data, dynamics, and amplitude? Second, is there potential for a farmer or agronomist to use SAR-derived indicators for spatially varying crop management and inputs between fields? The growth and development of a wheat crop depends on the interaction between the weather (e.g., solar radiation, rainfall, and temperature), crop genetics, field characteristics (e.g., aspect and soil type), and field management (e.g., previous cropping history and the timing, type of cultivation, and drilling). The results from the SAR data analysis indicates that it is possible to identify seasonal differences in the timing of crop growth, and this may be useful in identifying the most appropriate schedule of fertilizer and agrochemical application between fields. The analysis also suggests that it may be possible to improve the prediction of the eventual crop yield from an analysis of SAR characteristics.

A potential limitation with the method is that some of the parameters are determined retrospectively from a full temporal dataset, for example, the calculation of TZmax, the timing of the maximum value of VV/VH, requires data that occur after TZmax. Likewise, the value of S_steep, which occurs during the maturation of the crop, requires data that occur after that point. Developing a really effective predictive tool that can be used to modify field management instantaneously will require parameters that can be estimated in real-time. Further work is needed to improve the estimation of parameters, as well as the application of the methodology on other farms to validate the robustness of the method. The correlation of the productivity indicators with field measurements also has the potential to assist in better model simulations of crop development and growth. At the same time, temporal characteristics, as well as connection with the biophysical properties of the field, could be examined to improve the calibration and its incorporation into a data assimilation framework to estimate more efficiently the field production [1,7,8].

## 5. Conclusions

This paper describes a new methodology to derive wheat productivity indicators from Sentinel-1 VH/VV time series. The temporal curves, their first and second derivatives, and logistic curve fitting for growth and maturation periods were used to define 12 phenomorphological parameters. These provide information about the growth, development, and yield of wheat crop at field scale, offering an alternative approach to ground survey or yield estimation.

The analysis of the VH/VV ratio time series and the correlation matrix indicates that the time-based parameters appear to be related to biophysical changes in the field. In particular, the time period of "full" vegetation (Duration) was positively correlated with yield ($r = 0.61$), and a delay in the timing of the maximum VH/VV value (TZmax) was negatively correlated with yield ($r = -0.56$).

Automation of the SAR image analysis was possible, as the only inputs required were backscatter data from VH and VV polarisations. Future work will explore the use of this method to estimate commercial wheat production in other farming landscapes in the UK. It could also be adapted to monitor growth and development, as well as yield prediction for other arable crops, potentially allowing the remote quantitative assessment of environment and management impacts.

**Author Contributions:** Conceptualization, N.-C.V., T.W.W., J.M., and G.M.R.; methodology, N.-C.V.; software, N.-C.V.; formal analysis, N.-C.V.; data curation, N.-C.V. and G.F.; writing—original draft preparation, N.-C.V.; writing—review and editing, T.W.W., J.M., P.J.B., G.F., and G.M.R.; visualization, N.-C.V.; supervision, T.W.W., J.M., P.J.B., and G.M.R.; project administration, T.W.W. and G.M.R.; funding acquisition, T.W.W. and G.M.R. All authors have read and agreed to the published version of the manuscript.

**Funding:** Lead Author N.V. was funded by a studentship provided by the Soil Agricultural Research Innovation Accelerator (AgRIA) and jointly funded PhD programme between Cranfield University and Rothamsted Research. Contributions by coauthors G.F. and G.M.R. were funded by the NERC/UKRI Innovate-UK Project (NE/P008852/1) and G.M.R. acknowledges his support by the Natural Environment Research Council (NERC) and the Biotechnology and Biological Sciences Research Council (BBSRC) under the research programme NE/N018125/1 LTS-M ASSIST—Achieving Sustainable Agricultural Systems.

**Acknowledgments:** We thank Iain Cameron (Environment Systems Ltd.; Aberystwyth) for providing preprocessed satellite images and Stephen Hayward for setting up the server-based database within the Innovate-UK project mentioned above. N.C.V. is grateful to the Rothamsted statistic team for the insightful discussions.

**Conflicts of Interest:** The authors declare no conflict of interest.

## References

1. Jin, X.; Kumar, L.; Li, Z.; Feng, H.; Xu, X.; Yang, G.; Wang, J. A review of data assimilation of remote sensing and crop models. *Eur. J. Agron.* **2018**, *92*, 141–152. [CrossRef]
2. Veloso, A.; Mermoz, S.; Bouvet, A.; Le Toan, T.; Planells, M.; Dejoux, J.-F.; Ceschia, E. Understanding the temporal behavior of crops using Sentinel-1 and Sentinel-2-like data for agricultural applications. *Remote Sens. Environ.* **2017**, *199*, 415–426. [CrossRef]
3. De Sy, V.; Herold, M.; Achard, F.; Asner, G.P.; Held, A.; Kellndorfer, J.; Verbesselt, J. Synergies of multiple remote sensing data sources for REDD+ monitoring. *Curr. Opin. Environ. Sustain.* **2012**, *4*, 696–706. [CrossRef]
4. Kansakar, P.; Hossain, F. A review of applications of satellite earth observation data for global societal benefit and stewardship of planet earth. *Space Policy* **2016**, *36*, 46–54. [CrossRef]
5. Steele-Dunne, S.C.; McNairn, H.; Monsivais-Huertero, A.; Judge, J.; Liu, P.-W.; Papathanassiou, K. Radar Remote Sensing of Agricultural Canopies: A Review. *IEEE J. Sel. Top. Appl. Earth Obs. Remote Sens.* **2017**, *10*, 2249–2273. [CrossRef]
6. Kasampalis, D.; Alexandridis, T.; Deva, C.; Challinor, A.; Moshou, D.; Zalidis, G. Contribution of Remote Sensing on Crop Models: A Review. *J. Imaging* **2018**, *4*, 52. [CrossRef]
7. Betbeder, J.; Fieuzal, R.; Baup, F. Assimilation of LAI and Dry Biomass Data From Optical and SAR Images Into an Agro-Meteorological Model to Estimate Soybean Yield. *IEEE J. Sel. Top. Appl. Earth Obs. Remote Sens.* **2016**, *9*, 2540–2553. [CrossRef]
8. Huang, J.; Gómez-Dans, J.L.; Huang, H.; Ma, H.; Wu, Q.; Lewis, P.E.; Liang, S.; Chen, Z.; Xue, J.H.; Wu, Y.; et al. Assimilation of remote sensing into crop growth models: Current status and perspectives. *Agric. For. Meteorol.* **2019**, *276*, 107609. [CrossRef]
9. Blaes, X.; Defourny, P.; Wegmuller, U.; Della Vecchia, A.; Guerriero, L.; Ferrazzoli, P. C-band polarimetric indexes for maize monitoring based on a validated radiative transfer model. *IEEE Trans. Geosci. Remote Sens.* **2006**, *44*, 791–800. [CrossRef]
10. Casanova, J.J.; Judge, J.; Jang, M. Modeling Transmission of Microwaves Through Dynamic Vegetation. *IEEE Trans. Geosci. Remote Sens.* **2007**, *45*, 3145–3149. [CrossRef]
11. Van Emmerik, T.; Steele-Dunne, S.C.; Judge, J.; van de Giesen, N. Impact of Diurnal Variation in Vegetation Water Content on Radar Backscatter From Maize During Water Stress. *IEEE Trans. Geosci. Remote Sens.* **2015**, *53*, 3855–3869. [CrossRef]
12. Vicente-Guijalba, F.; Martinez-Marin, T.; Lopez-Sanchez, J.M. Crop Phenology Estimation Using a Multitemporal Model and a Kalman Filtering Strategy. *IEEE Geosci. Remote Sens. Lett.* **2014**, *11*, 1081–1085. [CrossRef]
13. Aschbacher, J.; Milagro-Pérez, M.P. The European Earth monitoring (GMES) programme: Status and perspectives. *Remote Sens. Environ.* **2012**, *120*, 3–8. [CrossRef]
14. Harfenmeister, K.; Spengler, D.; Weltzien, C. Analyzing Temporal and Spatial Characteristics of Crop Parameters Using Sentinel-1 Backscatter Data. *Remote Sens.* **2019**, *11*, 1569. [CrossRef]
15. Friesen, J.; Steele-Dunne, S.C.; van de Giesen, N. Diurnal Differences in Global ERS Scatterometer Backscatter Observations of the Land Surface. *IEEE Trans. Geosci. Remote Sens.* **2012**, *50*, 2595–2602. [CrossRef]
16. Paget, A.C.; Long, D.G.; Madsen, N.M. RapidScat Diurnal Cycles Over Land. *IEEE Trans. Geosci. Remote Sens.* **2016**, *54*, 3336–3344. [CrossRef]
17. Ndikumana, E.; Ho Tong Minh, D.; Dang Nguyen, H.; Baghdadi, N.; Courault, D.; Hossard, L.; El Moussawi, I. Estimation of Rice Height and Biomass Using Multitemporal SAR Sentinel-1 for Camargue, Southern France. *Remote Sens.* **2018**, *10*, 1394. [CrossRef]
18. Mattia, F.; Le Toan, T.; Picard, G.; Posa, F.I.; D'Alessio, A.; Notarnicola, C.; Gatti, A.M.; Rinaldi, M.; Satalino, G.; Pasquariello, G. Multitemporal c-band radar measurements on wheat fields. *IEEE Trans. Geosci. Remote Sens.* **2003**, *41*, 1551–1560. [CrossRef]
19. Alexakis, D.D.; Mexis, F.D.K.; Vozinaki, A.E.K.; Daliakopoulos, I.N.; Tsanis, I.K. Soil Moisture Content Estimation Based on Sentinel-1 and Auxiliary Earth Observation Products. A Hydrological Approach. *Sensors* **2017**, *17*, 1455. [CrossRef]
20. Vreugdenhil, M.; Wagner, W.; Bauer-Marschallinger, B.; Pfeil, I.; Teubner, I.; Rüdiger, C.; Strauss, P. Sensitivity of Sentinel-1 Backscatter to Vegetation Dynamics: An Austrian Case Study. *Remote Sens.* **2018**, *10*, 1396. [CrossRef]

21. Dobson, M.; Ulaby, F.; Hallikainen, M.; El-Rayes, M. Microwave Dielectric Behavior of Wet Soil-Part II: Dielectric Mixing Models. *IEEE Trans. Geosci. Remote Sens.* **1985**, *GE-23*, 35–46. [CrossRef]
22. Mironov, V.L.; Dobson, M.C.; Kaupp, V.H.; Komarov, S.A.; Kleshchenko, V.N. Generalized refractive mixing dielectric model for moist soils. *IEEE Trans. Geosci. Remote Sens.* **2004**, *42*, 773–785. [CrossRef]
23. Snapir, B.; Waine, T.W.; Corstanje, R.; Redfern, S.; De Silva, J.; Kirui, C. Harvest Monitoring of Kenyan Tea Plantations With X-Band SAR. *IEEE J. Sel. Top. Appl. Earth Obs. Remote Sens.* **2018**, *11*, 930–938. [CrossRef]
24. Bargiel, D. A new method for crop classification combining time series of radar images and crop phenology information. *Remote Sens. Environ.* **2017**, *198*, 369–383. [CrossRef]
25. McNairn, H.; Champagne, C.; Shang, J.; Holmstrom, D.; Reichert, G. Integration of optical and Synthetic Aperture Radar (SAR) imagery for delivering operational annual crop inventories. *ISPRS J. Photogramm. Remote Sens.* **2009**, *64*, 434–449. [CrossRef]
26. Van Tricht, K.; Gobin, A.; Gilliams, S.; Piccard, I. Synergistic use of radar Sentinel-1 and optical Sentinel-2 imagery for crop mapping: A case study for Belgium. *Remote Sens.* **2018**, *10*, 1642. [CrossRef]
27. De Bernardis, C.; Vicente-Guijalba, F.; Martinez-Marin, T.; Lopez-Sanchez, J.M. Contribution to Real-Time Estimation of Crop Phenological States in a Dynamical Framework Based on NDVI Time Series: Data Fusion With SAR and Temperature. *IEEE J. Sel. Top. Appl. Earth Obs. Remote Sens.* **2016**, *9*, 3512–3523. [CrossRef]
28. Fieuzal, R.; Baup, F.; Marais-Sicre, C. Monitoring Wheat and Rapeseed by Using Synchronous Optical and Radar Satellite Data—From Temporal Signatures to Crop Parameters Estimation. *Adv. Remote Sens.* **2013**, *2*, 162–180. [CrossRef]
29. Zhou, T.; Pan, J.; Zhang, P.; Wei, S.; Han, T. Mapping Winter Wheat with Multi-Temporal SAR and Optical Images in an Urban Agricultural Region. *Sensors* **2017**, *17*, 1210. [CrossRef]
30. Gao, Q.; Zribi, M.; Escorihuela, M.; Baghdadi, N.; Segui, P. Irrigation Mapping Using Sentinel-1 Time Series at Field Scale. *Remote Sens.* **2018**, *10*, 1495. [CrossRef]
31. Baghdadi, N.; El Hajj, M.; Zribi, M.; Bousbih, S. Calibration of the Water Cloud Model at C-Band for Winter Crop Fields and Grasslands. *Remote Sens.* **2017**, *9*, 969. [CrossRef]
32. Bériaux, E.; Waldner, F.; Collienne, F.; Bogaert, P.; Defourny, P. Maize Leaf Area Index Retrieval from Synthetic Quad Pol SAR Time Series Using the Water Cloud Model. *Remote Sens.* **2015**, *7*, 16204–16225. [CrossRef]
33. Li, J.; Wang, S. Using SAR-derived vegetation descriptors in a water cloud model to improve soil moisture retrieval. *Remote Sens.* **2018**, *10*, 1370. [CrossRef]
34. Clauss, K.; Ottinger, M.; Leinenkugel, P.; Kuenzer, C. Estimating rice production in the Mekong Delta, Vietnam, utilizing time series of Sentinel-1 SAR data. *Int. J. Appl. Earth Obs. Geoinf.* **2018**, *73*, 574–585. [CrossRef]
35. Mandal, D.; Kumar, V.; Ratha, D.; Dey, S.; Bhattacharya, A.; Lopez-Sanchez, J.M.; McNairn, H.; Rao, Y.S. Dual polarimetric radar vegetation index for crop growth monitoring using sentinel-1 SAR data. *Remote Sens. Environ.* **2020**, *247*, 111954. [CrossRef]
36. Chang, J.G.; Shoshany, M.; Oh, Y. Polarimetric Radar Vegetation Index for Biomass Estimation in Desert Fringe Ecosystems. *IEEE Trans. Geosci. Remote Sens.* **2018**, *56*, 7102–7108. [CrossRef]
37. Attema, E.P.W.; Ulaby, F.T. Vegetation modeled as a water cloud. *Radio Sci.* **1978**, *13*, 357–364. [CrossRef]
38. Brown, S.C.M.; Quegan, S.; Morrison, K.; Bennett, J.C.; Cookmartin, G. High-resolution measurements of scattering in wheat canopies—Implications for crop parameter retrieval. *IEEE Trans. Geosci. Remote Sens.* **2003**, *41*, 1602–1610. [CrossRef]
39. Huang, W.; Sun, G.; Ni, W.; Zhang, Z.; Dubayah, R. Sensitivity of multi-source SAR backscatter to changes in forest aboveground biomass. *Remote Sens.* **2015**, *7*, 9587–9609. [CrossRef]
40. Lahoz, W.A.; Schneider, P. Data assimilation: Making sense of Earth Observation. *Front. Environ. Sci.* **2014**, *2*, 1–28. [CrossRef]
41. Che, M.; Chen, B.; Zhang, H.; Fang, S.; Xu, G.; Lin, X.; Wang, Y. A New Equation for Deriving Vegetation Phenophase from Time Series of Leaf Area Index (LAI) Data. *Remote Sens.* **2014**, *6*, 5650–5670. [CrossRef]
42. Klosterman, S.T.; Hufkens, K.; Gray, J.M.; Melaas, E.; Sonnentag, O.; Lavine, I.; Mitchell, L.; Norman, R.; Friedl, M.A.; Richardson, A.D. Evaluating remote sensing of deciduous forest phenology at multiple spatial scales using PhenoCam imagery. *Biogeosciences* **2014**, *11*, 4305–4320. [CrossRef]
43. Son, N.-T.; Chen, C.-F.; Chang, L.-Y.; Chen, C.-R.; Sobue, S.-I.; Minh, V.-Q.; Chiang, S.-H.; Nguyen, L.-D.; Lin, Y.-W. A logistic-based method for rice monitoring from multitemporal MODIS-Landsat fusion data. *Eur. J. Remote Sens.* **2016**, *49*, 39–56. [CrossRef]

44. Østergaard, A.; Snoeij, P.; Navas Traver, I.; Ludwig, M.; Rostan, F.; Croci, R. C-band SAR for the GMES Sentinel-1 mission. In Proceedings of the European Microwave Week 2011 "Wave to Future" EuMW 2011,—8th European Radar Conference EuRAD 2011, Manchester, UK, 12–14 October 2011; pp. 234–240.
45. Khabbazan, S.; Vermunt, P.; Steele-Dunne, S.; Arntz, L.R.; Marinetti, C.; van der Valk, D.; Iannini, L.; Molijn, R.; Westerdijk, K.; van der Sande, C. Crop monitoring using Sentinel-1 data: A case study from The Netherlands. *Remote Sens.* **2019**, *11*, 1887. [CrossRef]
46. Cranfield University. The Soils Guide. 2019. Available online: www.landis.org.uk (accessed on 20 December 2019).
47. Avery, B.W.; Catt, J.A. The Soil-Map Units. In *The Soil at Rothamsted*; Lawes Agricultural Trust: Harpenden, UK, 1995; pp. 14–18.
48. Rothamsted Research. *Rothamsted Long-Term Monthly Rainfall*; Rothamsted Research: Harpenden, UK, 2018.
49. Mladenova, I.E.; Jackson, T.J.; Bindlish, R.; Hensley, S. Incidence Angle Normalization of Radar Backscatter Data. *IEEE Trans. Geosci. Remote Sens.* **2013**, *51*, 1791–1804. [CrossRef]
50. Vaudour, E.; Baghdadi, N.; Gilliot, J.M. Mapping tillage operations over a peri-urban region using combined SPOT4 and ASAR/ENVISAT images. *Int. J. Appl. Earth Obs. Geoinf.* **2014**, *28*, 43–59. [CrossRef]
51. Canisius, F.; Shang, J.; Liu, J.; Huang, X.; Ma, B.; Jiao, X.; Geng, X.; Kovacs, J.M.; Walters, D. Tracking crop phenological development using multi-temporal polarimetric Radarsat-2 data. *Remote Sens. Environ.* **2018**, *210*, 508–518. [CrossRef]
52. Molijn, R.A.; Iannini, L.; Mousivand, A.; Hanssen, R.F. Analyzing C-band SAR polarimetric information for LAI and crop yield estimations. In *Proceedings of the Remote Sensing for Agriculture, Ecosystems, and Hydrology XVI, Amsterdam, The Netherlands, 22–25 September 2014*; Neale, C.M.U., Maltese, A., Eds.; International Society for Optics and Photonics: Bellingham, WA, USA, 2014; Volume 9239, p. 92390V.
53. Zimmermann, B.; Kohler, A. Optimizing Savitzky–Golay Parameters for Improving Spectral Resolution and Quantification in Infrared Spectroscopy. *Appl. Spectrosc.* **2013**, *67*, 892–902. [CrossRef]
54. Savitzky, A.; Golay, M.J.E. Smoothing and Differentiation of Data by Simplified Least Squares Procedures. *Anal. Chem.* **1964**, *36*, 1627–1639. [CrossRef]
55. Biscoe, P.V.; Willington, V.B.A. Environmental effects on dry matter production. In *Nitrogen Requirement of Cereals: Proceedings of a Conference Organised by the Agricultural Development and Advisory Service, September 1982*; HMSO: London, UK, 1984.
56. Monteith, J.L. Climatic variation and the growth of crops. *Q. J. R. Meteorol. Soc.* **2007**, *107*, 749–774. [CrossRef]

remote sensing

MDPI

Article

# Rice-Field Mapping with Sentinel-1A SAR Time-Series Data

Lena Chang [1], Yi-Ting Chen [2], Jung-Hua Wang [3] and Yang-Lang Chang [4,*]

1 Department of Communications, Navigation and Control Engineering, National Taiwan Ocean University, Keelung 202301, Taiwan; lenachang@mail.ntou.edu.tw
2 Department of Electrical Engineering, National Taiwan Ocean University, Keelung 202301, Taiwan; 20953003@mail.ntou.edu.tw
3 Department of Electrical Engineering, AI Research Center, National Taiwan Ocean University, Keelung 202301, Taiwan; jhwang@mail.ntou.edu.tw
4 Department of Electrical Engineering, National Taipei University of Technology, Taipei 106344, Taiwan
* Correspondence: ylchang@mail.ntut.edu.tw; Tel.: +886-02-2771-2171 (ext. 2101)

**Abstract:** This study proposed a feature-based decision method for the mapping of rice cultivation by using the time-series C-band synthetic aperture radar (SAR) data provided by Sentinel-1A. In this study, a model related to crop growth was first established. The model was developed based on a cubic polynomial function which was fitted by the complete time-series SAR backscatters during the rice growing season. From the developed model, five rice growth-related features were introduced, including backscatter difference (BD), time interval (TI) between vegetative growth and maturity stages, backscatter variation rate (BVR), average normalized backscatter (ANB) and maximum backscatter (MB). Then, a decision method based on the combination of the five extracted features was proposed to improve the rice detection accuracy. In order to verify the detection performance of the proposed method, the test data set of this study consisted of 50,000 rice and non-rice fields which were randomly sampled from a research area in Taiwan for simulation verification. From the experimental results, the proposed method can improve overall accuracy in rice detection by 6% compared with the method using feature BD. Furthermore, the rice detection efficiency of the proposed method was compared with other four classifiers, including decision tree (DT), support vector machine (SVM), K-nearest neighbor (KNN) and quadratic discriminant analysis (QDA). The experimental results show that the proposed method has better rice detection accuracy than the other four classifiers, with an overall accuracy of 91.9%. This accuracy is 3% higher than fine SVM, which performs best among the other four classifiers. In addition, the consistency and effectiveness of the proposed method in rice detection have been verified for different years and studied regions.

**Keywords:** synthetic aperture radar; Sentinel-1A; rice detection; time-series data; rice growth-related features

**Citation:** Chang, L.; Chen, Y.-T.; Wang, J.-H.; Chang, Y.-L. Rice-Field Mapping with Sentinel-1A SAR Time-Series Data. *Remote Sens.* **2021**, *13*, 103. https://doi.org/10.3390/rs13010103

Received: 2 December 2020
Accepted: 26 December 2020
Published: 30 December 2020



## 1. Introduction

Rice is the primary staple food for a majority of the world's population, especially in Asia [1,2]. In 2018, the statistics from the FAO (Food and Agriculture Organization of the United Nations) show that rice is the second largest grain crop in the world and accounts for more than 12% of global cultivated land [3]. Global rice consumption has increased from approximately 437 million tons to 486 million tons in the past decade and average rice consumption per capita reached 53.9 kg in 2019 (https://www.statista.com/statistics/256002/global-per-capita-rice-use-since-2000/). Obviously, as the population has grown, the demand for rice has also increased substantially [4]. However, in recent years, global warming has greatly changed the temperature and rainfall [5–8], and has led to environmental and food security issues, such as land degradation and reduction of crop yields [9–12]. Considering the protection of the ecological environment and the development of rice crops, it is essential to monitor rice production and accurately detect the distribution of rice fields.

More than 90 percent of the world's rice is produced and consumed in the Asia-Pacific Region [3]. Effective monitoring of rice production and prices is also regarded as a key indicator of government performance in many Asian countries [12]. Especially Taiwan, with small land surface and dense population, is highly dependent on rice crops. Due to the narrow terrain and limited arable land, farmers often plant other crops on scattered rice fields. This farming method results in the complexity of crop varieties on small farmland (about 0.5 ha). Moreover, the complexity of food crops increases the difficulty and affects the accuracy of rice detection. In general, the government institutions spend a lot of time and labor on ground surveys and acquired aerial photos for annual rice-field mapping. Therefore, it is a challenge to accurately and efficiently monitor the distribution of rice fields in Taiwan.

In recent years, there are several studies on mapping paddy rice distribution using optical sensor data, such as MODIS and Landsat [13–17]. Some indexes have been extracted from optical remote sensing, such as the Land Surface Water Index (LSWI), Normalized Difference Vegetation Index (NDVI), Enhanced Vegetative Index (EVI) and Soil Adjusted Vegetation Index (SAVI) [18–25]. The study of Cheng Ru [20] combined the data of MODIS and SPOT by using the Spatial and Temporal Adaptive Reflectance Fusion Model (STARFM) to create high spatial-temporal images in Taiwan. Next, it constructed the smooth time-series vegetation indices using the empirical mode decomposition (EMD) and wavelet transform, respectively, regarding the scale of the study area and the application of the classifier. The rice crop classification was based on the phenology-based algorithm and artificial neural network (ANN) classifier. The overall accuracy of the classification was over 85% by the neural network classification. Jin et al. [21] proposed a phenology-based algorithm to map paddy rice extent using multi-temporal Landsat image at regional scale. In this study, three vegetation indexes, NDVI, EVI, and LSWI, were used to map rice-fields during the flooding/transplanting and ripening phases. The user and producer accuracies of paddy rice on the Landsat-based paddy rice map were 90% and 94%, respectively. Moreover, ref. [24] proposed a new normalized vegetation index based on the normalized EVI and SAVI to detect and map the paddy rice. The overall accuracy was 96.8% with a kappa coefficient of 0.97. Although optical images are useful in rice-field mapping, their practical application is limited due to the influence of climatic factors, such as rainfall. Especially in tropical areas, Asia included, the weather is cloudy more than 70% of the time during the rice growing season [26,27]. Due to the influence of cloud covering during the rice growing season, it is difficult to use optical sensors to acquire enough data for paddy rice mapping.

To alleviate the problem of optical sensor images being affected by clouds, synthetic aperture radar (SAR) has proved to be an effective tool for mapping paddy rice extent, because radar signals are less affected by cloud coverage compared to optical images [28,29]. In the past years, there have been several studies on the identification of rice-fields using SAR data. Studies [30–35] showed the effectiveness of SAR data in the X-band, C-band and L-band for rice monitoring. Researches [33,34,36–48] used Sentinel-1A time-series data to detect rice distribution. These studies explored the variation of temporal backscatter coefficients during the rice growth period. At seeding stage, the paddy field is flooded and the backscatter is low due to the specular reflection of water. During the growing period, the backscatter increases since the microwaves cannot penetrate rice. Thus, the temporal backscattering coefficient of rice changes from minimum at agronomic flooding to the maximum at tillering stage, after which the backscatter coefficient decreases. Since radar reflection of rice crops presents a large dynamic range during the growth period [47], Jyun-Bin [38] proposed the Normalized Difference Sigma-naught Index (NDSI) calculated from the time-series Sentinel-1A VH (vertical transmit and horizontal receive) and VV (vertical transmit and vertical receive) polarization data. The NDSI, backscatter difference between the maximum and minimum in the time-series data, was employed to perform rice-field mapping. The study [38] showed 92.1% overall accuracy and 0.85 kappa coefficient for VH polarization. Moreover, Lamin R. et al. [34] proposed a backscatter change threshold for paddy rice mapping. The threshold was determined by considering the spatial dynamics of rice backscattering, which was induced by inter-field variability from flooding to tillering and booting. The overall classification accuracy and kappa coefficient of [34] were 88.3% and 0.85, respectively. Nguyen [37] validated the performance of an existing phenology-based classification method for continental scale rice-field mapping. The classification method was based on the maximum backscatter, the difference between maximum backscatter and the backscatter from the beginning of the growing season, and the temporal interval between the beginning date and the date of maximum backscatter. The overall accuracy of [37] reaches more than 70% for all study areas. Additionally, Bazzi et al. [45] analyzed the temporal behavior of SAR backscattering coefficient of rice. The study proposed three matrices, derived from the Gaussian profile of the VV/VH time-series, the variance of the VV/VH time-series and the slope of the linear regression of the VH time-series. Based on the matrices, rice plots were mapped through decision tree and random forest classifiers. The study [45] showed high overall accuracy of 96.3% and 96.6% for decision tree and random forest classifier, respectively.

In this study, a feature-based decision approach was proposed to detect the mapping of rice cultivation using the multi-temporal SAR data provided by Sentinel-1A. The features related to rice growth were first extracted from fitting models which were established from the complete time-series data during the rice growing period. Then, based on the distribution of features, a threshold decision method was proposed for paddy rice mapping. The paper is organized as follows: Section 2 describes the study area and data used in this study, including the ground truth data and Sentinel-1A data. Then, the method of rice detection is introduced. The experimental results and discussions are presented in Section 3. Finally, some conclusions are drawn in Section 4.

## 2. Materials and Methods

### *2.1. Study Area*

The study area is composed of Yunlin and Changhua counties in central Taiwan, extending from 120°10′ to 120°40′E longitude and 23°30′ to 24°10′N latitude and at elevations up to 1760 m, as shown in Figure 1. About 90% of the terrains in this area are plains, except for the hills, tablelands and mountains on the east, with a total area of about 2365.23 km$^2$. These two counties are important rice growth regions in Taiwan, among which rice-planting areas rank as the top two in Taiwan and rice production accounts for about 35% of Taiwan's total rice production. Although rice is the main crop, many other non-rice food crops are planted in these regions, such as maize, peanut, wheat and onion. Most of the agricultural lands were scattered and the non-rice crops are planted among rice-fields. It belongs to the subtropical monsoon climate, with an average annual temperature of 22.4 °C. The annual rainfall is about 1723 mm, with a rainy season from May to June and a typhoon season from July to September. According to the rice phenology, rice cultivation can be divided into first-stage rice and second-stage rice. The sowing time of rice is from mid-February to late March and mid-July to early August for first-stage and second-stage rice, respectively, as shown in Table 1. Due to temperature and rainfall factors, the rice growth cycle takes about 110–140 day for first-stage rice and 100–110 day for second-stage rice.

**Table 1.** Sowing and harvesting time of different rice crop seasons in Taiwan.

| Rice Crop Seasons. | Sowing | Harvesting |
|---|---|---|
| First-stage | Feb.–Mar. | Jun.–Jul. |
| Second-stage | Jul.–Aug. | Nov.–Dec. |

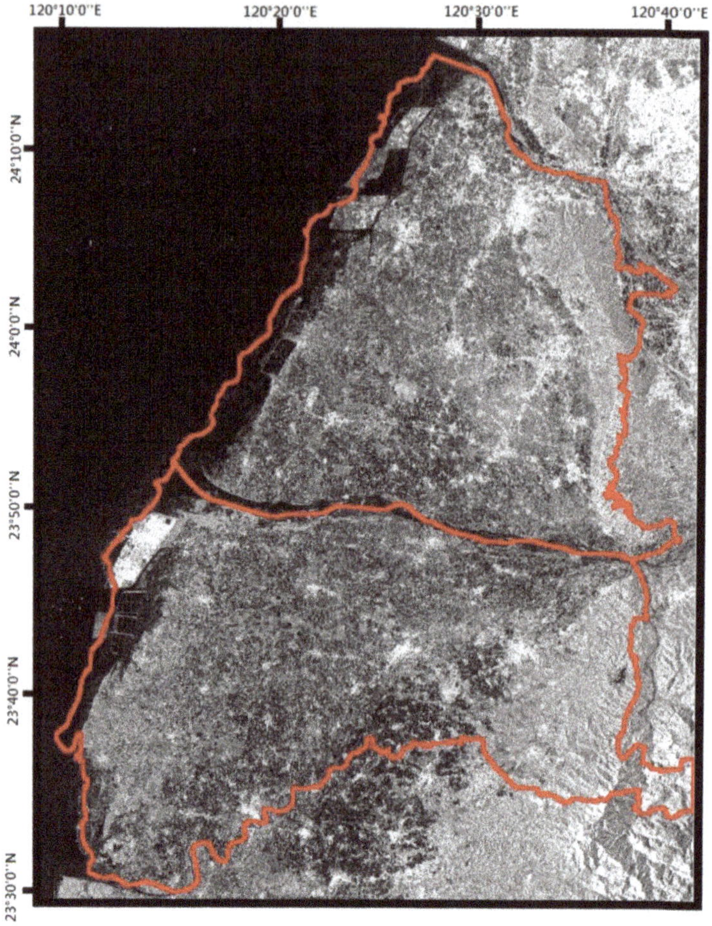


**Figure 1.** The study area consists of Yunlin and Changhua counties in central Taiwan. The red border represents the boundary of the study area. Background data received from Sentinel-1A on 8 February 2017.

## 2.2. Ground Data

Due to the narrow terrain of the study area, the rice planting area is relatively small and scattered. In the study area, the rice fields range from 0.2 to 0.9 ha, most of which are about 0.5 ha. In addition, rice crops are often cultivated mixed with other food crops. To evaluate the accuracy of rice detection, the experiment results of the proposed method were compared with the rice ground truth of first-stage rice provided by Agriculture and Food Agency (AFA) and Taiwan Agriculture Research Institute (TARI) in 2017. AFA and TARI are the agriculture related government institutions in Taiwan. The ground truth data provided by the AFA and TARI are identified through aerial photos and satellites (including Landsat-8 and Rapideye) with multiple periods and different spatial resolutions. These government institutions regularly collect high resolution satellite images, aerial photos and farmland parcel maps every year. First, agricultural lands are investigated by using census data and ground surveys. Then, aerial photos and satellite data from agricultural land are further applied in distinguishing and mapping the ground truth data of crops. The dataset used in the following experiments consists of rice fields and non-rice fields, and is established based on the intersection of ground truth data provided by both institutions. According to ground truth data, the rice-field distribution of the two study counties is shown in Figure 2. In the following experiments, 40,000 randomly selected rice fields were used in the training phase, and 10,000 sampled rice fields were used for validation.

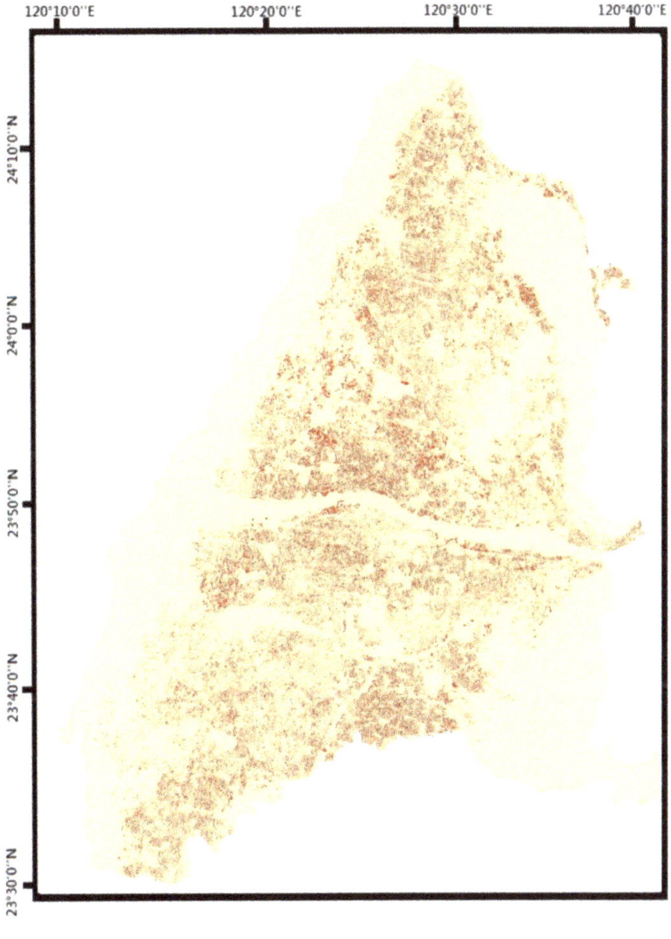


**Figure 2.** Rice ground data mapping from Agriculture and Food Agency (AFA) and Taiwan Agriculture Research Institute (TARI). Yellow indicates the study area, and red indicates the distribution of rice fields.

## 2.3. Sentinel-1A Data Preprocessing and Smoothing Processing

Sentinel-1A launched in April 2014 by the European Space Agency (ESA) is an earth observation satellite for the Copernicus Initiative. Copernicus, previously known as global monitoring for environment and security (GMES), is a European initiative for the implementation of information services dealing with environment and security. The revisit time of this satellite is 12 day at the equator. The Sentinel-1A satellite operates at the C-band (central frequency = 5.404 GHz), containing VH and VV polarizations with a spatial resolution of 5 m by 20 m in the range and azimuth directions. In this study, the Sentinel-1A data was acquired in the interferometric wide swath (IW) acquisition mode with a swath width of 250 km. Besides, data utilized level-1 ground range detected (GRD) format with a pixel spacing of 10 m by 10 m (https://sentinel.esa.int/). The Sentinel-1A IW level-1 GRD data, with VH and VV polarizations, ascending and descending orbit modes, was downloaded from https://search.asf.alaska.edu/#/. In addition, these acquired Sentinel-1A data are open access and free from the website. Since this study mainly focused on the first-stage rice-field detection, the data were downloaded from early February to late July in 2017. The complete acquisition dates of research data from Sentinel-1A were shown in Table 2. The incidence angles of the ascending and descending orbital modes in the study area are from 31.5 to 36.3° and 31.8 to 36.5°, respectively.

**Table 2.** The complete acquisition dates of research data from Sentinel-1A with ascending and descending orbit modes in 2017, respectively.

| Acquisition Dates \ Orbit Modes | Ascending Orbit Mode | Descending Orbit Mode |
|---|---|---|
| Feb. | 8th, 20th | 22nd |
| Mar. | 4th, 16th, 28th | |
| Apr. | 9th, 21st | 11th, 23rd |
| May | 3rd, 15th, 27th | 5th, 17th, 29th |
| Jun. | 8th | 10th, 22nd |
| Jul. | 2nd, 14th, 26th | 4th, 16th, 28th |

The Sentinel-1A data were pre-processed using the Sentinel Application Platform (SNAP) Sentinel-1 Toolbox software developed by ESA. Pretreatment includes three main steps. First, the radiometric calibration process was performed to convert the pixel data to actual backscattering values of sigma naught (dB). Thus, the pixel values in the imagery can be directly related to the radar backscatter of the scene. Then range Doppler terrain correction process corrected the geometric distortion in the range and projected range to Taiwan Datum 1997 (TWD97) earth ellipsoid model. Finally, the refined Lee filter [49] was applied to remove speckle noise in SAR data.

In this study, the proposed rice detection method is based on the rice-growth related features, which are extracted from the time-series backscatters during rice growth. Due to different farming behavior, such as different sowing dates, the heading or maturity dates of rice growth are also different. The times corresponding to minimum and maximum backscatter values are different for each rice-field. Therefore, instead of directly using backscatters, a model corresponding to backscattering change is established first, and then the rice features are extracted from the model.

However, the fluctuation of backscattering coefficients will cause model distortion and affect the extracted features, for example the maximum backscatter values and the growth time period. Therefore, it is difficult to model the rice growth curve effectively based on the acquired temporal data. To reduce the influence of the fluctuation of the backscattering coefficients, the SAR data are preprocessed by a smoothing approach [50]. Since the cultivation of crops was the block-based and clustered, a spatial mask was applied to perform pixel convolution for spatial smoothing. For example, the image in Figure 3a is the original SAR image in the study area. After the spatial smoothing, the output image was smoother, as shown in Figure 3c. Subsequently, temporal smoothing is performed on each pixel by a convolution mask to alleviate the randomness of temporal evolution. As shown in Figure 4a, the original SAR data fluctuates greatly. The temporal backscatter coefficient is smoother and the randomness has been reduced after the temporal convolution.

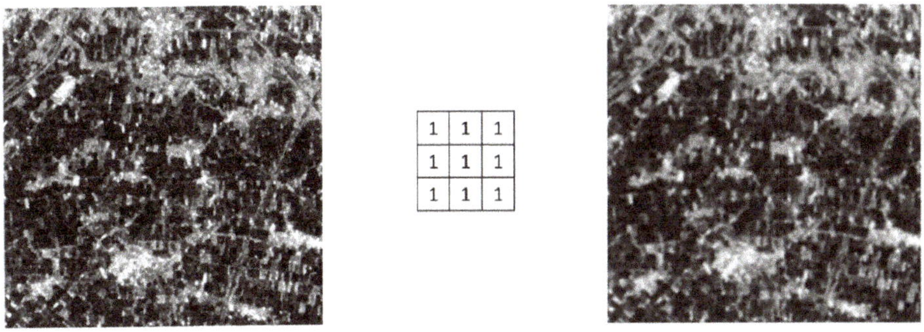

(**a**) original image (**b**) convolution mask (**c**) image after spatial smoothing

**Figure 3.** Example of spatial smoothing. (**a**) is the original input image, (**b**) is the convolution mask and (**c**) is the image after spatial smoothing.

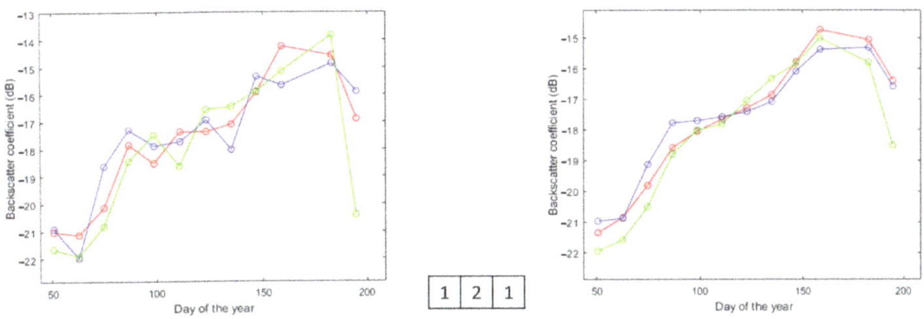


(**a**) original SAR data (**b**) convolution mask (**c**) SAR data after temporal smoothing

**Figure 4.** Example of temporal smoothing. (**a**) is the original data, (**b**) is the convolution mask and (**c**) is the data after temporal smoothing. (Each point in Figure 4 is the mean of all pixels within one rice field randomly selected from the study area).

### *2.4. Modeling and Feature Extraction*

After smoothing processing, a model of rice growth curve was established based on the complete time-series data of the rice growth period. Observing the temporal data in Figure 4, the time corresponding to minimum and maximum backscattering values can represent the sowing and heading dates of rice growing, respectively. Therefore, the date corresponding to the minimum backscatter value of the time-series data can be used to locate the starting date of the rice growth period and complete time-series data were collected from this date.

Due to the different sowing dates, the length of the complete time-series data of each rice-field is different. To overcome this problem in the study, a cubic polynomial function was used to fit the collected time-series data, and a model of the rice growth trend curve was established. Figure 5 shows the fitting curves of rice and non-rice crops. In terms of electromagnetic interaction mechanisms between radar waves, vegetation canopy, soil and water, there was high correlation between paddy rice backscatter coefficient and its specific growth period [32,51,52]. In the sowing, agricultural land was underwater, so the backscatter from rice-field was dominated by the double bounce volume scattering. Therefore, the backscatter energy of water was greater than that of paddy rice. Water layers are smooth and homogenous, causing reflected radar pulses to be weak and the backscatter coefficient is low in the sowing of rice growth period. During the vegetative to heading perio, the backscatter coefficient showed a significant increase, due to the volume scattering from within the rice canopy and multiple reflections between the plants and water surface. The backscatter then decreases slightly during the reproductive to harvest period, due to the fact that the water content of the plant decreases, and so do stem and leaf densities [41,42,46]. In the study area, the non-rice food crops include corn, peanut, onion and wheat. Figure 5 shows the temporal variation of backscatter coefficient of rice and the above non-rice food crops under VH and VV polarizations. These backscatter variation curves were obtained by using 50 crop fields randomly selected from the ground truth data of each crop. In order to show the deviation between the backscattering curve and the real data, the curve in Figure 5 was represented by the mean value and the corresponding standard deviation bar for each food crop. Obviously, the backscattering coefficient of rice changes greatly during the growing period, while the backscattering changes of other crops (such as peanuts and onions) are smoother. Thus, the backscatter coefficient of paddy rice changes significantly during the rice growth period compared with other non-rice crops.

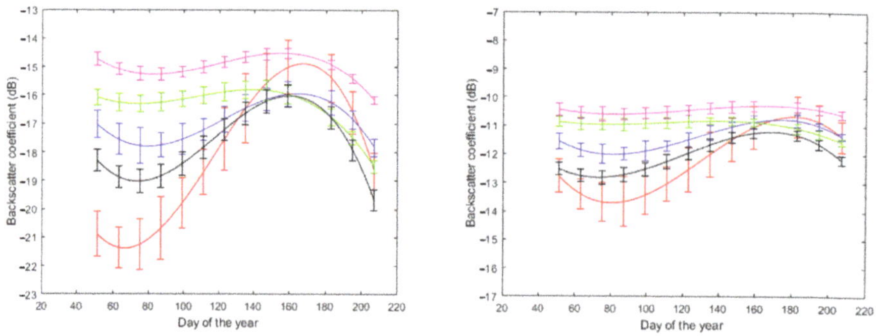


**Figure 5.** Backscattering curves (with mean and standard deviation bars) of food crops for VH and VV polarizations. Rice is shown in red, maize in blue, peanut in green, wheat in black and onion in magenta.

According to the growth and the cycle of rice crops, five features are extracted from the constructed rice growth trend model, as shown in Figure 5.

(1) **Backscatter Difference (BD)**

Compared with other non-rice crops in Figure 5, due to the obvious changes of rice plants during the growth period, the backscattering coefficient of rice also correspondingly changes significantly [30,33,34,36–38,40–44,46,47]. From the fitting curves of rice and non-rice in Figure 5, it can be observed that the backscatter change between the maximum and minimum values of rice is greater than that of other non-rice crops during the rice growing season (early Feb. to late Jul.). Thus, the backscatter difference (BD) between maximum and minimum values of time-series data during the rice growing season (from early Feb. to late Jul. in the study) was further examined for rice and non-rice crops, as shown in Figure 6. Figure 7a shows the probability density functions (pdf) of BD for different crops. It can be observed that the BD values of rice are distributed in 4–7 dB, while the BD values of maize and wheat are distributed in 1–5 dB and 2–5 dB, respectively, and the others are distributed in 1–4 dB. These results indicate that BD can help distinguish rice from other non-rice crops in the study area. In this study, BD was selected as one feature in rice-field detection. In following rice detection, the BD threshold was set based on the 95% confidence interval of pdf from the training data (i.e., 4.3 dB $\leq$ BD $\leq$ 6.6 dB in experiments for VH polarization with ascending orbit mode).

(2) **Time Interval (TI)**

In addition to the characteristics of BD, the growth period of crops is also an important feature [31,33,36,37,40,41]. Each crop has its own specific growth time period. The growth cycle of rice is carried out in the order of sowing, vegetative growth, heading and maturity, which takes about four months in Taiwan. According to the rice growth model in

Figure 6, the minimum backscatter point, the midpoint a, the maximum backscatter point and midpoint b are applied to represent the above four rice growth stages, respectively. The midpoints, a and b indicate the positions where the average value of the maximum and minimum backscatter appear in the rice growth curve. $T_a$ and $T_b$ are the corresponding dates of midpoint a and b, as shown in Figure 6. In the study, the time difference between two midpoints, $T_b - T_a$, of the rice growth curve, which denotes the time interval (TI) between vegetative growth and maturity, was used as one of the characteristics of rice growth. The TI distributions of rice and non-rice crops are given in Figure 7b. It can be observed that the average TI of non-rice crops are 54, 58, 62 and 65 day for onion, peanut, maize and wheat, respectively, which are relatively smaller than the 76 day average TI of rice. In the experiment, the TI decision interval is between 69.0 day and 82.5 day based on the 95% confidence interval of TI distributions.

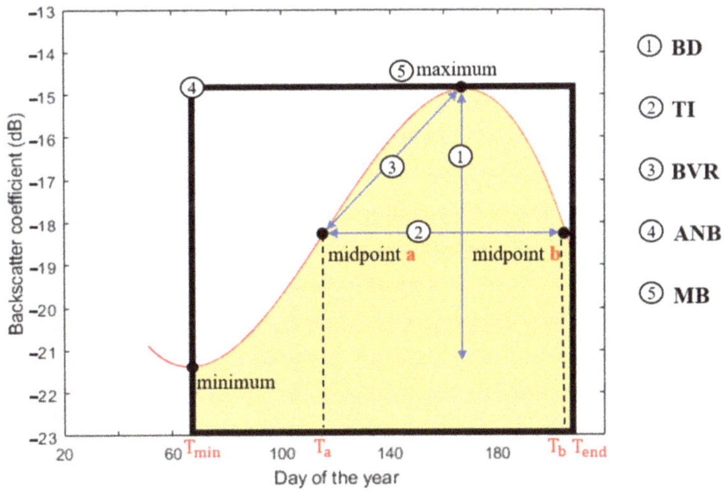


**Figure 6.** Extracted crop growth features. ① Backscatter difference (BD), ② time interval (TI), ③ backscatter variation rate (BVR), ④ average normalized backscatter (ANB) and ⑤ maximum backscatter (MB).

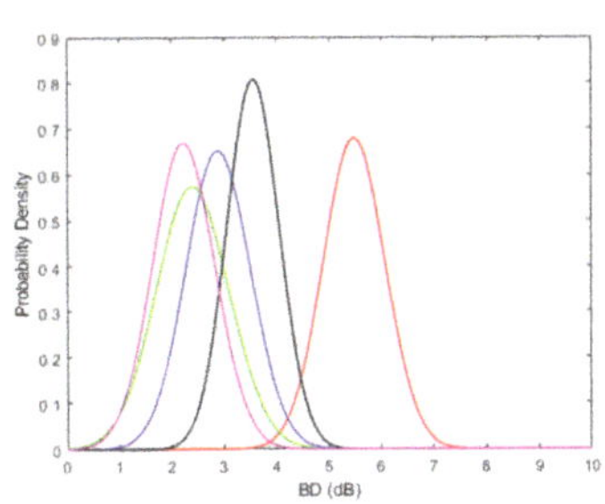


(a)

**Figure 7.** *Cont.*

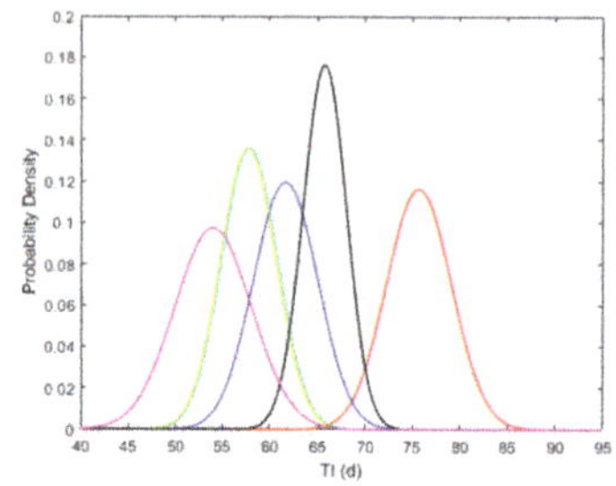


**(b)**

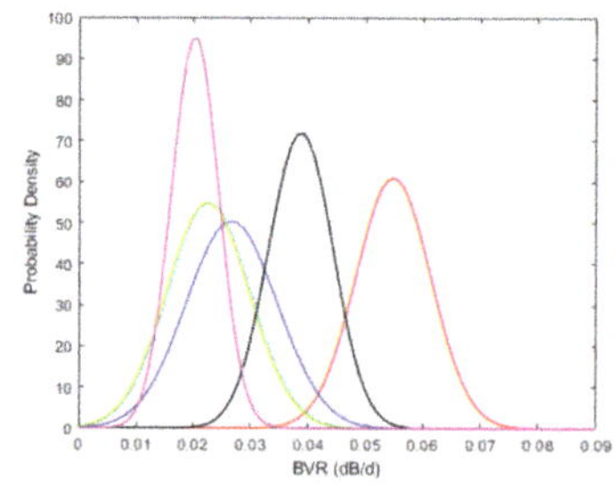


**(c)**

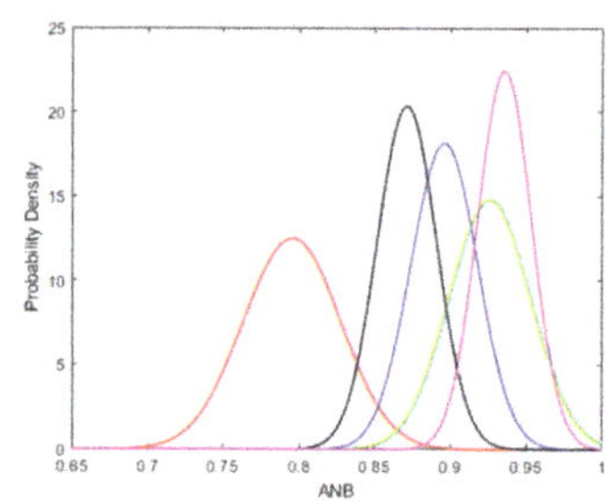


**(d)**

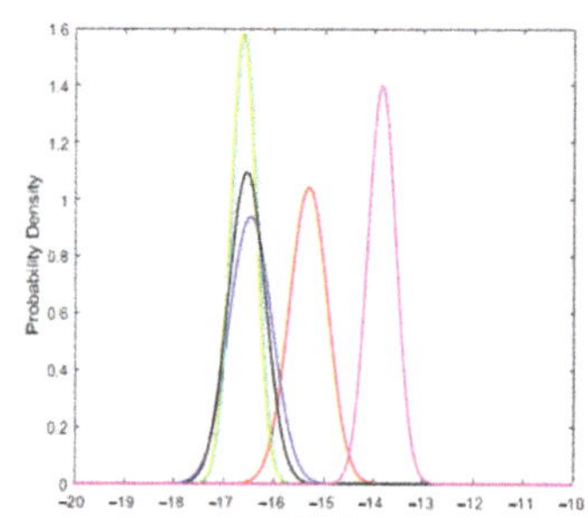


**(e)**

**Figure 7.** The probability density functions for VH polarization with ascending orbit mode of proposed features of (**a**) backscatter difference (BD), (**b**) time interval (TI), (**c**) backscatter variation rate (BVR), (**d**) average normalized backscatter (ANB) and (**e**) maximum backscatter (MB). Rice is shown in red, maize in blue, peanut in green, wheat in black and onion in magenta.

**(3) Backscatter Variation Rate (BVR)**

During the tillering, the variation rate of backscattering in rice is obviously accelerated. After the time corresponding to midpoint a of the curve in Figure 6, the backscatter of rice increases and reaches the peak at heading. Therefore, the slope from midpoint a to maximum backscatter point was used to represent the variation rate of backscattering at the tillering in the study, which is calculated by

$$\mathrm{BVR} = \frac{\sigma_{max} - \sigma_a}{T_{max} - T_a} \tag{1}$$

$\sigma_{max}$ and $\sigma_a$ are the backscatter values at the maximum backscatter point and midpoint a, respectively. $T_{max}$ and $T_a$ represent the corresponding dates of maximum backscatter point and midpoint a, respectively. Figure 7c shows the pdfs of backscatter variation rate (BVR) for rice and non-rice crops. It can be observed that the BVR values of rice are larger than those of non-rice crops. According to the 95% confidence interval of the pdf, the values between $4.2 \times 10^{-2}$ dB/d and $6.7 \times 10^{-2}$ dB/d were chosen as the decision interval of BVR in the subsequent rice detection experiments.

**(4) Average Normalized Backscatter (ANB)**

The trend of rice growth formed by the backscatter coefficients is consistent with the rice growth cycle from sowing, heading to maturity. When the maximum backscatter value is normalized to one for each crop in Figure 5, it can be observed that the value of average normalized backscatter is close to one for a flatter curve, while this value becomes smaller for a curve with greater variation. The ANB is calculated by:

$$\mathrm{ANB} = \frac{1}{T_{end} - T_{min}} \int_{T_{min}}^{T_{end}} \frac{\sigma(t)}{\sigma_{max}}\, dt \tag{2}$$

$T_{min}$ and $T_{end}$ represent the corresponding dates of the minimum backscatter point and end point in the rice growth season, respectively. In the study, the average normalized backscatter (ANB) is the ratio of the integrated area of the normalized growth curve (with the maximum value normalized to 1) to the growth period, as shown in Figure 6. According to the pdfs of ANB shown in Figure 7d, the ANB of rice is much smaller than that of other crops. Therefore, ANB was also selected as an indicator of rice detection. The decision interval of ANB was chosen to be from $7.3 \times 10^{-1}$ to $8.6 \times 10^{-1}$ in the rice detection experiment.

**(5) Maximum Backscatter (MB)**

Each crop has its backscatter distribution range, and the corresponding maximum backscatter (MB) value is also different, as shown in Figure 5. The pdfs of the MB values are given in Figure 7e, which shows that the MB distribution of rice is different from other crops. In order to distinguish rice from other non-rice objects, MB is used as one of rice identification features in the study [33,36,37,40,45]. In the experiment, the decision interval of MB was chosen between −16.0 dB and −14.5 dB for rice detection.

The above five features can be extracted from the fitting models. Then the distribution of each feature is estimated from the training data of rice and non-rice, respectively. To detect rice crops, a feature-based decision method is introduced based on five extracted features. The decision threshold is determined by the 95% confidence interval of rice feature distribution, shown in Figure 7. Moreover, it can be observed from Figure 7 that the distributions of the extracted features of rice overlap with those corresponding to other non-rice crops. These reasons make rice detection difficult. If only one or two features are used for rice detection, some non-rice crops will be misclassified as rice. This leads to lower detection accuracy of non-rice. Therefore, this study proposed a decision method based on all five features, BD, TI, BVR, ANB and MB, to detect the mapping of rice-fields. When all five extracted features of the test field meet the decision conditions, this field is classified as rice, otherwise it is classified as non-rice. In experiments, the rice field is identified by the following decision rules:

$$\text{Rice field conditions}:\begin{cases} 4.3\text{ dB} < BD < 6.6\text{ dB} \\ 69.0\text{ day} < TI < 82.5\text{ day} \\ 4.2\times10^{-2}\text{ dB/day} < BVR < 6.7\times10^{-2}\text{ dB/day} \\ 7.3\times10^{-1} < ANB < 8.6\times10^{-1} \\ -16.0\text{ dB} < MB < -14.5\text{ dB} \end{cases} \quad (3)$$

### 2.5. Classification Algorithms

To evaluate the performance of the proposed method, the detection results were compared with four other classification algorithms: decision tree (DT) [53], support vector machine (SVM) [54], K-nearest neighbor (KNN) [55] and quadratic discriminant analysis (QDA) [56]. DT, SVM and KNN are non-parametric supervised learning methods for classification. The DT classifier infers the decision rules from data features and divides the input dataset into categorical classes by recursive partitioning based on the splitting rules. SVM classifier constructs a hyperplane, through which a good separation can be achieved. That is the constructed hyperplane has the largest distance to the nearest training-data point of any class. The KNN classifier predicts the target label by finding the nearest neighbor category which is determined based on the distance measures. QDA is a statistical classifier that uses a quadratic decision surface to separate measurements of two or more classes. The decision boundary is generated by fitting class conditional densities to the data based on Bayes' rule.

Moreover, for DT, SVM and KNN classifiers, there are three scales to be considered in the experiment, namely Fine, Medium and Coarse, provided by MATLAB Machine Learning Toolbox. Rice detection performance was verified by 5-fold cross-validation [57] in MATLAB. The 5-fold cross-validation splits the dataset into five equal parts. In experiments, four parts were used as training data, and the remaining parts were used as test data. This process was repeated five times and the results averaged, each time using one different part as the testing data.

In the rice-field mapping experiment in Section 3.3, all image pixels will be used instead of only the samples of the dataset. Due to the clustering characteristics of crop planting, the image was first partitioned into clusters by the fuzzy c-means algorithm [58]. Therefore, pixels with similar temporal characteristics were grouped into a cluster. The corresponding model and rice-growth features of each cluster will be utilized in the subsequent rice detection.

## 3. Results

In this section, experiments were performed to verify the efficiency of the proposed rice-field detection method. Yunlin and Changhua counties, the most important rice growing areas in Taiwan, were selected as the study areas. The topographical characteristics of rice growing areas in these two counties are similar. There are no crops in the east, because these areas are the highlands and hills of the two counties. The coastal area is not suitable for rice growth due to the northeast monsoon. The planting area near the coast is scattered and small, and some short-term crops such as peanuts and garlic are grown. Compared with the eastern and coastal areas, the rice planting area is more extensive and denser in the inland areas of these two counties. In experiments, we have selected five regions from the inland area of Yunlin and Changhua counties and the image size of each region is 1000 × 1000 pixels, as shown in Figures 8 and 9. Regions 1 and 2 are both located in the Changhua County. Regions 3 and 4 are in Yunlin County. Region 5 is located in Yunlin County, except for the bottom.

For region 1, the farmland in this area is mainly for growing rice, but the rice-fields are scattered. Most of the non-rice areas are covered by buildings sparsely distributed in farmland.

In region 2, the agricultural land is intensive and rice is the main crop. According to the ground data of TARI, a few of the croplands are covered by peanut, garlic and watermelon. Similar to region 1, most of the buildings are scattered in this area.

Region 3 is the main agricultural land in Yunlin. Most farmland grows rice, but other non-rice fields, such as peanut, garlic, maize and cabbage, are scattered in this region.

There are diverse crops in the area of region 4. The crops contain rice, peanut, garlic, cabbage, potato and edamame. Although rice is the main crop in region 4, the rice-fields in this region are scattered and relatively small (less than 0.5 ha) compared with the average area of rice fields in other regions, about 0.5 ha.

In region 5, agricultural land is densely distributed and only rice is grown. Some buildings are sparsely located in this area. Besides, the bottom of region 5 is outside of the Yunlin County and there is no ground truth data for rice mapping, so it is marked as a non-interesting area shown in black in Figure 9e.

In the experiment, a dataset consisting of rice and non-rice samples was prepared. According to the ground truth data, 10,000 polygon fields of rice and non-rice were randomly selected from the above five regions, separately. There are a total of 50,000 rice and non-rice polygon fields, respectively. To evaluate the efficiency of the proposed method, the dataset was divided into training and testing samples in which 80 percent of samples (40,000 rice fields and non-paddy fields, respectively) were regarded as training data and 20 percent of samples (10,000 rice fields and non-paddy fields, respectively) were regarded as testing data. In the following, the features of rice were first extracted from the training data and approximated by using Gaussian distributions. Then, the decision criterion of each feature was determined according to the 95% confidence intervals of the corresponding feature distribution. Finally, the test data was used to evaluate the detection accuracy for the proposed feature-based rice decision method.

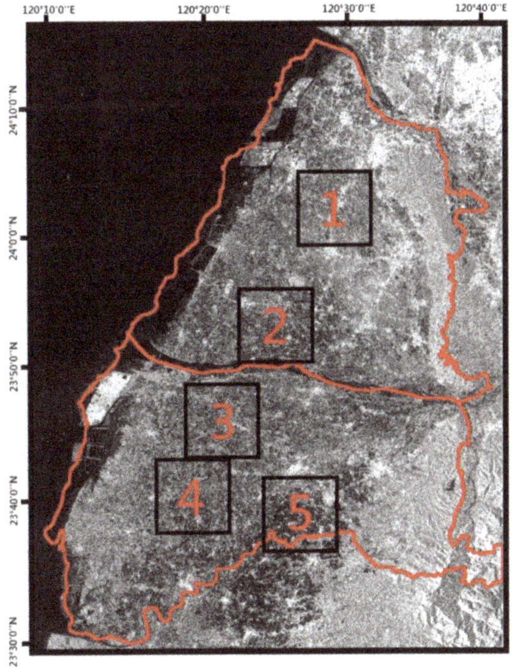


**Figure 8.** Five selected regions from the study area.

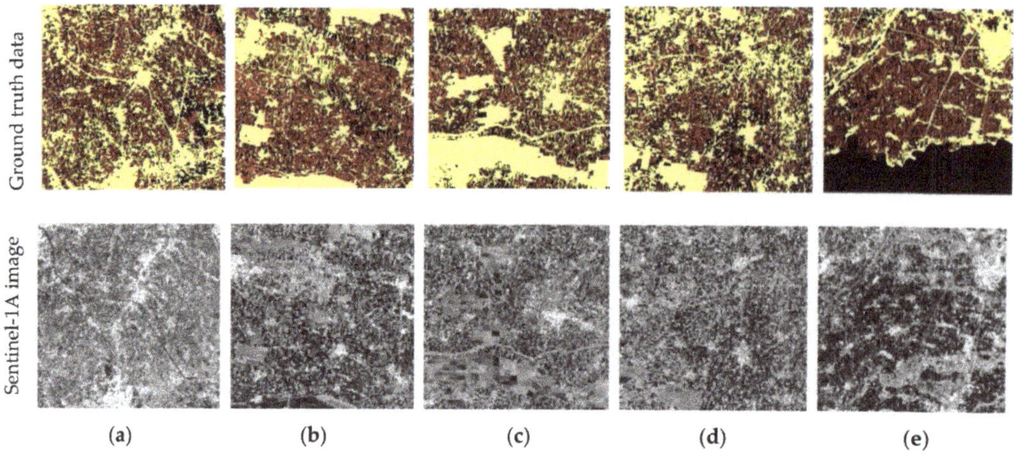


**Figure 9.** Sentinel-1A image and corresponding ground truth data of the five selected regions. (**a**–**e**) corresponding to regions 1–5 in Figure 8, respectively. Rice pixels are shown in red, non-rice pixels in yellow and non-interesting pixels in black.

### *3.1. Accuracy Assessment*

In this experiment, we compared the rice detection performance of the extracted rice growth features, including backscatter difference (BD), time interval (TI), backscatter variation rate (BVR), and average normalized backscatter (ANB). In addition, the simulation scenario contains two polarizations (VH and VV) and two orbit modes (ascending and descending).

First, rice detection experiments were conducted using only one extracted feature at a time. The maximum backscatter (MB) feature was not included in the simulation, because MB is suitable for use with other features. Table 3 shows the detection results under the experimental situation of VH polarization and ascending orbit mode. The classification overall accuracy (O.A.), user accuracy (U.A.) and producer accuracy (P.A.) were calculated from the confusion matrix [59]. It can be observed that the P.A. and O.A. of rice achieve over 90% and 85%, respectively, for each feature. These results validate the efficiency of the extracted features in rice detection. Moreover, the detection efficiency of feature TI is superior to other features. The effectiveness of TI in rice identification is due to the rice growing period being about 120 days, which is significantly different from other crops in this area. For example, the growing period of garlic is about 90 days in Yunlin County.

**Table 3.** Detection accuracy of the proposed features for VH polarization with ascending orbit mode. The P.A., U.A. and O.A. represent the producer accuracy, user accuracy and overall accuracy, respectively.

| | **Features** | **Rice** | | **Non-Rice** | | **O.A. (%)** |
|---|---|---|---|---|---|---|
| | | **P.A. (%)** | **U.A. (%)** | **P.A. (%)** | **U.A. (%)** | |
| One feature | BD | 91.1 | 82.4 | 80.6 | 90.0 | 85.8 |
| | TI | 92.3 | 85.7 | 84.6 | 91.6 | 88.4 |
| | BVR | 90.8 | 83.0 | 81.5 | 89.8 | 86.1 |
| | ANB | 90.3 | 82.5 | 80.9 | 89.2 | 85.6 |
| Combined features | BD + TI | 91.9 | 89.7 | 89.5 | 91.7 | 90.7 |
| | BD + TI + BVR | 91.5 | 91.2 | 91.2 | 91.4 | 91.3 |
| | BD + TI + BVR + ANB | 91.3 | 91.7 | 91.8 | 91.3 | 91.5 |
| | All | 90.9 | 92.1 | 92.3 | 91.0 | 91.6 |

The results in Table 3 also show that the accuracy of non-rice detection is much lower than that of rice detection. To improve the performance of non-rice detection, we performed the rice detection next by using the combined features. Five experiments were conducted here, including one feature (BD), two features combined (BD and TI), three features combined (BD, TI and BVR), four features combined (BD, TI, BVR and ANB) and all five features combined. Table 3 shows that the simulation scheme with two combined features can improve O.A. by about 5% compared with the case of using only one feature. Regarding the case with three or more features, the O.A. of rice slightly increases by about 1% and the P.A. of non-rice crops increases by about 2–3%, compared with the case using two features. With the increase in the number of combined features, the O.A. of rice and the P.A. of non-rice crop detection are still improved. Compared with only using feature BD, the P.A. of non-rice crops can be improved by about 12% by the combined use of all combined features, although the P.A. of rice crops is slightly reduced, it still remains above 90%. In general, the proposed method with combined features can achieve more than 90% O.A. in rice detection. These experimental results validate the efficiency of the proposed method based on the combined rice growth features.

Then, we examined the influence of polarizations and orbital modes on rice detection. Table 4 shows the results of the proposed method with all combined features for VH and VV polarizations with ascending and descending orbit modes. The O.A. is 91.6% and 91.3% for VH polarization with ascending and descending orbit modes, respectively. The orbit mode does not significantly affect the detection results. This result may come from the fact that incidence angles for the ascending and descending orbital modes are very close in the study area, which are from 31.5 to 36.3° and 31.8 to 36°, respectively. However, the O.A. is 84.2% and 86.4% for VV polarization with ascending and descending orbit modes, respectively. Thus, the rice detection accuracy of VH polarization is much better than that of VV polarization. This result is the same as the studies [33,43,51]. Compared with VV, VH is more representative of the actual rice growth structure and can provide more information in paddy rice distinguishing. Based on the above results, the simulation scheme of VH polarization with ascending orbit mode will be considered in the following experiments.

**Table 4.** Detection accuracy of all combined features for different orbit modes and polarizations.

| Orbit Modes | Polarizations | Rice | | Non-Rice | | O.A. (%) |
|---|---|---|---|---|---|---|
| | | P.A. (%) | U.A. (%) | P.A. (%) | U.A. (%) | |
| Ascending | VH | 90.9 | 92.1 | 92.3 | 91.0 | 91.6 |
| | VV | 82.1 | 85.7 | 86.3 | 82.8 | 84.2 |
| Descending | VH | 90.4 | 92.0 | 92.2 | 90.5 | 91.3 |
| | VV | 84.0 | 88.3 | 88.9 | 84.7 | 86.4 |

Next, an experiment was conducted to compare the accuracy of rice detection with and without smoothing preprocessing. In addition, most studies in the literature use the maximum and minimum backscatter coefficients to detect rice fields. Therefore, they can apply the raw backscatter coefficient data without smoothing. In the experiment, we compared the performance of all five features (with and without smoothing). As shown in Table 5, the O.A. of the proposed method using all five features is 90.2% and 80.1% with and without smoothing, respectively. The reason for the lower O.A. of the proposed method without smoothing processing is that the P.A. of rice is only 67.3%. This is due to model distortion caused by fluctuations in the backscattering coefficient and certain features such as TI and BVR are highly dependent on built models. Thus, the smoothing processing is important to establish the backscatter variation model for the proposed method.

**Table 5.** Detection accuracy of the combined features with and without smoothing processing for VH polarization with ascending orbit mode.

| Features | Smoothing Processing | Rice | | Non-Rice | | O.A. (%) |
|---|---|---|---|---|---|---|
| | | P.A. (%) | U.A. (%) | P.A. (%) | U.A. (%) | |
| All | Yes | 87.5 | 83.2 | 91.5 | 93.7 | 90.2 |
| All | No | 67.3 | 70.4 | 86.4 | 84.5 | 80.1 |

### 3.2. Classifier Comparison

In the experiment, the rice detection performance of the proposed method was compared with four other classification algorithms: DT, SVM, KNN and QDA, by using the dataset as in Section 3.1. The dataset consists of the rice and non-rice samples from five regions in Figure 9. Rice detection performance was verified by 5-fold validation in MATLAB and O.A., U.A. and P.A. were evaluated by using 100 Monte Carlo trials for all classifiers. The experiments were conducted using all extracted features as input data for all classification algorithms. In addition, three scales, Fine, Medium and Coarse, were considered for DT, SVM and KNN classifiers, provided by MATLAB Machine Learning Toolbox. Table 6 summarizes the detection results of all classifiers. SVM classifiers are one of the favorite methods in remote sensing research. In fact, they provide better performance than the other three classifiers, in addition to the proposed method. Among all SVM classifiers, the best one is SVM with a fine Gaussian kernel. The overall classification accuracy and kappa coefficient are 88.6% and 0.77, respectively. In addition, the fine-scaled KNN classifier achieves a high rice P.A. of 96.9%, but O.A. is only 69.2% which is due to the poor non-rice detection ability of KNN. Among the three scaled KNN classifiers, only coarse-scaled KNN has better classification performance, with 80.3% O.A. and 0.6 kappa. Compared with the other four classifiers, the proposed method can achieve higher classification accuracy with 91.9% O.A. and 0.83 kappa.

**Table 6.** Comparison of the proposed method with the other four classifiers for VH polarization with ascending orbit mode, including O.A., U.A., P.A. and kappa.

| Classifiers | | Rice | | Non-Rice | | O.A. (%) | Kappa |
|---|---|---|---|---|---|---|---|
| | | P.A. (%) | U.A. (%) | P.A. (%) | U.A. (%) | | |
| Proposed Method | | 91.2 | 92.4 | 92.6 | 91.3 | 91.9 | 0.83 |
| DT | Fine | 85.3 | 87.2 | 87.5 | 85.6 | 86.4 | 0.72 |
| | Medium | 82.2 | 85.7 | 86.4 | 82.9 | 84.3 | 0.68 |
| | Coarse | 88.1 | 73.2 | 67.7 | 85.0 | 77.9 | 0.55 |
| Gaussian SVM | Fine | 88.3 | 88.9 | 88.9 | 88.3 | 88.6 | 0.77 |
| | Medium | 82.3 | 85.7 | 86.4 | 83.0 | 84.3 | 0.68 |
| | Coarse | 85.3 | 77.4 | 75.1 | 83.6 | 80.2 | 0.6 |
| KNN | Fine | 96.9 | 62.3 | 41.5 | 93.0 | 69.2 | 0.38 |
| | Medium | 92.6 | 68.5 | 57.4 | 88.5 | 75.0 | 0.50 |
| | Coarse | 86.5 | 77.0 | 74.1 | 84.5 | 80.3 | 0.60 |
| QDA | | 85.7 | 77.6 | 75.3 | 84.0 | 80.5 | 0.61 |

### 3.3. Mapping Rice-Field Distribution

In this experiment, rice-field mapping of the study area was detected by the pre-trained classifiers in Section 3.2. According to the results of Table 6, the best scale of DT, SVM and KNN was selected for the rice detection. Thus, five classifiers were considered in the experiment, including the proposed method, fine-scaled DT, SVM with fine Gaussian kernel, coarse-scaled KNN and QDA. Instead of samples of the dataset, all image pixels were exploited in the experiment.

The rice detection results of the five selected regions are given in Figure 10 and Table 7. Figure 10 shows the rice-field mapping in which rice and non-rice pixels are shown in red and yellow, respectively, and non-interesting area in black. The detection accuracy of all classifiers was summarized in Table 7. Due to the large disparity in the number of pixels between rice fields and non-rice fields in the selected regions, the detection accuracy was determined by P.A. and O.A. in Table 7. Comparing the detection results of the five selected regions with the ground truth data in Figure 10, it can be observed that most rice fields can be correctly detected by the classifiers. As shown in Table 7, the rice P.A. of five selected regions can achieve over 80% for all classifiers. One exception is the poor rice detection performance of the KNN classifier in region 1. In region 2, the performance of the proposed method is similar to SVM, with 84.1% and 84.2% O.A., respectively. Obviously, the proposed method provides a non-rice P.A. of 88.7%, higher than the other classifiers in region 3. Whereas in region 4, the O.A. of all classifiers are lower than 80%, due to the poor detection efficiency of non-rice. Almost all classifiers can provide effective rice detection in region 5, especially the proposed method which can reach a high O.A. of 90.2%. Compared with other regions, the cultivated lands in region 4 are relatively small (less than 0.5 ha), densely distributed and have more non-rice species planted. Therefore, the backscattering of rice is affected by nearby non-rice crops. However, in this experiment, the decision criterion was determined based on the 95% confidence interval from the feature distributions of all training data. The decision parameters were not adjusted for the case of more diverse crop species. Thus, non-rice detection accuracy in region 4 is lower than other regions, resulting in lower overall detection accuracy. In contrast, rice fields are densely distributed and the crop types are single in region 5, so overall detection accuracy is better than other regions. Therefore, the diversity of crops and the dispersion degree of farmlands may affect the detection results. However, in all experimental regions, the proposed method can consistently provide higher classification accuracy than the other four classifiers.

In addition, to evaluate the efficiency of the proposed method, the processing timing of rice detection was evaluated by using selected region 5 with image size $1000 \times 1000$ pixels. The execution time of proposed method is 0.6 s. As for other classifiers, SVM takes more than 20 s, DT takes about 1.5 s, QDA takes about 1.2 s and KNN classifier takes about 2 s.

In the following experiment, we further estimated the rice-fields of Yunlin and Changhua counties by the proposed method. The rice field mapping of the study area is shown in Figure 11 with rice pixels colored in red. In order to clearly present the predicted results of rice-fields, three regions selected from Figure 11 were compared with the corresponding ground truth in Figure 12. Most of rice-fields can be detected by the proposed method except for a few sporadic agricultural lands (shown in the black circles in Figure 12). Obviously, the farmlands in these areas are small and scattered. The detection errors may be caused by other planting of crops, such as peanuts and garlic, between rice fields. According to the statistical data from AFA in 2017 (http://210.69.71.166/Pxweb2007/Dialog/statfile9L.asp), the total area of rice-field cultivation is 58,829.6 ha in Yunlin and Changhua counties. By using the proposed method, the total rice-field area in the study counties was estimated to be 54,673.79 ha based on the results in Figure 11, and the rice detection accuracy reached 92.9%.

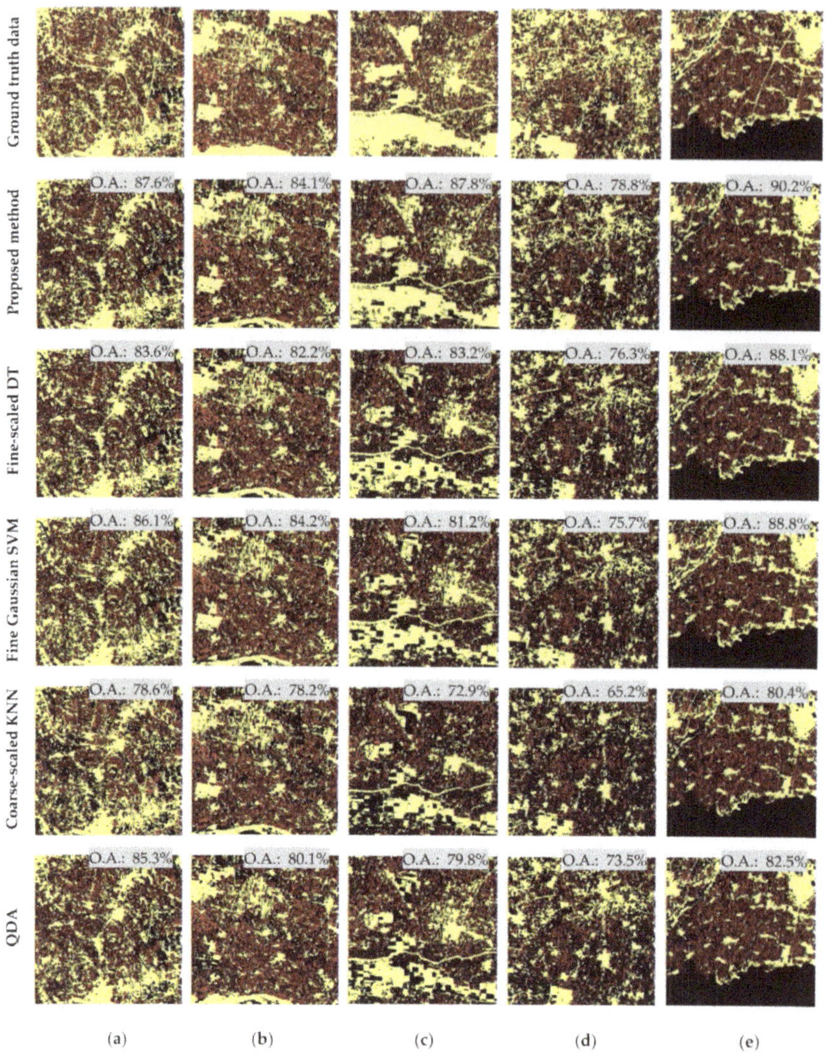


**Figure 10.** The rice-field detection of the five selected regions. (**a**–**e**) corresponding to regions 1–5 in Figure 8, respectively. The top row shows the ground truth data (corresponding to Figure 9), the remaining rows show the results from the five classifiers. Rice pixels are shown in red, non-rice pixels in yellow and non-interesting pixels in black.

**Table 7.** Detection accuracy of the five regions by the proposed method and the other four classifiers for VH polarization with ascending orbit mode.

| Regions | Classifiers | Rice P.A. (%) | Non-Rice P.A. (%) | O.A. (%) |
|---|---|---|---|---|
| | Proposed method | 84.7 | 88.4 | 87.6 |
| | Fine-scaled DT | 86.7 | 82.7 | 83.6 |
| 1 | Fine Gaussian SVM | 80.3 | 87.4 | 86.1 |
| | Coarse-scaled KNN | 70.4 | 80.9 | 78.6 |
| | QDA | 84.2 | 85.6 | 85.3 |
| | Proposed method | 82.4 | 84.8 | 84.1 |
| | Fine-scaled DT | 81.2 | 82.6 | 82.2 |
| 2 | Fine Gaussian SVM | 82.3 | 85.0 | 84.2 |
| | Coarse-scaled KNN | 85.1 | 75.2 | 78.2 |
| | QDA | 88.4 | 76.5 | 80.1 |
| | Proposed method | 83.8 | 88.7 | 87.8 |
| | Fine-scaled DT | 83.6 | 83.1 | 83.2 |
| 3 | Fine Gaussian SVM | 88.3 | 79.5 | 81.2 |
| | Coarse-scaled KNN | 91.5 | 68.4 | 72.9 |
| | QDA | 87.5 | 77.9 | 79.8 |
| | Proposed method | 90.1 | 76.1 | 78.8 |
| | Fine-scaled DT | 88.0 | 73.5 | 76.3 |
| 4 | Fine Gaussian SVM | 91.4 | 72.0 | 75.7 |
| | Coarse-scaled KNN | 89.6 | 59.4 | 65.2 |
| | QDA | 87.7 | 70.1 | 73.5 |
| | Proposed method | 87.5 | 91.5 | 90.2 |
| | Fine-scaled DT | 86.8 | 88.8 | 88.1 |
| 5 | Fine Gaussian SVM | 87.5 | 89.4 | 88.8 |
| | Coarse-scaled KNN | 90.0 | 75.7 | 80.4 |
| | QDA | 92.3 | 77.7 | 82.5 |

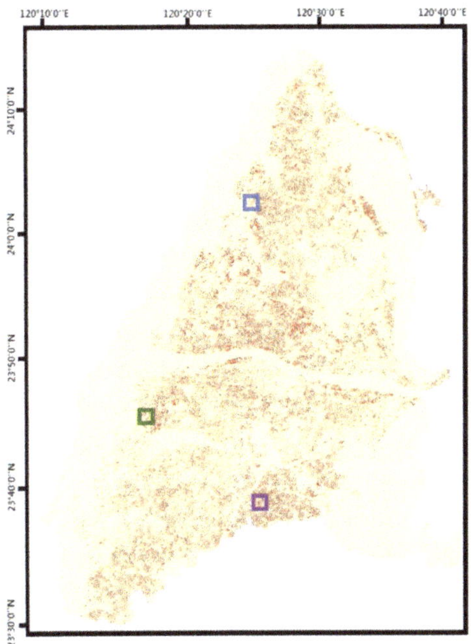


**Figure 11.** The rice-field mapping of the study area. The yellow color is the study area and the rice pixels are shown in red. The blue, green and purple squares represent the three selected regions.

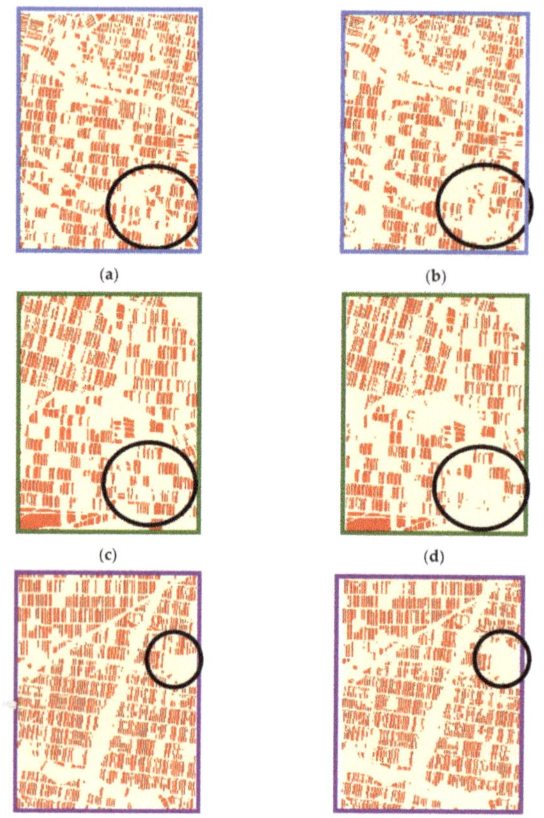

(a) (b) (c) (d) (e) (f)

**Figure 12.** Comparing the detection results of selected regions with ground truth data of the selected regions from Figure 11. (**a**,**c**,**e**) are ground truth data; (**b**,**d**,**f**) are rice-fields mapping.

## 4. Discussion

Rice growth has a specific time sequence, from transplanting, heading to maturity. The variation of rice plants at different growth stages is reflected in the SAR backscatter value. The study has made full use of the phenological characteristics of rice, and proposed the associated features of rice growth based on the temporal SAR backscatter values. All extracted features were combined for rice detection. Experiment results validate the efficiency of the proposed rice-field detection, with OA greater than 90%.

Other researches [38,60] have also detected the distribution of rice fields in the same study area as ours. Jyun-bin [38] utilized the Normalized Difference Sigma-naught Index (NDSI), which is calculated based on the maximum and minimum backscatter values of the Sentinel-1A time-series data. NDSI is the same as the proposed feature BD, which achieves 85.8% overall detection accuracy as shown in Table 3. Chia-Hao [60] proposed a temporal-local binary pattern (T-LBP) method which combined time-series SAR data and local binary patterns. T-LBP describes the temporal change pattern of rice pixel in time-series Sentinel-1 image by applying LBP operator into temporal SAR backscatter. The results of [60] showed that the overall accuracy, rice-fields' user accuracy and rice-fields' producer accuracy were 74.1%, 72.2% and 77.6%, respectively.

To demonstrate the flexibility of the proposed method with different training and testing data, an experiment was conducted based on the same dataset as in Section 3.1, but using the samples in one region as training data, and the remaining samples in other regions as test data. For example, the decision model was pre-trained by using the samples from region 1 and tested by the data in regions 2–5. Therefore, in the experiment, there are 20% training data (10,000 rice fields and non-paddy fields) and 80% test data (40,000 rice fields and non-paddy fields). The detection results for VH polarization with ascending orbit mode in 2017 were summarized in Table 8. The results show that the pre-trained model based on the data in region 5 has a better testing average O.A. than the models based on other regions, where the average O.A. is the average value of O.A. of the other four testing region. Whereas, the pre-trained model based on region 4 has a lowest average O.A. These results may come from the difference in the complexity of crops. As mentioned in Section 3.1, there are multiple crops planted in region 4 and a single rice crop grown in region 5. In addition, the mean of average O.A. from different training regions is 90.7% which was 1% lower than the O.A. in Table 3. The results verify the effectiveness of the proposed method under different training data.

**Table 8.** The overall accuracy for different training and testing data for VH polarization with ascending orbit mode in 2017.

| Train \ Test | Region 1 | Region 2 | Region 3 | Region 4 | Region 5 | Average O.A. |
|---|---|---|---|---|---|---|
| Region 1 | N/A | 92.3 | 91.6 | 90.4 | 89.8 | 91.0 |
| Region 2 | 91.9 | N/A | 91.2 | 91.1 | 89.6 | 91.0 |
| Region 3 | 89.7 | 90.3 | N/A | 91.0 | 88.9 | 90.0 |
| Region 4 | 89.6 | 90.1 | 89.8 | N/A | 88.4 | 89.5 |
| Region 5 | 92.1 | 91.9 | 91.8 | 91.4 | N/A | 91.8 |

N/A: not applicable.

Furthermore, in order to validate the stability of the proposed method in different years, the pre-trained classifier in Section 3.2 was applied to estimate the rice-fields of five selected regions, shown in Figure 8, in 2018. The detection results in 2017 and 2018 were summarized in Table 9. As shown in Table 9, rice P.A. of the five selected regions in 2018 were slightly lower than the detection results in 2017. For regions 2, 4 and 5, the O.A. values in 2018 were slightly lower than those in 2017, while for regions 1 and 3, the O.A. values in 2018 were slightly higher than those in 2017. Thus, by using the pre-trained classifier in 2017, the accuracy of rice detection can be maintained in 2018 for all selected regions.

**Table 9.** Detection accuracy in 2017 and 2018 by the proposed method for VH polarization with ascending orbit mode.

| Regions | Years | Rice | | Non-Rice | | O.A. (%) |
|---|---|---|---|---|---|---|
| | | P.A. (%) | U.A. (%) | P.A. (%) | U.A. (%) | |
| 1 | 2017 | 84.7 | 67.5 | 88.4 | 95.3 | 87.6 |
| | 2018 | 83.8 | 68.0 | 88.8 | 95.0 | 87.7 |
| 2 | 2017 | 82.4 | 70.3 | 84.8 | 91.7 | 84.1 |
| | 2018 | 80.7 | 67.6 | 83.1 | 90.8 | 82.4 |
| 3 | 2017 | 83.8 | 63.9 | 88.7 | 95.8 | 87.8 |
| | 2018 | 83.6 | 64.3 | 88.9 | 95.8 | 87.9 |
| 4 | 2017 | 90.1 | 46.9 | 76.1 | 97.0 | 78.8 |
| | 2018 | 89.8 | 45.0 | 74.2 | 96.8 | 77.1 |
| 5 | 2017 | 87.5 | 83.2 | 91.5 | 93.7 | 90.2 |
| | 2018 | 86.7 | 81.3 | 90.4 | 93.3 | 89.2 |

Then, the performance of the pre-trained classifier in Section 3.2 was further tested by using one experimental area in Hualien County, which is located in eastern Taiwan. The mountainous terrain of Hualien County is very different from the topographic features of Yunlin and Changhua counties. The selected experimental area was shown in Figure 13 with an image size of 500 $\times$ 500 pixels. The P.A. of rice and non-rice is 91.7% and 83.8%, respectively. The O.A. is 84.3% in Hualien County. These results validate the stable performance of the proposed method in different terrain areas.

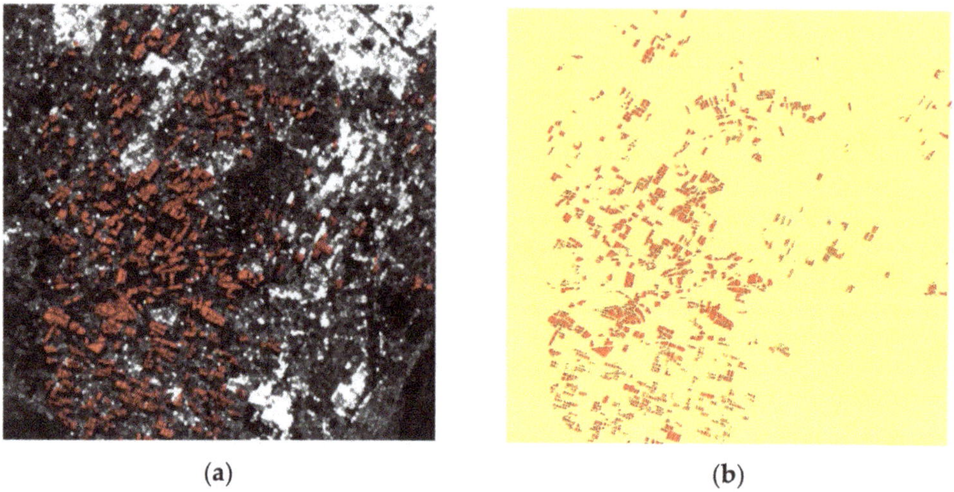

**Figure 13.** The selected area of Hualien County of (**a**) Sentinel-1 image with ground truth and (**b**) rice detection result. Rice is shown in red and non-rice in yellow.

Instead of using the SAR backscattering coefficients directly, the proposed method uses the SAR temporal variation of rice plots during rice growing season to extract rice features. Therefore, the proposed combined-feature method can accurately detect rice fields in different years and regions. In future research, combinations of different polarizations will be considered for rice-field detection. The combination of VH and VV polarizations should provide more features of rice, thereby improving the performance of the proposed method in rice detection.

## 5. Conclusions

In the study, a rice detection method based on the rice growth-related features was proposed by using the Sentinel-1A time-series SAR data. Five rice growth related features were introduced, including BD, TI, BVR, ANB and MB.

Experiments conducted using only one feature show that the detection accuracy of feature TI, with 88.4% O.A., is higher than other features, as shown in Table 3. In the study, the feature-based rice decision method with the combination of all five features was proposed. This method can achieve over 90% overall detection accuracy through test dataset verification. In addition, experiments have examined the effectiveness of the proposed method in paddy field mapping. The results show that the proposed method can reach the rice detection accuracy of 91.9% O.A. and 0.83 kappa. Compared with the best classifiers among the other four classifiers (DT, SVM, KNN and QDA), the detection accuracy O.A. of the proposed method was 3% higher than that of the fine SVM classifier, as shown in Table 6. Furthermore, the proposed method has been verified to have consistently high rice detection accuracy in different years and regions. In conclusion, the proposed method can improve the accuracy of rice detection and is effective in rice-field mapping for the experimental areas of Yunlin County and Changhua County in central Taiwan.

**Author Contributions:** Data curation, Y.-T.C.; funding acquisition, Y.-L.C.; methodology, L.C.; project administration, Y.-L.C.; software, Y.-T.C.; supervision, J.-H.W.; validation, L.C.; writing—original draft, Y.-T.C.; writing—review and editing, L.C.; Y.-T.C., L.C. and Y.-L.C. made equal contributions. All authors have read and agreed to the published version of the manuscript.

**Funding:** This research was funded by the National Space Organization, Taiwan.

**Institutional Review Board Statement:** Not applicable.

**Informed Consent Statement:** Not applicable.

**Data Availability Statement:** No new data were created or analyzed in this study. Data sharing not applicable.

**Acknowledgments:** This work was sponsored by National Taipei University of Technology (grant nos. USTP-NTUT-NTOU-107-02 and NTUT-USTB-108-02).

**Conflicts of Interest:** The authors declare no conflict of interest.

## References

1. Elert, E. Rice by the numbers: A good grain. *Nature* **2014**, *514*, S50–S51. [CrossRef] [PubMed]
2. Mohanty, S.; Wassmann, R.; Nelson, A.; Moya, P.; Jagadish, S.V. *Rice and Climate Change: Significance for Food Security and Vulnerability*; IRRI: Manila, Philippines, 2014; Volume 49, pp. 1–14.
3. FAO. FAO Rice Market Monitor. 2018, Volume 21. Available online: http://www.fao.org/economic/est/publications/rice-publications/rice-market-monitor-rmm/en/ (accessed on 30 April 2018).
4. Kubo, M.; Purevdorj, M. The future of rice production and consumption. *J. Food Distrib. Res.* **2004**, *35*, 128–142.
5. Sass, R.L.; Cicerone, R.J. Photosynthate allocations in rice plants: Food production or atmospheric methane? *Proc. Natl. Acad. Sci. USA* **2002**, *99*, 11993–11995. [CrossRef] [PubMed]
6. Wheeler, T.; von Braun, J. Climate change impacts on global food security. *Science* **2013**, *341*, 508–513. [CrossRef] [PubMed]
7. Metz, B.; Davidson, O.R.; Bosch, P.R.; Dave, R.; Meyer, L.A. Climate Change 2007: Impacts, Adaptation and Vulnerability. In Proceedings of the 8th Session of Working Group II of the IPCC, Brussels, Belgium, 13 April 2007; Volume 987.
8. Matthews, R.; Wassmann, R. Modelling the impacts of climate change and methane emission reductions on rice production: A review. *Eur. J. Agron.* **2003**, *19*, 573–598. [CrossRef]
9. Chipanshi, A.C.; Chanda, R.; Totolo, O. Vulnerability assessment of the maize and sorghum crops to climate change in Botswana. *Clim. Chang.* **2004**, *61*, 339–360. [CrossRef]
10. Peng, S.; Huang, J.; Sheely, J.E.; Laza, R.C.; Visperas, R.M.; Zhong, X.; Centeno, G.S.; Khush, G.S.; Cassman, K.G. Rice yields decline with higher night temperature from global warming. *Proc. Natl. Acad. Sci. USA* **2005**, *101*, 9971–9975. [CrossRef]
11. Furuya, J.; Kobayashi, S. Impact of global warming on agricultural product markets: Stochastic world food model analysis. *Sustain. Sci.* **2009**, *4*, 71–79. [CrossRef]
12. GRiSP (Global Rice Science Partnership). *Rice Almanac*, 4th ed.; International Rice Research Institute: Manila, Philippines, 2013; Volume 283.
13. Son, N.T.; Chen, C.F.; Chen, C.R.; Duc, H.N.; Chang, L.Y. A phenology-based classification of time-series MODIS data for rice crop monitoring in Mekong Delta, Vietnam. *Remote Sens.* **2013**, *6*, 135–156. [CrossRef]
14. Kontgis, C.; Schneider, A.; Ozdogan, M. Mapping rice paddy extent and intensification in the Vietnamese Mekong River Delta with dense time stacks of Landsat data. *Remote Sens. Environ.* **2015**, *169*, 255–269. [CrossRef]
15. Su, T. Efficient paddy field mapping using Landsat-8 imagery and object-based image analysis based on advanced fractal net evolution approach. *GISci. Remote Sens.* **2016**, *54*, 354–380. [CrossRef]
16. Onojeghuo, A.O.; Blackburn, G.A.; Wang, Q.; Atkinson, P.M.; Kindred, D.; Miao, Y. Rice crop phenology mapping at high spatial and temporal resolution using downscaled MODIS time-series. *GISci. Remote Sens.* **2018**, *55*, 659–677. [CrossRef]
17. Sakamoto, T.; Sprague, D.S.; Okamoto, K.; Ishitsuka, N. Semi-automatic classification method for mapping the rice-planted areas of Japan using multi-temporal Landsat images. *Remote Sens. Appl. Soc. Environ.* **2018**, *10*, 7–17. [CrossRef]
18. Pena-Barragan, J.M.; Ngugi, M.K.; Plant, R.E.; Six, J. Object-based crop identification using multiple vegetation indices, textural features and crop phenology. *Remote Sens. Environ.* **2011**, *115*, 1301–1316. [CrossRef]
19. Son, N.T.; Chen, C.F.; Chen, C.R.; Chang, L.Y.; Duc, H.N.; Nguyen, L.D. Prediction of rice crop yield using MODIS EVI-LAI data in the Mekong Delta, Vietnam. *Int. J. Remote Sens.* **2013**, *34*, 7275–7292. [CrossRef]
20. Chen, C.R. Satellite-Based Information for Rice Crop Monitoring. Ph.D. Thesis, National Central University, Taoyuan City, Taiwan, 2014.
21. Jin, C.; Xiao, X.; Dong, J.; Qin, Y.; Wang, Z. Mapping paddy rice distribution using multi-temporal Landsat imagery in the Sanjiang Plain, northeast China. *Front. Earth Sci.* **2016**, *10*, 49–62. [CrossRef]
22. Yu, F.H.; Xu, T.Y.; Cao, Y.L.; Yang, G.J.; Du, W.; Wang, S. Models for estimating the leaf NDVI of japonica rice on a canopy scale by combining canopy NDVI and multisource environmental data in northeast China. *Int. J. Agric. Biol. Eng.* **2016**, *9*, 132–142.
23. Guan, X.D.; Huang, C.; Liu, G.H.; Meng, X.L.; Liu, Q.S. Mapping rice cropping systems in Vietnam using an NDVI-based time-series similarity measurement based on DTW distance. *Remote Sens.* **2016**, *8*, 19. [CrossRef]
24. Li, P.; Xiao, C.; Feng, Z. Mapping rice planted area using a new normalized EVI and SAVI (NVI) derived from Landsat-8 OLI. *IEEE Geosci. Remote Sens. Lett.* **2018**, *15*, 1822–1826. [CrossRef]

25. Gumma, M.K.; Thenkabail, P.S.; Deevi, K.C.; Mohammed, I.A.; Teluguntla, P.; Oliphant, A.; Xiong, J.; Aye, T.; Whitbread, A.M. Mapping cropland fallow areas in Myanmar to scale up sustainable intensification of pulse crops in the farming system. *GISci. Remote Sens.* **2018**, *55*, 926–949. [CrossRef]
26. Huke, R.E.; Huke, E.H. *Rice Area by Type of Culture: South, Southeast, and East Asia, A Revised and Updated Data Base*; IRRI: Manila, Philippines, 1997; Volume 61.
27. NASA. NASA Cloud Fraction (1 Month Terra/MODIS) 1 FAO FAOSTAT. Available online: http://faostat3.fao.org/faostatgateway/go/to/home/E (accessed on 1 June 2014).
28. Zhu, Z.; Woodcock, C.E.; Rogan, J.; Kellndorfer, J. Assessment of spectral, polarimetric, temporal, and spatial dimensions for urban and peri-urban land cover classification using Landsat and SAR data. *Remote Sens. Environ.* **2012**, *119*, 72–82. [CrossRef]
29. Niu, X.; Ban, Y. Multi-temporal RADARSAT-2 polarimetric SAR data for urban land-cover classification using an object-based support vector machine and a rule-based approach. *Int. J. Remote Sens.* **2013**, *34*, 1–26. [CrossRef]
30. Ribbes, F. Rice-field mapping and monitoring with RADARSAT data. *Int. J. Remote Sens.* **1999**, *20*, 745–765. [CrossRef]
31. Shao, Y.; Fan, X.; Liu, H.; Xiao, J.; Ross, S.; Brisco, B.; Brown, R.; Staples, G. Rice monitoring and production estimation using multitemporal RADARSAT. *Remote Sens. Environ.* **2001**, *76*, 310–325. [CrossRef]
32. Wang, L.-F.; Kong, J.A.; Ding, K.-H.; Le Toan, T.; Ribbes-Baillarin, F.; Floury, N. Electromagnetic scattering model for rice canopy based on Monte Carlo simulation. *Prog. Electromagn. Res.* **2005**, *52*, 153–171. [CrossRef]
33. Nguyen, D.B.; Gruber, A.; Wagner, W. Mapping rice extent and cropping scheme in the Mekong Delta using Sentinel-1A data. *Remote Sens. Lett.* **2016**, *7*, 1209–1218. [CrossRef]
34. Mansaray, L.R.; Zhang, D.; Zhou, Z.; Huang, J. Evaluating the potential of temporal Sentinel-1A data for paddy rice discrimination at local scales. *Remote Sens. Lett.* **2017**, *8*, 967–976. [CrossRef]
35. Lopez-Sanchez, J.M.; Vicente-Guijalba, F.; Erten, E.; Campos-Taberner, M.; Garcia-Haro, F.J. Retrieval of vegetation height in rice-fields using polarimetric SAR interferometry with TanDEM-X data. *Remote Sens. Environ.* **2017**, *192*, 30–44. [CrossRef]
36. Nguyen, D.B.; Clauss, K.; Cao, S.; Naeimi, V.; Kuenzer, C.; Wagner, W. Mapping Rice Seasonality in the Mekong Delta with Multi-Year Envisat ASAR WSM Data. *Remote Sens.* **2015**, *7*, 15868–15893. [CrossRef]
37. Nguyen, D.B.; Wagner, W. European Rice Cropland Mapping with Sentinel-1 Data: The Mediterranean Region Case Study. *Water* **2017**, *9*, 392. [CrossRef]
38. Chen, J.-B. Rice Crop Classification Using Sentinel-1A SAR Data in Central Taiwan. Master's Thesis, National Central University, Taoyuan City, Taiwan, 2016.
39. Mohite, J.; Sawant, S.; Kumar, A.; Prajapati, M.; Pusapati, S.; Singh, D.; Pappula, S. Operational near real time rice area mapping using multi-temporal Sentinel-1 SAR observations. *ISPRS* **2018**, *42*, 433–438. [CrossRef]
40. Clauss, K.; Ottinger, M.; Kuenzer, C. Mapping rice areas with Sentinel-1 time-series and superpixel segmentation. *Int. J. Remote Sens.* **2018**, *39*, 1399–1420. [CrossRef]
41. Phan, H.; Le Toan, T.; Bouvet, A.; Nguyen, L.D.; Pham Duy, T.; Zribi, M. Mapping of rice varieties and sowing date using X-band SAR data. *Sensors* **2018**, *18*, 316. [CrossRef] [PubMed]
42. Tian, H.; Wu, M.; Wang, L.; Niu, Z. Mapping early, middle and late rice extent using Sentinel-1A and Landsat-8 data in the Poyang Lake Plain, China. *Sensors* **2018**, *18*, 185. [CrossRef]
43. Lasko, K.; Prasad Vadrevu, K.; Tuan Tran, V.; Justice, C. Mapping double and single crop paddy rice with Sentinel-1A at varying spatial scales and polarizations in Hanoi, Vietnam. *IEEE J. Sel. Top Appl. Earth Obs. Remote Sens.* **2018**, *11*, 498–512. [CrossRef]
44. Park, S.; Im, J.; Park, S.; Yoo, C.; Han, H.; Rhee, J. Classification and mapping of paddy rice by combining Landsat and SAR time-series data. *Remote Sens.* **2018**, *10*, 447. [CrossRef]
45. Bazzi, H.; Baghdadi, N.; El Hajj, M.; Zribi, M.; Tong, H.; Ndikumana, E.; Dominique, C.; Belhouchette, H.; Ho Tong Minh, D. Mapping paddy rice using Sentinel-1 SAR time-series in Camargue, France. *Remote Sens.* **2019**, *11*, 887. [CrossRef]
46. Verma, A.K.; Nandan, R.; Verma, A. Knowledge based classifier based on backscattering coefficient for monitoring the crop growth analysis using multi-temporal images of space-borne Synthetic Aperture Radar (SAR) sensors. *ISPRS* **2019**, *42*, 643–647. [CrossRef]
47. Sudarmanian, N.S.; Pazhanivelan, S. Detection of SAR signature for rice crop using Sentinel 1A data. *Bull. Environ. Pharmacol. Life Sci.* **2019**, *8*, 32–39.
48. Courault, D.; Hossard, L.; Flamain, F.; Baghdadi, N.; Irfan, K. Assessment of agricultural practices from Sentinel 1 & 2 images applied on rice fields to develop a farm typology in the Camargue region. *IEEE J. Sel. Top. Appl. Earth Obs. Remote Sens.* **2020**, *13*, 5027–5035.
49. Lee, J.-S. Refined filtering of image noise using local statistics. *Comput. Graph. Image Proc.* **1981**, *15*, 380–389. [CrossRef]
50. George, S.; Geoff, W. Smoothing of three dimensional models by convolution. In Proceedings of the CGI '96 Proceedings, Pohang, Korea, 24–28 June 1996; pp. 184–190.
51. Bouvet, A.; Le Toan, T.; Lam Dao, N. Monitoring of the rice cropping system in the Mekong Delta using ENVISAT/ASAR dual polarization data. *IEEE Trans. Geosci. Remote Sens.* **2009**, *47*, 517–526. [CrossRef]
52. Le Toan, T.; Ribbes, F.; Wang, L.-F.; Floury, N.; Ding, K.-H.; Kong, J.A.; Fujita, M.; Kurosu, T. Rice crop mapping and monitoring using ERS-1 data based on experiment and modeling results. *IEEE Trans. Geosci. Remote Sens.* **1997**, *35*, 41–56. [CrossRef]
53. Quinlan, J.R. Induction of decision trees. *Mach. Learn.* **1986**, *1*, 81–106. [CrossRef]
54. Cortes, C.; Vapnik, V. Support-vector networks. *Mach. Learn.* **1995**, *20*, 273–297. [CrossRef]

55. Mucherino, A.; Papajorgji, P.J.; Pardalos, P.M. K-Nearest Neighbor Classification. In *Springer Optimization and Its Applications (SOIA)*; Springer: New York, NY, USA, 2009; Volume 34.
56. Santosh, S.; Maya, R.G.; Bela, A.F. Bayesian quadratic discriminant analysis. *Mach. Learn. Res.* **2007**, *8*, 1277–1305.
57. Stone, M. Cross-validatory choice and assessment of statistical predictions. *J. R. Stat. Soc.* **1974**, *36*, 111–147. [CrossRef]
58. James, C.B.; Robert, E.; Willian, F. FCM: The fuzzy c-means clustering algorithm. *Comput. Geosci.* **1984**, *10*, 191–203.
59. Congalton, R.G. Accuracy assessment and validation of remotely sense and other spatial information. *Int. J. Wildland Fire* **2001**, *10*, 321–328. [CrossRef]
60. Chang, C.-H.; Chu, T.-H.; Fuh, C.-S. Detecting Rice Crop Fields Distribution with Spatial-Temporal Remote Sensing Information during the First Rice Season in 2018. In *GLP 2018*; NTU: Taoyuan City, Taiwan, 2018; Available online: https://www.csie.ntu.edu.tw/~{}fuh/personal/DetectingRiceCropFieldsDistributionwithSpatio-TemporalRemoteSensingInformationDuringtheFirstRiceSeasonin2018.pdf (accessed on 30 September 2018).

remote sensing

MDPI

*Article*

# Crop Height Estimation of Corn from Multi-Year RADARSAT-2 Polarimetric Observables Using Machine Learning

**Qinghua Xie [1,2,3], Jinfei Wang [3,4], Juan M. Lopez-Sanchez [2], Xing Peng [1,*], Chunhua Liao [3], Jiali Shang [5], Jianjun Zhu [6], Haiqiang Fu [6] and J. David Ballester-Berman [2]**

1 School of Geography and Information Engineering, China University of Geosciences (Wuhan), Wuhan 430074, China; xieqh@cug.edu.cn
2 Institute for Computer Research (IUII), University of Alicante, E-03080 Alicante, Spain; juanma.lopez@ua.es (J.M.L.-S.); davidb@ua.es (J.D.B.-B.)
3 Department of Geography and Environment, The University of Western Ontario, London, ON N6A 5C2, Canada; jfwang@uwo.ca (J.W.); cliao33@uwo.ca (C.L.)
4 Institute for Earth and Space Exploration, The University of Western Ontario, London, ON N6A 3K7, Canada
5 Ottawa Research and Development Centre, Agriculture and Agri-Food Canada, Ottawa, ON K1A 0C6, Canada; jiali.shang@canada.ca
6 School of Geosciences and Info-Physics, Central South University, Changsha 410083, China; zjj@csu.edu.cn (J.Z.); haiqiangfu@csu.edu.cn (H.F.)
* Correspondence: pengxing@cug.edu.cn

**Abstract:** This study presents a demonstration of the applicability of machine learning techniques for the retrieval of crop height in corn fields using space-borne PolSAR (Polarimetric Synthetic Aperture Radar) data. Multi-year RADARSAT-2 C-band data acquired over agricultural areas in Canada, covering the whole corn growing period, are exploited. Two popular machine learning regression methods, i.e., Random Forest Regression (RFR) and Support Vector Regression (SVR) are adopted and evaluated. A set of 27 representative polarimetric parameters are extracted from the PolSAR data and used as input features in the regression models for height estimation. Furthermore, based on the unique capability of the RFR method to determine variable importance contributing to the regression, a smaller number of polarimetric features (6 out of 27 in our study) are selected in the final regression models. Results of our study demonstrate that PolSAR observables can produce corn height estimates with root mean square error (RMSE) around 40–50 cm throughout the growth cycle. The RFR approach shows better overall accuracy in corn height estimation than the SVR method in all tests. The six selected polarimetric features by variable importance ranking can generate better results. This study provides a new perspective on the use of PolSAR data in retrieving agricultural crop height from space.

**Keywords:** crop height; RADARSAT-2; corn; Synthetic Aperture Radar (SAR); PolSAR; machine learning; RFR; SVR; agriculture

**Citation:** Xie, Q.; Wang, J.; Lopez-Sanchez, J.M.; Peng, X.; Liao, C.; Shang, J.; Zhu, J.; Fu, H.; Ballester-Berman, J.D. Crop Height Estimation of Corn from Multi-Year RADARSAT-2 Polarimetric Observables Using Machine Learning. *Remote Sens.* **2021**, *13*, 392. https://doi.org/10.3390/rs13030392

Academic Editors: Olaniyi Ajadi and David McAlpin
Received: 31 December 2020
Accepted: 19 January 2021
Published: 23 January 2021



## 1. Introduction

Crop height is an important agronomic descriptor related to crop type, biomass estimation, phenological stage, potential yield, detection of growth anomalies (e.g., diseases, pests, weather disasters, and cereal lodging), and precision fertilization [1–3]. Traditional methods to monitor crop height by visual inspection require a huge workforce over large areas [4]. Synthetic Aperture Radar (SAR), with its capability of imaging in day and night and all weather conditions and its sensitivity to the geometric and physical properties of the target, has shown to be an effective remote sensing technique in crop biophysical parameter retrieval at regional and global scales. For crop height estimation with SAR observations, the approaches can be generally categorized into three types: backscattering model methods, interferometric SAR (InSAR) methods, and data-driven empirical model methods [1].

The backscattering models for crop height retrieval include both physical models and semi-empirical models. The physical model developed for crop height usually adopts a discrete approach, such as the Radiative Transfer Theory Model (RTM), which is able to simulate the backscattering coefficient for crop targets from the perspective of fine electromagnetic scattering as a function of various geometric and physical parameters of the plant, such as canopy height, dielectric constant, number of layers and leaves, leaf angle and size, stem width, and so on [5–12]. The physical scattering model usually depends on the polarization and crop type. Due to the complexity of the physical scattering model, crop height estimation may require computationally expensive Monte Carlo simulations to relate the SAR measurements to parameters describing the entire canopy's physical characteristics [1]. Moreover, the inversion process of model parameters often leads to ill-posed problems due to a high-dimensional parameter space [1,4,13,14]. Although the merging of a metamodel (e.g., the polynomial chaos expansion (PCE)) with the backscattering model enables a significant reduction of the computational cost and the complexity involved in the inversion scheme, the growth stage needs to be identified in advance to narrow the solution space [4,13,14]. For vegetation, the most popular semi-empirical model is the Water Cloud Model (WCM) proposed by Attema and Ulaby [15]. Due to its simplicity and practicability, the WCM has been extensively applied to soil moisture estimation and to the retrieval of various vegetation biophysical variables, such as leaf area index (LAI), aboveground biomass, and vegetation height [16–18]. However, the retrieval results from the original WCM often show low accuracies attributed to many assumptions and simplifications involved in the model. In past studies, many modifications of the WCM have been developed by considering more complex scattering mechanisms or more vegetation geometrical properties. Moreover, there were some studies reporting calibration of the model coefficients of the WCM for specific areas, vegetation types, or SAR sensors based on some ground measurements [19].

The InSAR method exploits the interferometric phase between two co-registered SAR images acquired in the same polarization to capture the height of the scattering phase center [20], which is approximately considered as the crop surface height in agricultural areas, like a digital surface model. Then, an external digital terrain model (DTM), also called vegetation-free digital elevation model (DEM), is required to derive the crop height itself [1]. In order to obtain accurate crop surface height, it needs some strict conditions in general. For example, the available SAR data is expected to work with short wavelengths and appropriate polarization to enable the scattering phase center to be located as close as possible to the top of the canopy. For the same purpose, a structurally dense crop is required as well. Since the crop height is usually very low with respect to forest height, a relatively large spatial baseline is required to reduce the height of ambiguity. Moreover, a short enough revisit time is expected to obtain high coherence since crop height is assumed unchanged within this time interval. In addition, the quality of the DTM data used to remove the underlying topography from crop surface height is also an unignorable factor. Although recently, single-polarization InSAR data at the L- and P-band have proven its capability to generate a comparable performance in DTM inversion in forest areas with respect to the traditional PolInSAR method with fully polarimetric SAR data [21], currently, the accurate DTM product over a vegetation area is mostly generated from other measurement technologies, such as light detection and ranging(LIDAR), polarimetric SAR interferometry (PolInSAR) [22–24], SAR Tomography (TomoSAR) [25], and field topographic mapping. The PolInSAR method combines the interferometric and polarimetric information to better separate the different scattering phase centers in the vegetation volume, which has been demonstrated to be useful for estimating vegetation structural parameters [26,27]. For vegetation height estimation, the PolInSAR technique has been validated in a variety of forest types with many airborne and few spaceborne datasets at different radar frequencies [22,23,28–39]. The performance is constrained by two key aspects: temporal decorrelation and spatial baseline. A shorter revisit time is expected to provide higher coherence, which is related to the quality of the interferometric phase [40]. A relatively large baseline is expected to provide enough sensitivity of height measurement. Since crops grow faster and have shorter heights than forests, the availability of PolInSAR data for crop height estimation is more constrained than for forest height inversion [1–3]. Until now, few successful examples of crop height retrieval with the PolInSAR method are restricted to data acquired in indoor experiments [41,42] and airborne campaigns [43,44]. With regard to spaceborne datasets, several authors have reported successful results with a dataset over paddy rice fields acquired from the science phase of the TanDEM-X mission (bistatic configuration) from April to September 2015, being the baselines especially adjusted to ten times the regular configuration, i.e., around 2–3 km [1,3]. Although the PolInSAR method shows the capability to produce accurate estimates of crop height, currently the available PolInSAR datasets for crop height monitoring are very limited.

Another available option to estimate crop height is the data-driven empirical model method, which in this study refers to the use of a regression approach to train an empirical model between some PolSAR observables and crop height. The unknown crop heights in a scene are predicted by the trained regression model and the corresponding PolSAR observables. A large number of previous studies have been reported to investigate the correlations between PolSAR observables (e.g., backscattering coefficients, polarimetric decomposition parameters, radar vegetation index, and correlation coefficients), and crop parameters (e.g., LAI, PAI, biomass, phenological stages, canopy coverage, and crop height) over different crop fields at different radar frequencies [45–54]. These research results have shown the potential of PolSAR observables for crop parameter retrieval. However, to date, there are few studies reporting crop height retrieval based on regression with PolSAR observables [55–57]. In these studies, due to the limitation of images available and field data collected, the volume of available observed samples for training is in general relatively small. The crop types studied are also limited, such as sunflower, wheat, and canola. Moreover, the number and types of selected PolSAR observables are limited, and an empirical relationship model or linear/polynomial regression model is usually chosen [55–57].

Due to the aforementioned limitations of the backscattering model methods and interferometric SAR methods in practice, this study is focused on data-driven empirical model methods. More specifically, the present study aims at providing a comprehensive demonstration and validation of crop height retrieval of corn by exploiting a large number of PolSAR observables and ground measurements with machine learning regression methods. A large dataset formed by multi-temporal C-band RADARSAT-2 (launched by the Canadian Space Agency) images and quasi-synchronous in situ measurements of crop height along three years over two geographically close study sites in Canada, with similar agricultural practices and climatic conditions, are exploited here. Corn has been chosen as the target crop for this experiment because it has significant socio-economic interest for humans' staple food, a raw material of ethanol and animal feed, and it is the cereal with the highest production worldwide. Moreover, the height range of corn during the whole growing period is relatively large, reaching over 3 m in our study at the final stage, which results in more radar signature differences within the time-series SAR acquisitions. In addition, radar response to corn with RADARSAT-2 data have been well studied in previous literature [42,45,46,48,50,58–60]. Two typical machine learning regression methods, i.e., Random Forest Regression (RFR) and Support Vector Regression (SVR), are adopted for model training instead of linear regression. Moreover, the unique ability of the RFR method to provide variable importance contributing to the regression can help us understand the results and further investigate the performance after filtering the selection of input PolSAR observables.

## 2. Materials and Methods

### 2.1. Study Site and Dataset

As shown in Figure 1, two geographically close sites both located in Southwestern Ontario, Canada, were selected. One site was located in the west of London, and the other one was near Stratford. Both study sites were agricultural areas including mainly crop fields, a few buildings, and forests. Corn, wheat, and soybean were the dominant crop types in both study sites. There was also some alfalfa, hay, and grass growing in these areas. Both study sites were suitable for cultivating crops because of abundant precipitation, mild weather, and productive soil, with relatively flat topography. In both study areas, corn and soybean were seeded in May and harvested in October of the same year. In contrast, the winter wheat growing period crossed over two years, seeded in October and harvested in July of the following year. It should be noted that in both study areas, crop rotation did not need to be done in the same field. The farmers' practice was to retain residuals for soil conservation. Therefore, one crop field may harvest residuals from another crop field from a previous year. For example, the cornfields may have residuals of wheat or soybean. All cornfields employed for height inversion and the locations of the sample points for collecting ground measurements are marked in green and red in Figure 1, respectively.

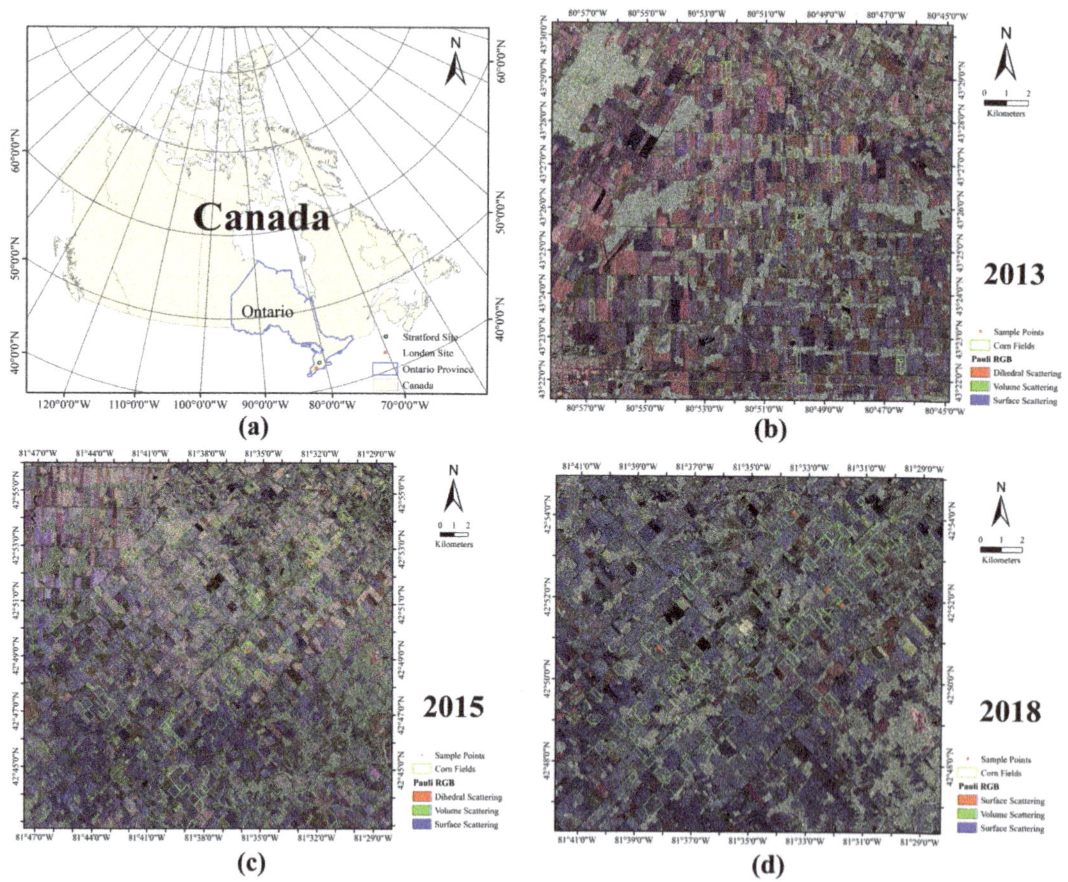


**Figure 1.** Locations and Pauli RGB images of the two study sites (Stratford and London). (**a**) Geographical locations of the two study sites; (**b**) Pauli RGB image acquired on 26 June 2013 at the Stratford site; (**c**) Pauli RGB image acquired on 23 June 2015 at the London site; (**d**) Pauli RGB image acquired on 1 July 2018 at the London site. All boundaries of cornfields and sample points for collecting ground measurements are highlighted in green and red, respectively. (RADARSAT-2 Data and Products © MacDonald, Dettwiler and Associates Ltd. (2013, 2015, 2018)—All Rights Reserved. RADARSAT is an official trademark of the Canadian Space Agency).

In total, 19 scenes of Fine Quad-Pol Wide (FQW) RADARSAT-2 data acquired in 2013, 2015, and 2018 were employed in this study. More specifically, eight scenes were acquired over the Stratford site in 2013 covering the whole corn growing period. Four scenes in 2015 and seven scenes in 2018 were acquired over the London site, which mainly covered the late growth stages. Table 1 shows the acquisition dates, beam modes, incidence angles, resolutions, and orbits of all RADARSAT-2 images.

**Table 1.** RADARSAT-2 Images and Ground Data Acquired for 2013, 2015, and 2018.

| Date | Mode | Incidence | Resolution | Orbit | Fieldwork Date | Number of Corn Sample Points | Average Corn Height (cm) | Study Site |
|---|---|---|---|---|---|---|---|---|
| 23 May 2013 | FQ9W | 27.2 ~ 30.5 | 5.1 × 4.7 | Ascending | 24 May 2013 | 4 | 5.75 | Stratford |
| 2 June 2013 | FQ19W | 37.7 ~ 40.4 | 4.7 × 4.7 | Ascending | 4 June 2013 | 16 | 10.06 | |
| 16 June 2013 | FQ9W | 27.2 ~ 30.5 | 5.1 × 4.7 | Ascending | 16 June 2013 | 17 | 25.13 | |
| 26 June 2013 | FQ19W | 37.7 ~ 40.4 | 4.7 × 4.7 | Ascending | 24 June 2013/ 25 June 2013 | 17 | 59.87 | |
| 10 July 2013 | FQ9W | 27.2 ~ 30.5 | 5.1 × 4.7 | Ascending | 10 July 2013 | 17 | 142.29 | |
| 20 July 2013 | FQ19W | 37.7 ~ 40.4 | 4.7 × 4.7 | Ascending | 21 July 2013 | 11 | 214.35 | |
| 3 August 2013 | FQ9W | 27.2 ~ 30.5 | 5.1 × 4.7 | Ascending | 3 August 2013 | 13 | 254.84 | |
| 13 August 2013 | FQ19W | 37.7 ~ 40.4 | 4.7 × 4.7 | Ascending | 13 August 2013/ 14 August 2013 | 17 | 260.78 | |
| 23 June 2015 | FQ10W | 28.4 ~ 31.6 | 5.5 × 4.7 | Ascending | 23 June 2015 | 25 | 88.44 | London |
| 10 August 2015 | FQ10W | 28.4 ~ 31.6 | 5.5 × 4.7 | Ascending | 11 August 2015 | 6 | 266.61 | |
| 3 September 2015 | FQ10W | 28.4 ~ 31.6 | 5.5 × 4.7 | Ascending | 3 September 2015 | 6 | 265.61 | |
| 13 September 2015 | FQ20W | 38.6 ~ 41.3 | 5.1 × 4.7 | Ascending | 13 September 2015 | 6 | 276.72 | |
| 1 July 2018 | FQ10W | 28.4 ~ 31.6 | 5.5 × 4.7 | Ascending | 4 July 2018 | 24 | 182.07 | |
| 25 July 2018 | FQ10W | 28.4 ~ 31.6 | 5.5 × 4.7 | Ascending | 25 July 2018 | 32 | 252.76 | |
| 1 August 2018 | FQ5W | 22.5 ~ 26.0 | 5.0 × 4.7 | Ascending | 2 August 2018 | 32 | 275.22 | |
| 18 August 2018 | FQ10W | 28.4 ~ 31.6 | 5.5 × 4.7 | Ascending | 18 August 2018 | 32 | 267.77 | |
| 25 August 2018 | FQ5W | 22.5 ~ 26.0 | 5.0 × 4.7 | Ascending | 25 August 2018 | 8 | 214.99 | |
| 1 September 2018 | FQ1W | 17.2 ~ 21.2 | 4.8 × 4.7 | Ascending | 1 September 2018 | 32 | 267.04 | |
| 15 September 2018 | FQ9W | 27.3 ~ 30.5 | 5.1 × 4.7 | Descending | 11 September 2018 | 32 | 267.22 | |

For each year, ground measurements including crop height, crop type, ground photos, soil moisture, and crop phenological stage were recorded. Due to the limitations of weather conditions, human resources, and other reasons, the in situ fieldwork was sometimes not conducted on the exact dates that the RADARSAT-2 satellite overpassed. Since the maximum offset of time was only three days (which occurred on 1 July 2018), it was reasonable to assume that the corn heights did not change from the dates of satellite acquisitions to the ground campaign dates. As it is shown in Table 1, the final number of sample points for each image was different because of limitations due to weather conditions, human resources, and other logistical reasons. For the field campaign in 2013, 17 sample points per day within a maximum of five corn fields were selected for collecting ground measurements. Finally, 112 corn sample points over the Stratford site were collected. For the ground campaign in 2015, 25 sample points per day at a maximum of four cornfields were selected to conduct measurements. In total, 43 corn sample points over the London site were collected. For the ground campaign in 2018, 32 sample points per day in a maximum of four cornfields were exploited. In total, 192 corn sample points over the London site were collected. In summary, a total of 347 corn sample points were collected from the multi-year fieldwork campaigns. Three replicate height readings per sample point were carried out in 2013 and 2015, while twelve readings were conducted in 2018. The measured corn height had a wide range, with values between 3.5 cm and 333.75 cm. The average value of corn height of the ground samples on each fieldwork date ranged from 5.75 cm to 276.72 cm.

### 2.2. Polarimetric Observables

For a fully polarimetric SAR system, the acquired single look complex (SLC) data in H-V polarization basis can be represented by a 2 × 2 scattering matrix, i.e., [61,62],

$$S = \begin{bmatrix} S_{HH} & S_{HV} \\ S_{VH} & S_{VV} \end{bmatrix} \tag{1}$$

where $S_{ij}$ ($i, j = H$ or $V$) represents the scattering coefficient from transmitted polarization $i$ and received polarization $j$. The scattering matrix $S$ is used to describe a "pure single target" or deterministic target. For distributed targets, typical in natural media, the second-order statistics (covariance matrix or coherency matrix) are usually exploited to conduct polarimetric analysis. Under the assumption of reciprocal scattering ($S_{HV} = S_{VH}$), the lexicographic basis vector and the Pauli basis vector, respectively, can be expressed as [61,62]

$$k_l = \begin{bmatrix} S_{HH} & \sqrt{2}S_{HV} & S_{VV} \end{bmatrix}^T \tag{2}$$

$$k_P = \frac{1}{\sqrt{2}}\begin{bmatrix} S_{HH}+S_{VV} & S_{HH}-S_{VV} & 2S_{HV} \end{bmatrix}^T \tag{3}$$

Then, the corresponding covariance matrix $C$ and the coherency matrix $T$ are given as [61,62]

$$C = \left\langle k_l k_l^{*T} \right\rangle = \frac{1}{2}\begin{bmatrix} \left\langle |S_{HH}|^2 \right\rangle & \sqrt{2}\langle S_{HH}S_{HV}^* \rangle & \langle S_{HH}S_{VV}^* \rangle \\ \sqrt{2}\langle S_{HV}S_{HH}^* \rangle & 2\left\langle |S_{HV}|^2 \right\rangle & \sqrt{2}\langle S_{HV}S_{VV}^* \rangle \\ \langle S_{VV}S_{HH}^* \rangle & \sqrt{2}\langle S_{VV}S_{HV}^* \rangle & \left\langle |S_{VV}|^2 \right\rangle \end{bmatrix} \tag{4}$$

$$\begin{aligned} T &= \langle k_P k_P^{*T} \rangle \\ &= \frac{1}{2}\begin{bmatrix} \left\langle |S_{HH}+S_{VV}|^2 \right\rangle & \langle (S_{HH}+S_{VV})(S_{HH}-S_{VV})^* \rangle & 2\langle (S_{HH}+S_{VV})S_{HV}^* \rangle \\ \langle (S_{HH}-S_{VV})(S_{HH}+S_{VV})^* \rangle & \left\langle |S_{HH}-S_{VV}|^2 \right\rangle & 2\langle (S_{HH}-S_{VV})S_{HV}^* \rangle \\ 2\langle S_{HV}(S_{HH}+S_{VV})^* \rangle & 2\langle S_{HV}(S_{HH}-S_{VV})^* \rangle & 4\left\langle |S_{HV}|^2 \right\rangle \end{bmatrix} \end{aligned} \tag{5}$$

Based on these polarimetric observation matrices, a large number of polarimetric parameters can be extracted for crop monitoring applications [47,48,63,64]. According to their wide usage in crop monitoring studies, a total of 27 polarimetric observables were chosen in this study, as listed in Table 2. First, as the basic products provided by a fully polarimetric SAR system, radar backscattering coefficients in co-polar channels (HH, VV) and cross-polar channel (HV) were chosen, which corresponded to the diagonal elements (C11, C22, C33) in the covariance matrix. Due to their clear physical interpretation in terms of scattering mechanisms, radar backscattering coefficients in the Pauli channels were considered, which could be extracted from the coherency matrix (T11, T22). The widely used total backscattering power, SPAN, was also selected, which was extracted from the sum of the diagonal elements in either the covariance matrix or the coherency matrix (C11 + C22 + C33 or T11 + T22 + T33). The correlation and phase between polarimetric channels (in both the linear and the Pauli basis) were also exploited, which provided additional information about the scattering from the scene. In past studies, these observables have proven very useful for crop phenology monitoring and crop state detection based on multi-temporal analysis of the radar polarimetric response [48,63,65–67]. They provided four complex correlation coefficients, which resulted in eight real observables (amplitudes and phases). Moreover, three backscattering ratios between different linear channels (HH/VV, HV/HH, and HV/VV) were also considered, which had proven to be sensitive to target characteristics. A widely used approach for generating polarimetric features is polarimetric target decomposition, which can be generally categorized as either coherent polarimetric decomposition or incoherent decomposition [61,62]. Due to the capacity to describe distributed targets, incoherent polarimetric decomposition approaches are more suitable for interpreting most nature targets [61,62]. They can be further divided into model-based decomposition and eigenvector-eigenvalue based decomposition. As the pioneering and one of the most popular model-based decomposition methods, the Freeman-Durden three-component decomposition generates three scattering power parameters representing surface, double-bounce, and volume scattering mechanisms, respectively (Freeman and Durden, 1998), which were selected in this study. Additionally, three polarimetric parameters with clear physical meanings from the representative eigenvector-eigenvalue based decomposition proposed by Cloude and Pottier were also used, including the polarimetric scattering entropy H (the degree of scattering randomness in the scattering medium), the alpha angle $\alpha$ (the average dominant scattering mechanism), and the polarimetric anisotropy A (the relative importance between the second and the third scattering mechanism) [68].

**Table 2.** List of 27 polarimetric observables selected in this study.

| Polarimetric Observable | Description |
|---|---|
| C11, C22, C33, | Backscattering coefficients in the linear polarization channels |
| T11, T22 | Backscattering coefficients in the Pauli polarization channels |
| SPAN | Total backscattering power |
| $\lvert\rho_{HHVV}\rvert$, $\lvert\rho_{HVVV}\rvert$, $\lvert\rho_{HHHV}\rvert$, $\lvert\rho_{HH+VV,HH-VV}\rvert$ | Correlation between polarimetric channels |
| $\phi_{HHVV}, \phi_{HVVV}, \phi_{HHHV}, \phi_{HH+VV,HH-VV}$ | Phase difference between polarimetric channels |
| HH/VV, HV/HH, HV/VV | Backscattering ratios |
| $P_s, P_d, P_v$ | Scattering Power from different scattering mechanisms derived from Freeman-Durden decomposition |
| H, A, $\alpha$ | Entropy, anisotropy, alpha angle from Cloude-Pottier decomposition |
| $\lvert\delta\rvert, \phi_\delta, \tau$ | Magnitude and phase of the particle scattering anisotropy, the degree of orientation randomness derived from Neumann decomposition |
| RVI | Radar Vegetation Index |

Another model-based decomposition, Neumann decomposition, is aimed at describing vegetation scattering by considering the morphological characteristics of vegetation in scattering modeling [36,69,70]. It has shown the potential advantage of identifying more types of volume scenes [71]. Moreover, two of its output parameters have proven to provide physical meanings similar to Cloude-Pottier decomposition outputs [36,69,70]. A recent study has shown that the third parameter in the Neumann decomposition, the phase of the particle scattering anisotropy, is more effective in improving the classification accuracy with respect to the Cloude-Pottier decomposition [72]. Therefore, the three output parameters from the Neumann decomposition were employed. In addition, the radar vegetation index (RVI) [73] has shown high sensitivity to crop morphological features and thus was also considered in this study.

For each RADARSAT-2 image, a series of preprocessing steps including calibration, speckle filter, and geocoding was conducted. Sigma naught values were obtained by the calibration. A 9 $\times$ 9 boxcar filter was applied to reduce the speckle noise. Then, a coherency matrix was generated at each pixel. A Digital Elevation Model (PDEM) of the Ontario province, Canada, with a spatial resolution of 30 m was used to geocode the coherency matrix of each image in the Universal Transverse Mercator (UTM) geographic reference. Afterward, 27 polarimetric features listed in Table 2 were extracted for each image. The spatial resolution of all the polarimetric features was the same because they followed the same processing. The final pixel spacing in the geocoded products was 10 m. Furthermore, the corresponding feature vectors, for a total of 347 sample points during the whole RADARSAT-2 acquisition period, were obtained based on the geolocation records of sample points.

### 2.3. *Machine Learning Method Used in This Study*

Due to the high capacity of prediction, machine learning methods are frequently used for classification and regression problems in remote sensing studies [74,75]. In particular, support vector regression (SVR) and random forest regression (RFR) are two representative examples in this domain and thus were considered for crop height retrieval in this study.

#### 2.3.1. Support Vector Regression (SVR)

SVR is the application of a well-known support vector machine model in regression, which has been used in agricultural biophysical parameter estimation with remote sensing data [76–78]. The core idea of SVR is to find an optimal approximating hyperplane to distinguish the input vectors and the predictor variables based on training data, which could be determined by solving a convex optimization problem. Theoretically, it is designed to establish an optimal linear separator and hence is suitable for linear data distribution. However, SVR can also handle nonlinear data distributions after embedded into a kernel framework. With a kernel function, the training data are nonlinearly transformed from the original space to a higher dimensional feature space. In this new space, the new data are better conditioned to make use of a linear separator.

In this study, an IDL-based tool, named imageSVM developed as a non-commercial product at the Geomatics Lab of Humboldt-Universität zu Berlin, specifically designed for support vector machine classification and regression analysis of remote sensing data, was used to carry out the SVR analysis. The radial basis function kernel, called the RBF kernel or the Gaussian kernel, was adopted in the regression model. The kernel parameter, the regularization parameter, and the Epsilon loss function parameter were required to be set to parameterize the SVR. A cross-validation strategy to tune these three parameters was used to reduce model overfitting.

#### 2.3.2. Random Forest Regression (RFR)

Unlike SVR, RFR is an ensemble learning method, which uses the subset of the training data to construct a set of decision trees and adopts various non-parametric predictive models [79]. Like the morphological structure of a real tree, a decision tree includes a root node, multiple internal nodes (splits), and various terminal nodes (leaves). Starting from the root to a leaf, a set of decision rules is applied to subdivide the training data into smaller subsets. The predictor variable is assigned as the leaf node. A strategy of bootstrap sampling with replacement is used in RFR to create each individual decision tree. The excluded samples, called out-of-bag samples, are used for model validation. The final prediction is generated by averaging the results from individual decision trees to obtain better prediction performance. In addition, a beneficial property of RFR is that it can also provide estimates of variable importance in the regression, which allows for a better understanding of the sensitivity of the input features to the predictor variable.

In this study, an IDL-based tool, imageRF [80], particularly designed for random forest classification and regression analysis of remote sensing image data, was used to carry out the RFR analysis. The number of decision trees was set to 200 based on the overall consideration of the prediction accuracy and computation time. In the bootstrap sampling for each decision tree, one-third of the training data were set as out-of-bag samples for independent validation. The number of randomly selected features at each split node was determined by the square root of all features.

#### 2.3.3. Experimental Design

From Table 1, it is evident that the corn was very short on 23 May, 2 June, and 16 June in 2013. The main backscattering contribution came from the soil, influenced by soil residual and tillage on these dates. The ground photos corresponding to these dates are shown in Figure 2. For this reason, we first carried out the tests by excluding these three observations. Consequently, 16 RADARSAT-2 images and 310 corn sample points were used. In addition, we performed tests involving these three images and all sample points (i.e., 19 RADARSAT-2 images and 347 sample points) and compared the results. In order to construct a regression model, training samples were required for model calibration. For this purpose, the sample points collected from multi-year fieldwork campaigns were randomly divided into two parts. A portion of the samples was dedicated to training samples for model calibration, while the remaining samples were assigned as testing samples for model validation. In our study, 80% of samples were used for calibration and 20% for validation.

It is a common fact that the calibration and prediction accuracies of the regression model can be affected by features of training and testing samples including their distribution and numbers. In order to simulate more random scenarios and investigate the uncertainty of the accuracy, a strategy of bootstrap sampling with replacement was used. As some training samples may have been recycled using this strategy, samples were generated multiple times by random sampling to reduce bias in height estimation [81]. The entire sample points including both short and tall corn heights were considered in the bootstrap sampling method. In total, 10 realizations of random sampling (hereafter named as scenarios) were carried out for generating different datasets for training and testing.

**Figure 2.** Field photos of corn in 2013. (**a**) 25 May; (**b**) 4 June; (**c**) 16 June.

## 3. Results.

### *3.1. Comparison between SVR and RFR*

For each dataset, SVR and RFR were conducted. The statistical analysis of model calibration and validation for each dataset with both regression algorithms is shown in Table 3. As expected, results appear varied in different scenarios, which can be attributed to the dependence of the regression model accuracies on the training and testing sets. The results of RFR were generally better than the ones of SVR. RFR obtained overall lower values of root mean square error (RMSE) and higher values of Pearson correlation coefficient (R), despite an opposite behavior for scenario 8. For model calibration, the differences between the performances of both regression algorithms were notable (average RMSE = 22.36 cm for RFR and RMSE = 44.12 cm for SVR), whereas small differences were observed for model validation (average RMSE = 50.40 cm for RFR and RMSE = 54.69 cm for SVR).

**Table 3.** Statistics of model calibration and validation for 10 different datasets with Support Vector Regression (SVR) and Random Forest Regression (RFR) in case of using all 27 polarimetric variables.

| Scenario | Model Calibration | | | | Model Validation | | | |
|---|---|---|---|---|---|---|---|---|
| | SVR | | RFR | | SVR | | RFR | |
| | RMSE (cm) | R | RMSE (cm) | R | RMSE (cm) | R | RMSE (cm) | R |
| 1 | 42.05 | 0.84 | 22.15 | 0.98 | 56.61 | 0.64 | 52.81 | 0.74 |
| 2 | 43.04 | 0.82 | 23.01 | 0.98 | 49.64 | 0.76 | 48.73 | 0.85 |
| 3 | 51.14 | 0.75 | 22.65 | 0.98 | 51.35 | 0.75 | 49.27 | 0.82 |
| 4 | 43.10 | 0.82 | 22.53 | 0.98 | 49.62 | 0.76 | 49.83 | 0.82 |
| 5 | 41.27 | 0.84 | 21.95 | 0.98 | 58.49 | 0.64 | 51.73 | 0.78 |
| 6 | 41.28 | 0.84 | 21.98 | 0.98 | 58.49 | 0.64 | 51.82 | 0.78 |
| 7 | 50.37 | 0.75 | 22.19 | 0.98 | 54.17 | 0.75 | 50.93 | 0.80 |
| 8 | 42.59 | 0.83 | 22.35 | 0.98 | 48.92 | 0.81 | 51.10 | 0.80 |
| 9 | 43.08 | 0.82 | 22.28 | 0.98 | 49.63 | 0.76 | 48.82 | 0.84 |
| 10 | 43.30 | 0.82 | 22.51 | 0.98 | 49.56 | 0.76 | 48.99 | 0.84 |
| Average | 44.12 | 0.81 | 22.36 | 0.98 | 54.69 | 0.73 | 50.40 | 0.81 |

In addition to the statistical indices shown in Table 3, Figure 3 illustrates the scatter plots of measured and predicted corn height obtained with both SVR and RFR methods in scenario 2, in which the regression produced the overall best accuracies. The RFR results exhibited a higher correlation than the results from SVR in both model calibration (see Figure 3a,c) and model validation (see Figure 3b,d). In detail, the model of RFR, in general, yielded overestimated values for lower corn height, while underestimation was observed for higher corn height (taller than around 225 cm). The SVR model generated overestimation and underestimation results for either lower corn height or for higher corn height, while a larger underestimation appeared for higher corn height.

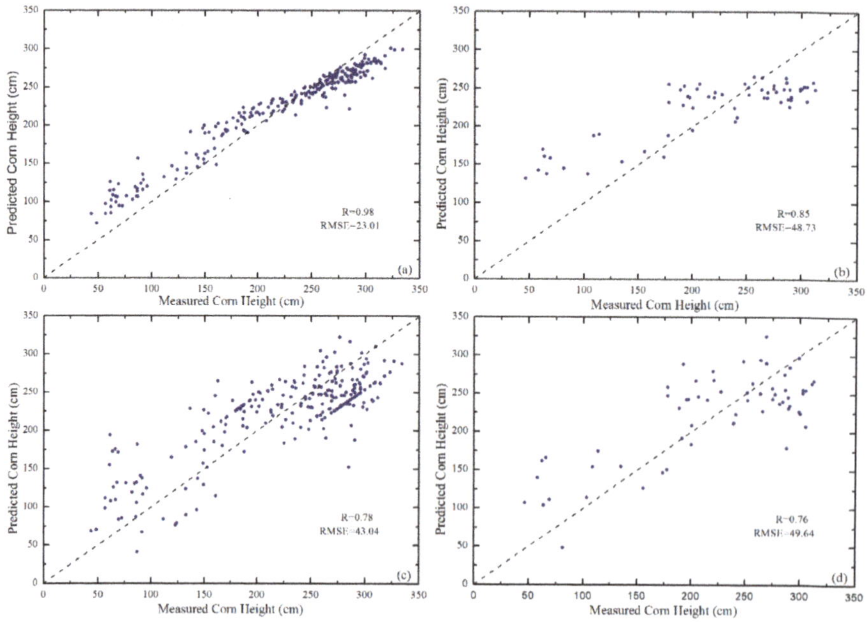


**Figure 3.** Comparison between measured and estimated corn height with scenario 2 in Table 3: (**a**) calibration of the Random Forest Regression (RFR) model; (**b**) validation of the RFR model; (**c**) calibration of the Support Vector Regression (SVR) model; (**d**) validation of the SVR model.

### 3.2. Normalized Variable Importance of RFR

As explained above, the Random Forest algorithm can provide the relative importance of different input variables to the classification or regression accuracy. Therefore, this interesting property of RFR was been also used for further analysis.

Since RFR was carried out under 10 different scenarios and 27 input polarimetric observables were selected in our study, the output values of normalized variable importance of RFR regression for each scenario were variable. It is difficult and unfair to analyze the variable importance using values from one specific scenario. For this reason, Figure 4 shows the ranking of the importance of each polarimetric observables in the regression based on the average values of output normalized variable importance for the 10 scenarios. For better visualization, the parameters belonging to the same or similar group are marked with the same color. The contribution of the double-bounce component (Pd) from the Freeman-Durden decomposition was the most important variable among the selected 27 polarimetric observables. It demonstrated a strong correlation between double-bounce scattering and crop height. Moreover, considering the polarimetric decompositions used in this study, parameters from the Freeman-Durden decomposition appeared to be more relevant than parameters from the Neumann and Cloude-Pottier decompositions. The reasons may be attributed to the nature of the Freeman-Durden decomposition models, which are physically based rather than purely mathematical as in the Cloude-Pottier decomposition. Hence, its applicability is more general than the Neumann decomposition that assumes the dominance of volume scattering, which is not always the case for crops. In particular, the contribution of the volume scattering component (Pv) from the Freeman-Durden decomposition took second place in the variable importance ranking, just after Pd. Notably, the polarimetric anisotropy contributed more entropy and alpha angle from Cloude-Pottier decomposition.

The magnitude of the degree of orientation randomness of the particle scattering anisotropy was the most important variable from the Neumann decomposition. Among the backscattering coefficients, C22 played the most important role, which was highly related to the volume scattering component Pv, and took third place in the ranking. C33, which represented the VV polarization, showed less contribution and was placed in the last position among the linear backscattering coefficients. T22 was the second most important, and was highly related to the double-bounce scattering component Pd, which took fourth place. The total scattering power SPAN ranked behind T22 but was still at the front position. Among the backscattering ratios, HV/HH was the most important parameter, and the corresponding contribution was significantly larger than the other two linear ratios. Although co-polar correlation magnitude $|\rho_{HHVV}|$ took up a middle position similar to RVI in the ranking, most correlation magnitudes and polarimetric phases contributed less to the regression and were found at the end of the importance list. From the aforementioned analysis, it was clear to see that SAR parameters related to the double bounce and volume scattering components (e.g., Pd, Pv, C22, T22) showed high sensitivity to crop height and strongly drove height estimation for the RFR method. The reasons may be attributed to the nature of the scattering mechanisms which interact differently with different plant structures (e.g., stem, flower, leaf, tassel) as the crop development advances. For example, double-bounce and volume scattering components vary significantly with crop growth stages, which generally are low in the early stage and high in later development stages (e.g., stem elongation, tassel, and stigmata emergence).

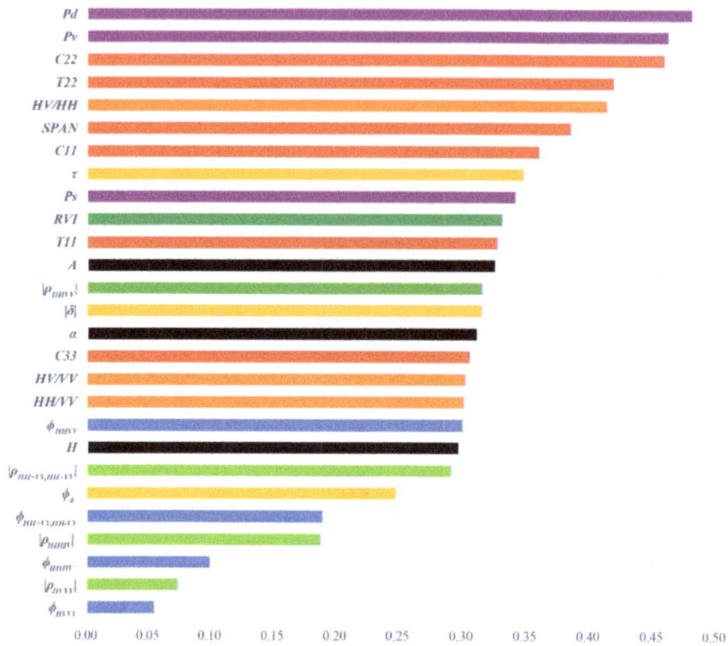


**Figure 4.** Normalized variable importance of RFR in regression for corn height based on averaged output values of 10 scenarios. Different types of polarimetric variables are represented as follows: Backscattering coefficients and SPAN (red), Correlation magnitude (light green), Polarimetric phase (blue), Backscattering ratios (orange), Freeman-Durden decomposition (purple), Cloude-Pottier decomposition (black), Neumann Decomposition (yellow) and RVI (green). A description of how variables were named can be found in Table 2.

## 4. Discussion

### *4.1. Tests with Fewer Polarimetric Observables*

It is well known that using more features does not always generate better regression accuracies because of redundant or correlated information. Therefore, it is interesting to further check the regression accuracies in case fewer polarimetric observables can be used. From the diagram of variable importance ranking shown in Figure 3, the contributions of the first six polarimetric variables are obviously larger than the rest of the variables. Therefore, additional RFR tests were conducted by using only the first six polarimetric variables (i.e., Pd, Pv, C22, T22, HV/HH, and SPAN). At the same time, corresponding SVR tests were also carried out for comparison purposes. The statistical analyses of model calibration and validation for 10 different scenarios with SVR and RFR are presented in Table 4. It is apparent that regression estimation accuracies of SVR and RFR improve with respect to the previous results for each scenario in Table 3, even though sometimes model calibration accuracies decrease slightly. As in the previous results, RFR provides better accuracies than SVR in most scenarios. The differences between the two machine learning regression methods are smaller in the case of using fewer polarimetric variables. The average values of RMSE and R with RFR prediction are 47.76 cm and 0.79, while the corresponding values are 47.90 cm and 0.78 with SVR prediction, respectively. The best regression estimation results were produced in scenario 1 with RFR. The corresponding values of RMSE and R reach 42.69 cm and 0.84, respectively. To further analyze the results in scenario 1, the scatterplots of measured and estimated corn height with SVR and RFR methods are presented in Figure 5. The patterns of model calibration are similar to the ones shown in Figure 3. The overall distributions of RFR and SVR estimation results show similar patterns, in which overestimation is mostly observed for lower corn height, and underestimation is observed for taller corn height. To analyze the SVR and RFR results at different stages of corn growth and maintain sufficient samples for statistics, the simple division method for growth stages of corn proposed in [48] was adopted. The corn height of less than 150 cm is addressed as the early stage, and height that is taller than 150 cm is defined as the later stage [48]. After calculation of values of RMSE and R, statistics of model validation at an early stage and later stage for 10 different scenarios with SVR and RFR are presented in Table 5. It is clear to see that the results of SVR and RFR are better at the later stage than the ones at the early stage.

**Table 4.** Statistics of model calibration and validation for 10 different datasets with SVR and RFR in case of using the top six polarimetric variables in the variable importance ranking.

| Scenario | Model Calibration | | | | Model Validation | | | |
|---|---|---|---|---|---|---|---|---|
| | SVR | | RFR | | SVR | | RFR | |
| | RMSE (cm) | R | RMSE (cm) | R | RMSE (cm) | R | RMSE (cm) | R |
| 1 | 46.42 | 0.79 | 22.33 | 0.97 | 45.45 | 0.79 | 42.69 | 0.84 |
| 2 | 49.14 | 0.76 | 22.69 | 0.97 | 48.04 | 0.78 | 47.81 | 0.79 |
| 3 | 45.84 | 0.79 | 21.91 | 0.97 | 50.75 | 0.74 | 48.07 | 0.78 |
| 4 | 45.73 | 0.79 | 22.04 | 0.97 | 46.20 | 0.80 | 48.12 | 0.79 |
| 5 | 45.34 | 0.81 | 21.53 | 0.97 | 53.89 | 0.72 | 50.68 | 0.74 |
| 6 | 48.69 | 0.77 | 21.28 | 0.97 | 51.57 | 0.73 | 51.29 | 0.73 |
| 7 | 45.81 | 0.80 | 22.42 | 0.97 | 45.38 | 0.81 | 46.14 | 0.82 |
| 8 | 45.82 | 0.80 | 22.44 | 0.97 | 45.38 | 0.81 | 47.07 | 0.81 |
| 9 | 45.71 | 0.79 | 22.31 | 0.97 | 46.14 | 0.80 | 47.72 | 0.79 |
| 10 | 45.74 | 0.79 | 22.36 | 0.97 | 46.22 | 0.80 | 48.01 | 0.79 |
| Average | 46.42 | 0.79 | 22.13 | 0.97 | 47.90 | 0.78 | 47.76 | 0.79 |

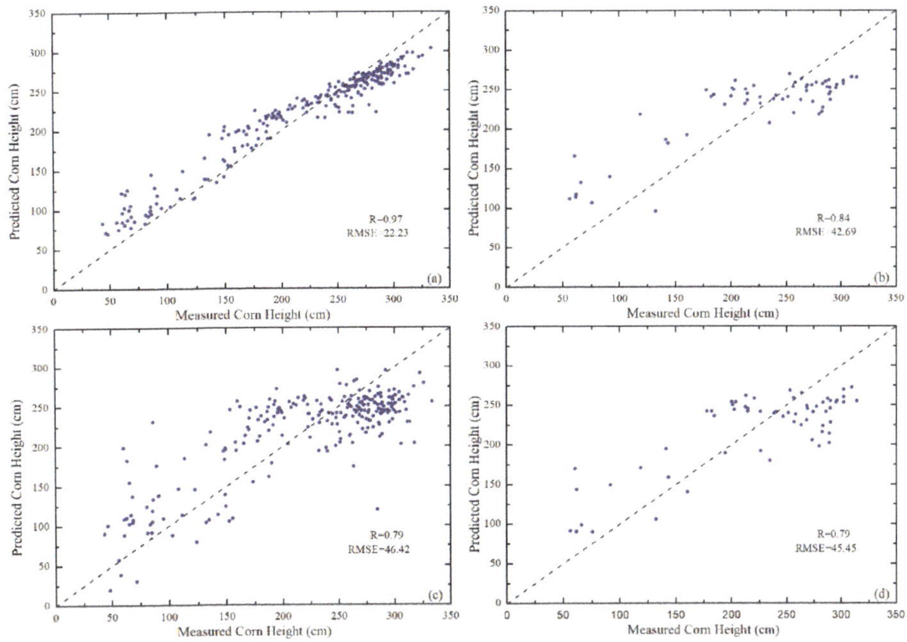


**Figure 5.** Comparison between measured and estimated corn height with scenario 1 in Table 4: (**a**) calibration of the RFR model; (**b**) validation of the RFR model; (**c**) calibration of the SVR model; (**d**) validation of the SVR model.

**Table 5.** Statistics of model validation at two different growing stages for 10 different datasets with SVR and RFR in case of using the top six polarimetric variables in the variable importance ranking.

| Scenario | Model Validation | | | | | | | |
|---|---|---|---|---|---|---|---|---|
| | SVR (Height < 150 cm) | | SVR (Height > 150 cm) | | RFR (Height < 150 cm) | | RFR (Height > 150 cm) | |
| | RMSE (cm) | R | RMSE (cm) | R | RMSE (cm) | R | RMSE (cm) | R |
| 1 | 53.91 | 0.52 | 43.41 | 0.27 | 62.02 | 0.51 | 37.23 | 0.40 |
| 2 | 59.61 | 0.16 | 45.16 | 0.32 | 49.19 | 0.64 | 47.51 | 0.26 |
| 3 | 44.46 | 0.80 | 52.01 | 0.10 | 48.47 | 0.92 | 47.98 | 0.10 |
| 4 | 51.75 | 0.34 | 44.91 | 0.35 | 50.71 | 0.58 | 47.55 | 0.26 |
| 5 | 55.13 | 0.76 | 53.49 | 0.31 | 68.37 | 0.72 | 43.55 | 0.30 |
| 6 | 60.29 | 0.67 | 48.46 | 0.35 | 66.06 | 0.71 | 45.58 | 0.26 |
| 7 | 59.99 | 0.37 | 41.99 | 0.47 | 64.46 | 0.48 | 41.71 | 0.41 |
| 8 | 59.99 | 0.37 | 41.99 | 0.47 | 64.59 | 0.49 | 42.89 | 0.38 |
| 9 | 51.76 | 0.34 | 44.84 | 0.35 | 49.00 | 0.60 | 47.44 | 0.28 |
| 10 | 51.75 | 0.34 | 44.94 | 0.35 | 51.06 | 0.63 | 47.33 | 0.28 |
| Average | 54.86 | 0.47 | 46.12 | 0.33 | 57.39 | 0.63 | 44.88 | 0.29 |

Based on the regression model, a corn height map can be generated on each date. For example, the map of estimated corn height on 15 September 2018 is presented in Figure 6. Locations of four cornfields and eight sample points in each field are marked on the map. The estimated corn height ranges between 80 and 295 cm. Considering most cornfields are at the late crop growing stage on that date, the results of estimated height are underestimated which is consistent with the previous analysis. In addition, the values of estimated height somehow show variation among different fields, which can be attributed to the diversity of cornfield conditions, such as soil moisture and roughness, topography, precipitation, and fertilization. Table 6 shows the measured and estimated corn heights in 32 sample points. The difference is very small at some points and large at other points. The calculated RMSE and R values are 32.61 cm and 0.59, respectively.

**Table 6.** The measured and estimated corn heights of 32 sample points on 15 September 2018 based on RFR model prediction.

| Field Name | Corn Height (cm) | 1 | 2 | 3 | 4 | 5 | 6 | 7 | 8 |
|---|---|---|---|---|---|---|---|---|---|
| C1 | Measured | 274.83 | 251.5 | 295.17 | 286.83 | 290.67 | 285.5 | 235.75 | 271.41 |
| | Estimated | 252.55 | 228.26 | 283.05 | 275.69 | 279.13 | 274.72 | 244.21 | 263.61 |
| C2 | Measured | 273.92 | 301 | 371.58 | 260.42 | 241.08 | 283.17 | 298.75 | 283.92 |
| | Estimated | 255.66 | 259.90 | 255.99 | 255.32 | 244.74 | 273.18 | 282.48 | 227.11 |
| C3 | Measured | 255.75 | 251.67 | 219.83 | 248.08 | 162.5 | 220.67 | 253.5 | 204.33 |
| | Estimated | 258.35 | 261.41 | 228 | 232.20 | 199.91 | 254.53 | 248.51 | 261.60 |
| C4 | Measured | 276.25 | 288.50 | 290.75 | 288.08 | 296.58 | 298.92 | 306 | 284.25 |
| | Estimated | 270.46 | 280.46 | 246.22 | 252.87 | 255.87 | 284.98 | 268.49 | 262.39 |
| RMSE (cm) | | 32.61 | | | | | | | |
| R | | 0.59 | | | | | | | |

### 4.2. Tests with All Images Including Very Short Corn Height

In previous tests, the first three images in 2013 are excluded due to their very short corn height. It is interesting to check the performance when all images, including very short corn height, are used. Ten realizations of SVR and RFR tests were carried out again, respectively. As in previous tests, 80% of samples were used for calibration and 20% for validation. However, 19 RADARSAT-2 images and 347 corn samples were used this time. The statistical analysis of model calibration and validation for each dataset with both regression algorithms is shown in Table 7. As expected, the results still show variation in different scenarios and the results of RFR are better than SVR. However, the values of RMSE and R are worse than the results in Table 3 (using 16 RADARSAT-2 images with

310 corn samples). The average values of RMSE and R for RFR in model validation are 54.55 cm and 0.83, and the values for SVR are 56.75 cm and 0.80, respectively. It somehow indicates the limitation of this kind of regression methods, i.e., very short height will affect the accuracy of estimation.

**Figure 6.** Map of corn height on 15 September 2018 based on RFR model prediction.

**Table 7.** Statistics of model calibration and validation for 10 different datasets with SVR and RFR in case of using all 27 polarimetric variables and all 19 RADARSAT-2 images.

| Scenario | Model Calibration | | | | Model Validation | | | |
|---|---|---|---|---|---|---|---|---|
| | SVR | | RFR | | SVR | | RFR | |
| | RMSE (cm) | R | RMSE (cm) | R | RMSE (cm) | R | RMSE (cm) | R |
| 1 | 48.63 | 0.87 | 22.38 | 0.99 | 58.81 | 0.80 | 54.41 | 0.84 |
| 2 | 48.53 | 0.86 | 22.63 | 0.99 | 58.75 | 0.82 | 55.62 | 0.86 |
| 3 | 50.05 | 0.86 | 22.22 | 0.99 | 59.87 | 0.75 | 54.56 | 0.83 |
| 4 | 50.45 | 0.87 | 22.78 | 0.99 | 54.37 | 0.76 | 52.98 | 0.78 |
| 5 | 49.09 | 0.87 | 22.20 | 0.99 | 55.16 | 0.79 | 54.73 | 0.84 |
| 6 | 48.27 | 0.87 | 22.18 | 0.99 | 57.64 | 0.83 | 57.62 | 0.84 |
| 7 | 49.13 | 0.86 | 23.05 | 0.99 | 53.44 | 0.85 | 53.15 | 0.88 |
| 8 | 49.95 | 0.86 | 22.43 | 0.99 | 54.86 | 0.76 | 50.65 | 0.82 |
| 9 | 48.51 | 0.87 | 22.14 | 0.99 | 56.53 | 0.81 | 55.59 | 0.82 |
| 10 | 49.23 | 0.86 | 22.47 | 0.99 | 58.05 | 0.78 | 56.20 | 0.83 |
| Average | 49.18 | 0.87 | 22.45 | 0.99 | 56.75 | 0.80 | 54.55 | 0.83 |

### *4.3. Limitations and Future Research*

In this study, two common machine learning techniques used for scientific purposes, i.e., RFR and SVR, were evaluated for crop height estimation of corn from multi-year RADARSAT-2 polarimetric observables. There are some limitations to these methods. Firstly, the methods depend on the availability of a large number of sample data and a good sample distribution. In our case, although the number of samples is large, corn samples have more values of heights taller than 2 m and fewer values between 1 m and 2 m. This might affect the accuracy of model calibration and estimation. Secondly, the output results somehow show overestimation and underestimation. The results show worse performance at the early growth stages, especially in the case of estimating very short crop height. Applying piecewise regression methods at different growing stages may improve the results. Third, the PolSAR data are acquired in different modes with different incidence angles, which might also influence the accuracy of estimation. Additionally, field conditions such as soil moisture, irrigation, and fertilization, may also affect the estimation results.

Future work will focus on testing these techniques for height estimation of other crop types, such as wheat, soybean, and rice. Moreover, it is worth investigating how changes in crop physiology associated with crop development and precipitation (or irrigation regimes) could contribute to changes in the priority of different PolSAR parameters as the crop matures along the full season. Tests and analysis with PolSAR data at other radar frequencies (such as TerraSAR-X at X band, ALOS-2 at L band) over different sites will also be investigated in future research.

## 5. Conclusions

This study presents a demonstration of crop height retrieval based on space-borne PolSAR data with machine learning techniques. The techniques have been tested with RADARSAT-2 data in cornfields covering the whole corn growing period. The potential of two popular machine learning regression algorithms (SVR and RFR) was investigated, including an identification of the relevant features by means of the normalized variable importance of RFR. A set of 27 representative PolSAR observables was initially selected and analyzed in this work. The results show a satisfactory prediction performance for corn height mapping at a large scale, with RMSE around 40–50 cm considering the whole growth cycle, with corn height over 3 m at late stages. The RFR approach overall outperforms the SVR method in all tests. The best result is generated by the RFR method when selecting a subset of six polarimetric features, with an RMSE of 42.8 cm, which indicates that fewer selected polarimetric features can generate better results than using all features. Regarding the analysis of the relative importance of all polarimetric features considered, results highlight that parameters related to double-bounce and volume scattering are the most important polarimetric features for corn height estimation. In addition, the HV/HH ratio appears to be the most representative among all three backscattering ratios. Compared with past studies on crop height retrieval with SAR data, this research provides a potentially efficient method and a new perspective on the use of PolSAR data.

**Author Contributions:** Conceptualization, Q.X. and J.W.; methodology, Q.X.; software, Q.X.; validation, Q.X. and C.L.; formal analysis, Q.X., J.W. and X.P.; investigation, Q.X. and J.W.; resources, J.W.; data curation, Q.X., C.L., J.W. and J.S.; writing—original draft preparation, Q.X.; writing—review and editing, J.W., J.M.L.-S., J.Z., J.S., X.P. and J.D.B.-B.; visualization, Q.X.; supervision, J.W.; project administration, Q.X. and J.W.; funding acquisition, Q.X., J.W., J.Z., H.F. and J.M.L.-S. All authors have read and agreed to the published version of the manuscript.

**Funding:** This research was funded in part by the National Natural Science Foundation of China (Grant No. 41804004, 41820104005, 41531068, 41904004), the Canadian Space Agency SOAR-E program (Grant No. SOAR-E-5489), the Fundamental Research Funds for the Central Universities, China University of Geosciences (Wuhan) (Grant No. CUG190633), and the Spanish Ministry of Science, Innovation and Universities, State Research Agency (AEI) and the European Regional Development Fund under project TEC2017-85244-C2-1-P.

**Institutional Review Board Statement:** Not applicable.

**Informed Consent Statement:** Not applicable.

**Data Availability Statement:** No new data were created or analyzed in this study. Data sharing is not applicable to this article.

**Acknowledgments:** RADARSAT-2 Data and Products © MacDonald, Dettwiler and Associates Ltd. (2013, 2015, 2018)—All Rights Reserved. RADARSAT is an official trademark of the Canadian Space Agency. The authors would like to thank the Canadian Space Agency and Agriculture and Agri-Food Canada for providing the RADARSAT-2 data, and Yang Song, Yu Bo, Xiaodong Huang from GITA lab, UWO for their help with field work. In addition, the authors acknowledge A&L Canada Inc. for the access to the corn fields.

**Conflicts of Interest:** The authors declare no conflict of interest.

## References

1. Erten, E.; Lopez-Sanchez, J.M.; Yuzugullu, O.; Hajnsek, I. Retrieval of agricultural crop height from space: A comparison of SAR techniques. *Remote Sens. Environ.* **2016**, *187*, 130–144. [CrossRef]
2. Lopez-Sanchez, J.M.; Ballester-Berman, J.D. Potentials of polarimetric SAR interferometry for agriculture monitoring. *Radio Sci.* **2009**, *44*, 1–20. [CrossRef]
3. Lopez-Sanchez, J.M.; Vicente-Guijalba, F.; Erten, E.; Campos-Taberner, M.; Garcia-Haro, F.J. Retrieval of vegetation height in rice fields using polarimetric SAR interferometry with TanDEM-X data. *Remote Sens. Environ.* **2017**, *192*, 30–44. [CrossRef]
4. Yuzugullu, O.; Erten, E.; Hajnsek, I. Assessment of Paddy Rice Height: Sequential Inversion of Coherent and Incoherent Models. *IEEE J. Sel. Top. Appl. Earth Obs. Remote Sens.* **2018**, *11*, 3001–3013. [CrossRef]
5. Karam, M.; Amar, F.; Fung, A.K.; Mougin, E.; Lopes, A.; Le Vine, D.M.; Beaudoin, A. A microwave polarimetric scattering model for forest canopies based on vector radiative transfer theory. *Remote Sens. Environ.* **1995**, *53*, 16–30. [CrossRef]
6. Karam, M.; Fung, A.; Lang, R.; Chauhan, N. A microwave scattering model for layered vegetation. *IEEE Trans. Geosci. Remote Sens.* **1992**, *30*, 767–784. [CrossRef]
7. Karam, M.; Fung, A.; Antar, Y. Electromagnetic wave scattering from some vegetation samples. *IEEE Trans. Geosci. Remote Sens.* **1988**, *26*, 799–808. [CrossRef]
8. Liu, Y.; Chen, K.-S.; Xu, P.; Li, Z.-L. Modeling and Characteristics of Microwave Backscattering From Rice Canopy over Growth Stages. *IEEE Trans. Geosci. Remote Sens.* **2016**, *54*, 6757–6770. [CrossRef]
9. Toure, A.; Thomson, K.; Edwards, G.; Brown, R.; Brisco, B. Adaptation of the MIMICS backscattering model to the agricultural context-wheat and canola at L and C bands. *IEEE Trans. Geosci. Remote Sens.* **1994**, *32*, 47–61. [CrossRef]
10. Ulaby, F.T.; Sarabandi, K.; McDonald, K.; Whitt, M.; Dobson, M.C. Michigan microwave canopy scattering model. *Int. J. Remote Sens.* **1990**, *11*, 1223–1253. [CrossRef]
11. Ni, W.; Sun, G.; Ranson, K.J.; Zhang, Z.; He, Y.; Huang, W.; Guo, Z. Model-Based Analysis of the Influence of Forest Structures on the Scattering Phase Center at L-Band. *IEEE Trans. Geosci. Remote Sens.* **2013**, *52*, 3937–3946. [CrossRef]
12. Zhang, Y.; Liu, X.; Su, S.; Wang, C. Retrieving canopy height and density of paddy rice from Radarsat-2 images with a canopy scattering model. *Int. J. Appl. Earth Obs. Geoinf.* **2014**, *28*, 170–180. [CrossRef]
13. Yuzugullu, O.; Marelli, S.; Erten, E.; Sudret, B.; Hajnsek, I. Determining Rice Growth Stage with X-Band SAR: A Metamodel Based Inversion. *Remote Sens.* **2017**, *9*, 460. [CrossRef]
14. Yuzugullu, O.; Erten, E.; Hajnsek, I. Estimation of Rice Crop Height from X- and C-Band PolSAR by Metamodel-Based Optimization. *IEEE J. Sel. Top. Appl. Earth Obs. Remote Sens.* **2016**, *10*, 194–204. [CrossRef]
15. Attema, E.P.W.; Ulaby, F.T. Vegetation modeled as a water cloud. *Radio Sci.* **1978**, *13*, 357–364. [CrossRef]
16. Yang, Z.; Li, K.; Shao, Y.; Brisco, B.; Liu, L. Estimation of Paddy Rice Variables with a Modified Water Cloud Model and Improved Polarimetric Decomposition Using Multi-Temporal RADARSAT-2 Images. *Remote Sens.* **2016**, *8*, 878. [CrossRef]
17. Homayouni, S.; McNairn, H.; Hosseini, M.; Jiao, X.; Powers, J. Quad and compact multitemporal C-band PolSAR observations for crop characterization and monitoring. *Int. J. Appl. Earth Obs. Geoinf.* **2019**, *74*, 78–87. [CrossRef]
18. Mandal, D.; Kumar, V.; McNairn, H.; Bhattacharya, A.; Rao, Y. Joint estimation of Plant Area Index (PAI) and wet biomass in wheat and soybean from C-band polarimetric SAR data. *Int. J. Appl. Earth Obs. Geoinf.* **2019**, *79*, 24–34. [CrossRef]
19. Baghdadi, N.; El Hajj, M.; Zribi, M.; Bousbih, S. Calibration of the Water Cloud Model at C-Band for Winter Crop Fields and Grasslands. *Remote Sens.* **2017**, *9*, 969. [CrossRef]
20. Bamler, R.; Hartl, P. Synthetic aperture radar interferometry. *Inverse Probl.* **1998**, *14*, R1–R54. [CrossRef]
21. Fu, H.; Zhu, J.; Wang, C.; Zhao, R.; Xie, Q. Underlying Topography Estimation Over Forest Areas Using Single-Baseline InSAR Data. *IEEE Trans. Geosci. Remote Sens.* **2018**, *57*, 2876–2888. [CrossRef]
22. Fu, H.; Wang, C.; Zhu, J.; Xie, Q.; Zhang, B. Estimation of Pine Forest Height and Underlying DEM Using Multi-Baseline P-Band PolInSAR Data. *Remote Sens.* **2016**, *8*, 820. [CrossRef]
23. Liao, Z.; He, B.; Van Dijk, A.I.; Bai, X.; Quan, X. The impacts of spatial baseline on forest canopy height model and digital terrain model retrieval using P-band PolInSAR data. *Remote Sens. Environ.* **2018**, *210*, 403–421. [CrossRef]
24. Fu, H.; Zhu, J.; Wang, C.; Li, Z. Underlying topography extraction over forest areas from multi-baseline PolInSAR data. *J. Geod.* **2018**, *92*, 727–741. [CrossRef]
25. Peng, X.; Li, X.; Wang, C.; Fu, H.; Du, Y. A Maximum Likelihood Based Nonparametric Iterative Adaptive Method of Synthetic Aperture Radar Tomography and Its Application for Estimating Underlying Topography and Forest Height. *Sensors* **2018**, *18*, 2459. [CrossRef]

26. Papathanassiou, K.; Cloude, S. Single-baseline polarimetric SAR interferometry. *IEEE Trans. Geosci. Remote Sens.* **2001**, *39*, 2352–2363. [CrossRef]
27. Cloude, S.R.; Papathanassiou, K.P. Polarimetric SAR interferometry. *IEEE Trans. Geosci. Remote Sens.* **1998**, *36*, 1551–1565. [CrossRef]
28. Xie, Q.; Zhu, J.; Wang, C.; Fu, H.; Lopez-Sanchez, J.M.; Ballester-Berman, J.D. A Modified Dual-Baseline PolInSAR Method for Forest Height Estimation. *Remote Sens.* **2017**, *9*, 819. [CrossRef]
29. Fu, H.; Wang, C.; Zhu, J.; Xie, Q.; Zhao, R. Inversion of vegetation height from PolInSAR using complex least squares adjustment method. *Sci. China Earth Sci.* **2015**, *58*, 1018–1031. [CrossRef]
30. Garestier, F.; Dubois-Fernandez, P.C.; Papathanassiou, K.P. Pine Forest Height Inversion Using Single-Pass X-Band PolInSAR Data. *IEEE Trans. Geosci. Remote Sens.* **2007**, *46*, 59–68. [CrossRef]
31. Garestier, F.; Dubois-Fernandez, P.; Champion, I. Forest Height Inversion Using High-Resolution P-Band Pol-InSAR Data. *IEEE Trans. Geosci. Remote Sens.* **2008**, *46*, 3544–3559. [CrossRef]
32. Hajnsek, I.; Kugler, F.; Lee, S.-K.; Papathanassiou, K.P. Tropical-Forest-Parameter Estimation by Means of Pol-InSAR: The INDREX-II Campaign. *IEEE Trans. Geosci. Remote Sens.* **2009**, *47*, 481–493. [CrossRef]
33. Garestier, F.; Dubois-Fernandez, P.; Champion, I.; Le Toan, T. Pine forest investigation using high resolution P-band Pol-InSAR data. *Remote Sens. Environ.* **2011**, *115*, 2897–2905. [CrossRef]
34. Kugler, F.; Schulze, D.; Hajnsek, I.; Pretzsch, H.; Papathanassiou, K.P. TanDEM-X Pol-InSAR Performance for Forest Height Estimation. *IEEE Trans. Geosci. Remote Sens.* **2014**, *52*, 6404–6422. [CrossRef]
35. Praks, J.; Kugler, F.; Papathanassiou, K.P.; Hajnsek, I.; Hallikainen, M. Height Estimation of Boreal Forest: Interferometric Model-Based Inversion at L- and X-Band versus HUTSCAT Profiling Scatterometer. *IEEE Geosci. Remote Sens. Lett.* **2007**, *4*, 466–470. [CrossRef]
36. Neumann, M.; Ferro-Famil, L.; Reigber, A. Estimation of Forest Structure, Ground, and Canopy Layer Characteristics from Multibaseline Polarimetric Interferometric SAR Data. *IEEE Trans. Geosci. Remote Sens.* **2009**, *48*, 1086–1104. [CrossRef]
37. Zhang, B.; Fu, H.; Zhu, J.; Peng, X.; Xie, Q.; Lin, D.; Liu, Z. A Multibaseline PolInSAR Forest Height Inversion Model Based on Fourier-Legendre Polynomials. *IEEE Geosci. Remote Sens. Lett.* **2020**, 1–5. [CrossRef]
38. Xie, Y.; Fu, H.; Zhu, J.; Wang, C.; Xie, Q. A LiDAR-Aided Multibaseline PolInSAR Method for Forest Height Estimation: With Emphasis on Dual-Baseline Selection. *IEEE Geosci. Remote Sens. Lett.* **2020**, *17*, 1807–1811. [CrossRef]
39. Kumar, S.; Govil, H.; Srivastava, P.K.; Thakur, P.K.; Kushwaha, S.P.S. Spaceborne Multifrequency PolInSAR-Based Inversion Modelling for Forest Height Retrieval. *Remote Sens.* **2020**, *12*, 4042. [CrossRef]
40. Zhang, Z.; Wang, C.; Zhang, H.; Tang, Y.; Liu, X. Analysis of Permafrost Region Coherence Variation in the Qinghai–Tibet Plateau with a High-Resolution TerraSAR-X Image. *Remote Sens.* **2018**, *10*, 298. [CrossRef]
41. Sagues, L.; Lopez-Sanchez, J.M.; Fortuny, J.; Fabregas, X.; Broquetas, A.; Sieber, A. Indoor experiments on polarimetric SAR interferometry. *IEEE Trans. Geosci. Remote Sens.* **2000**, *38*, 671–684. [CrossRef]
42. Ballester-Berman, J.; Lopez-Sanchez, J.M.; Fortuny-Guasch, J. Retrieval of biophysical parameters of agricultural crops using polarimetric SAR interferometry. *IEEE Trans. Geosci. Remote Sens.* **2005**, *43*, 683–694. [CrossRef]
43. Lopez-Sanchez, J.M.; Hajnsek, I.; Ballester-Berman, J.D. First Demonstration of Agriculture Height Retrieval with PolInSAR Airborne Data. *IEEE Geosci. Remote Sens. Lett.* **2012**, *9*, 242–246. [CrossRef]
44. Pichierri, M.; Hajnsek, I.; Papathanassiou, K.P. A Multibaseline Pol-InSAR Inversion Scheme for Crop Parameter Estimation at Different Frequencies. *IEEE Trans. Geosci. Remote Sens.* **2016**, *54*, 4952–4970. [CrossRef]
45. Nasirzadehdizaji, R.; Balik Sanli, F.; Abdikan, S.; Cakir, Z.; Sekertekin, A.; Ustuner, M. Sensitivity Analysis of Multi-Temporal Sentinel-1 SAR Parameters to Crop Height and Canopy Coverage. *Appl. Sci.* **2019**, *9*, 655. [CrossRef]
46. Jiao, X.; McNairn, H.; Shang, J.; Pattey, E.; Liu, J.; Champagne, C. The sensitivity of RADARSAT-2 polarimetric SAR data to corn and soybean leaf area index. *Can. J. Remote Sens.* **2011**, *37*, 69–81. [CrossRef]
47. Erten, E.; Taskin, G.; Lopez-Sanchez, J.M. Selection of PolSAR Observables for Crop Biophysical Variable Estimation with Global Sensitivity Analysis. *IEEE Geosci. Remote Sens. Lett.* **2019**, *16*, 766–770. [CrossRef]
48. Liao, C.; Wang, J.; Shang, J.; Huang, X.; Liu, J.; Huffman, T. Sensitivity study of Radarsat-2 polarimetric SAR to crop height and fractional vegetation cover of corn and wheat. *Int. J. Remote Sens.* **2017**, *39*, 1475–1490. [CrossRef]
49. Sarti, M.; Migliaccio, M.; Nunziata, F.; Mascolo, L.; Brugnoli, E. On the sensitivity of polarimetric SAR measurements to vegetation cover: The Coiba National Park, Panama. *Int. J. Remote Sens.* **2017**, *38*, 6755–6768. [CrossRef]
50. Hosseini, M.; McNairn, H.; Merzouki, A.; Pacheco, A. Estimation of Leaf Area Index (LAI) in corn and soybeans using multi-polarization C- and L-band radar data. *Remote Sens. Environ.* **2015**, *170*, 77–89. [CrossRef]
51. Wiseman, G.; McNairn, H.; Homayouni, S.; Shang, J. RADARSAT-2 Polarimetric SAR Response to Crop Biomass for Agricultural Production Monitoring. *IEEE J. Sel. Top. Appl. Earth Obs. Remote Sens.* **2014**, *7*, 4461–4471. [CrossRef]
52. Reisi-Gahrouei, O.; Homayouni, S.; McNairn, H.; Hosseini, M.; Safari, A. Crop biomass estimation using multi regression analysis and neural networks from multitemporal L-band polarimetric synthetic aperture radar data. *Int. J. Remote Sens.* **2019**, *40*, 6822–6840. [CrossRef]
53. Mandal, D.; Kumar, V.; Lopez-Sanchez, J.M.; Bhattacharya, A.; McNairn, H.; Rao, Y.S. Crop biophysical parameter retrieval from Sentinel-1 SAR data with a multi-target inversion of Water Cloud Model. *Int. J. Remote Sens.* **2020**, *41*, 5503–5524. [CrossRef]

54. Mandal, D.; Hosseini, M.; McNairn, H.; Kumar, V.; Bhattacharya, A.; Rao, Y.; Mitchell, S.; Robertson, L.D.; Davidson, A.; Dabrowska-Zielinska, K. An investigation of inversion methodologies to retrieve the leaf area index of corn from C-band SAR data. *Int. J. Appl. Earth Obs. Geoinf.* **2019**, *82*, 101893. [CrossRef]
55. Fieuzal, R.; Baup, F. Estimation of leaf area index and crop height of sunflowers using multi-temporal optical and SAR satellite data. *Int. J. Remote Sens.* **2016**, *37*, 2780–2809. [CrossRef]
56. Yang, H.; Chen, E.; Li, Z.; Yang, G.; Xu, X.; Yuan, L.; Feng, Q.; Zhao, L. Capability of multi-temporal Radarsat-2 data in monitoring canola crop and its plant height. In Proceedings of the 2014 the Third International Conference on Agro-Geoinformatics, Beijing, China, 11–14 August 2014; pp. 1–5.
57. Canisius, F.; Shang, J.; Liu, J.; Huang, X.; Ma, B.; Jiao, X.; Geng, X.; Kovacs, J.M.; Walters, D. Tracking crop phenological development using multi-temporal polarimetric Radarsat-2 data. *Remote Sens. Environ.* **2018**, *210*, 508–518. [CrossRef]
58. El Hajj, M.; Baghdadi, N.; Bazzi, H.; Zribi, M. Penetration Analysis of SAR Signals in the C and L Bands for Wheat, Maize, and Grasslands. *Remote Sens.* **2018**, *11*, 31. [CrossRef]
59. Ballester-Berman, J.D.; Lopez-Sanchez, J.M.; Fortuny-Guasch, J. Retrieval of height and topography of corn fields by polarimetric SAR interferometry. In Proceedings of the IEEE International Geoscience and Remote Sensing Symposium, IGARSS'04, Anchorage, AK, USA, 20–24 September 2004; Volume 2, pp. 1228–1231.
60. Liu, C.; Shang, J.; Vachon, P.W.; McNairn, H. Multiyear Crop Monitoring Using Polarimetric RADARSAT-2 Data. *IEEE Trans. Geosci. Remote Sens.* **2012**, *51*, 2227–2240. [CrossRef]
61. Cloude, S.R. *Polarisation: Applications in Remote Sensing Polarisation*; Oxford University Press: New York, NY, USA, 2010; ISBN 978-0-19-956973-1.
62. Lee, J.; Pottier, E. *Polarimetric Radar Imaging: From Basics to Applications*; CRC Press: Boca Raton, FL, USA, 2009; ISBN 142005497X.
63. Vicente-Guijalba, F.; Martinez-Marin, T.; Lopez-Sanchez, J.M. Dynamical Approach for Real-Time Monitoring of Agricultural Crops. *IEEE Trans. Geosci. Remote Sens.* **2014**, *53*, 3278–3293. [CrossRef]
64. Huapeng, L.; Zhang, C.; Zhang, S.; Atkinson, P.M. Full year crop monitoring and separability assessment with fully-polarimetric L-band UAVSAR: A case study in the Sacramento Valley, California. *Int. J. Appl. Earth Obs. Geoinf.* **2019**, *74*, 45–56. [CrossRef]
65. Lopez-Sanchez, J.M.; Ballester-Berman, J.D.; Hajnsek, I. First Results of Rice Monitoring Practices in Spain by Means of Time Series of TerraSAR-X Dual-Pol Images. *IEEE J. Sel. Top. Appl. Earth Obs. Remote Sens.* **2010**, *4*, 412–422. [CrossRef]
66. Lopez-Sanchez, J.M.; Cloude, S.R.; Ballester-Berman, J.D. Rice Phenology Monitoring by Means of SAR Polarimetry at X-Band. *IEEE Trans. Geosci. Remote Sens.* **2012**, *50*, 2695–2709. [CrossRef]
67. Lopez-Sanchez, J.M.; Vicente-Guijalba, F.; Ballester-Berman, J.D.; Cloude, S.R. Influence of Incidence Angle on the Coherent Copolar Polarimetric Response of Rice at X-Band. *IEEE Geosci. Remote Sens. Lett.* **2014**, *12*, 249–253. [CrossRef]
68. Cloude, S.; Pottier, E. An entropy based classification scheme for land applications of polarimetric SAR. *IEEE Trans. Geosci. Remote Sens.* **1997**, *35*, 68–78. [CrossRef]
69. Neumann, M.; Ferro-Famil, L.; Jager, M.; Reigber, A.; Pottier, E. A polarimetric vegetation model to retrieve particle and orientation distribution characteristics. In Proceedings of the 2009 IEEE International Geoscience and Remote Sensing Symposium, Cape Town, South Africa, 12–17 July 2009; Volume 4, p. IV-145.
70. Neumann, M.; Saatchi, S.S.; Ulander, L.M.H.; Fransson, J.E.S. Assessing Performance of L- and P-Band Polarimetric Interferometric SAR Data in Estimating Boreal Forest Above-Ground Biomass. *IEEE Trans. Geosci. Remote Sens.* **2012**, *50*, 714–726. [CrossRef]
71. Xie, Q.; Zhu, J.; Lopez-Sanchez, J.M.; Wang, C.; Fu, H. A Modified General Polarimetric Model-Based Decomposition Method with the Simplified Neumann Volume Scattering Model. *IEEE Geosci. Remote Sens. Lett.* **2018**, *15*, 1229–1233. [CrossRef]
72. Xie, Q.; Wang, J.; Liao, C.; Shang, J.; Lopez-Sanchez, J.M.; Fu, H.; Liu, X. On the Use of Neumann Decomposition for Crop Classification Using Multi-Temporal RADARSAT-2 Polarimetric SAR Data. *Remote Sens.* **2019**, *11*, 776. [CrossRef]
73. Kim, Y.; Van Zyl, J. Comparison of forest parameter estimation techniques using SAR data. In Proceedings of the IGARSS 2001, Scanning the Present and Resolving the Future, IEEE 2001 International Geoscience and Remote Sensing Symposium (Cat. No.01CH37217), Sydney, NSW, Australia, 9–13 July 2001; Volume 3, pp. 1395–1397.
74. Zhu, X.X.; Tuia, D.; Mou, L.; Xia, G.-S.; Zhang, L.; Xu, F.; Fraundorfer, F. Deep Learning in Remote Sensing: A Comprehensive Review and List of Resources. *IEEE Geosci. Remote Sens. Mag.* **2017**, *5*, 8–36. [CrossRef]
75. Zhang, L.; Zhang, L.; Du, B. Deep Learning for Remote Sensing Data: A Technical Tutorial on the State of the Art. *IEEE Geosci. Remote Sens. Mag.* **2016**, *4*, 22–40. [CrossRef]
76. Tuia, D.; Verrelst, J.; Alonso, L.; Perez-Cruz, F.; Camps-Valls, G. Multioutput Support Vector Regression for Remote Sensing Biophysical Parameter Estimation. *IEEE Geosci. Remote Sens. Lett.* **2011**, *8*, 804–808. [CrossRef]
77. Karimi, Y.; Prasher, S.O.; Madani, A.; Kim, S. Application of support vector machine technology for the estimation of crop biophysical parameters using aerial hyperspectral observations. *Can. Biosyst. Eng.* **2008**, *50*, 713–720.
78. Siegmann, B.; Jarmer, T. Comparison of different regression models and validation techniques for the assessment of wheat leaf area index from hyperspectral data. *Int. J. Remote Sens.* **2015**, *36*, 4519–4534. [CrossRef]
79. Breiman, L. Random Forests. *Mach. Learn.* **2001**, *45*, 5–32. [CrossRef]
80. Waske, B.; Van Der Linden, S.; Oldenburg, C.; Jakimow, B.; Rabe, A.; Hostert, P. Imager—A user-oriented implementation for remote sensing image analysis with Random Forests. *Environ. Model. Softw.* **2012**, *35*, 192–193. [CrossRef]
81. Austin, P.C.; Tu, J.V. Bootstrap Methods for Developing Predictive Models. *Am. Stat.* **2004**, *58*, 131–137. [CrossRef]

*remote sensing*

MDPI

*Article*

# Using Time Series Sentinel-1 Images for Object-Oriented Crop Classification in Google Earth Engine

**Chong Luo [1], Beisong Qi [1], Huanjun Liu [1,2,*], Dong Guo [1], Lvping Lu [3], Qiang Fu [1] and Yiqun Shao [3]**

1 Northeast Institute of Geography and Agroecology, Chinese Academy of Sciences, Changchun 130102, China; luochong@iga.ac.cn (C.L.); qibeisong@iga.ac.cn (B.Q.); guodong@iga.ac.cn (D.G.); fuqaing@iga.ac.cn (Q.F.)
2 School of Pubilc Adminstration and Law, Northeast Agricultural University, Harbin 150030, China
3 School of Resources and Environment, Northeast Agricultural University, Harbin 150030, China; yifei97518@gmail.com (L.L.); shao31945@gmail.com (Y.S.)
* Correspondence: liuhuanjun@neigae.ac.cn

**Abstract:** The purpose of this study was to evaluate the feasibility and applicability of object-oriented crop classification using Sentinel-1 images in the Google Earth Engine (GEE). In this study, two study areas (Keshan farm and Tongnan town) with different average plot sizes in Heilongjiang Province, China, were selected. The research time was two consecutive years (2018 and 2019), which were used to verify the robustness of the method. Sentinel-1 images of the crop growth period (May to September) in each study area were composited with three time intervals (10 d, 15 d and 30 d). Then, the composite images were segmented by simple noniterative clustering (SNIC) according to different sizes and finally, the training samples and processed images were input into a random forest classifier for crop classification. The results showed the following: (1) the overall accuracy of using the object-oriented classification method combined composite Sentinel-1 image represented a great improvement compared with the pixel-based classification method in areas with large average plots (increase by 10%), the applicable scope of the method depends on the plot size of the study area; (2) the shorter time interval of the composite Sentinel-1 image was, the higher the crop classification accuracy was; (3) the features with high importance of composite Sentinel-1 images with different time intervals were mainly distributed in July, August and September, which was mainly due to the large differences in crop growth in these months; and (4) the optimal segmentation size of crop classification was closely related to image resolution and plot size. Previous studies usually emphasize the advantages of object-oriented classification. Our research not only emphasizes the advantages of object-oriented classification but also analyzes the constraints of using object-oriented classification, which is very important for the follow-up research of crop classification using object-oriented and synthetic aperture radar (SAR).

**Keywords:** Sentinel-1; temporal composite; object-oriented; crop classification; Google Earth Engine

**Citation:** Luo, C.; Qi, B.; Liu, H.; Guo, D.; Lu, L.; Fu, Q.; Shao, Y. Using Time Series Sentinel-1 Images for Object-Oriented Crop Classification in Google Earth Engine. *Remote Sens.* **2021**, *13*, 561. https://doi.org/10.3390/rs13040561

Academic Editor: Olaniyi Ajadi
Received: 27 December 2020
Accepted: 2 February 2021
Published: 4 February 2021



## 1. Introduction

With continuous global population growth, the problem of food security is becoming increasingly serious [1–3]. To meet the increase in global demand for food in the future, improving the efficiency of food production so that it can be increased is the future governmental focus [4–6]. The rational distribution of grain production is key to improving the efficiency of grain production [7,8]. The accurate identification of the distribution of different crops on cultivated land is the premise of the rational distribution of food production; thus, accurate identification of the distribution of different crops on cultivated land is the basic condition needed to achieve regional sustainable development and ensure food security [9,10].

Remote sensing technology is the most commonly used technology in crop classification [11,12]. Optical images have always been the main data used in crop classification.

Many studies have used various machine learning methods to classify single-date or multi-temporal optical images, usually using MODIS, which has a moderate spatial resolution and Sentinel-2 and Landsat-8, which have medium spatial resolutions [13–16]. However, optical images are more vulnerable to the influence of clouds, especially in some areas with hot and rainy seasons and there are a lack of available optical images in the critical period of crop growth [17,18]. With an increasing number of synthetic aperture radar (SAR) platforms, SAR images show great potential for mapping crop distribution [19–21].

SAR is an active Earth observation system that can be installed on aircraft, satellites, spaceships and other flight platforms. It can observe the earth on all days and in all weather and it has a certain surface penetration ability [22–24]. SAR mainly records all kinds of backscatter responses produced by crop canopy biophysical structure. Many studies have shown that using multi-temporal SAR data can obtain better classification results than those obtained using single-date SAR data and using multi-polarization SAR data can obtain better classification results than can single polarization SAR data [25,26].

Although there are many studies on crop classification using SAR data, most are pixel-based classification. Speckle noise caused by the coherence of SAR data will seriously affect the accuracy of pixel-based classification results [27,28]. Some researchers have proven that the combination of object-oriented methods and SAR data can greatly improve the accuracy of crop classification but the general research is only in study areas with uniform plot sizes and do not consider the applicability of object-oriented methods combined with SAR data in different plot sizes [29,30]. In addition, some studies have proven that when using composite optical images for crop classification, composite images with shorter time interval can obtain higher accuracy [16]. In general, the relevant research shows that the object-oriented method is better than the traditional pixel-based method in crop classification using SAR [26]. However, there is little research on whether the object-oriented classification is applicable to all regions and what constraints it has. Therefore, clarifying the impact of the time interval and spatial resolution of SAR data on the accuracy of crop classification and evaluating the applicability of object-oriented methods in crop classification can provide a basis for extending the application of SAR data in crop classification.

The Google Earth Engine (GEE) is a Google cloud-based platform for processing satellite images and other geographic data. The GEE platform stores Pb-level processing-ready data and researchers can process several images quickly in parallel tasks, which greatly improves the efficiency of image processing [31]. GEE has been applied to various scales of geospatial mapping, such as rice distribution mapping, fallow land mapping, tidal flats mapping, land cover mapping and so on [32–35]. C-band Sentinel-1 is considered to be the most promising radar data for crop classification because it has medium temporal and spatial resolutions and is provided free of charge to the public [36]. These conditions provide support for crop classification based on SAR data.

The main purposes of this study were as follows: (1) to evaluate the impact of different time intervals composite and different segmentation size object-oriented methods on the accuracy of Sentinel-1 crop classification; (2) to study the key period of using Sentinel-1 images for crop classification; and (3) to compare the classification accuracy of two study areas with different plot sizes and evaluate the applicability of this method.

## 2. Materials and Methods

To study the potential of using object-oriented and time series Sentinel-1 in crop classification in GEE, two study areas with large differences in plot size were selected in this study. First, the Sentinel-1 images with different time intervals were obtained by GEE processing, the images were segmented by the simple noniterative clustering (SNIC) image segmentation method and then the processed images were classified using a random forest classifier. Finally, the classification accuracies of different scenarios were compared to evaluate the effectiveness and applicability of the methods. The flowchart of this study is shown in Figure 1.

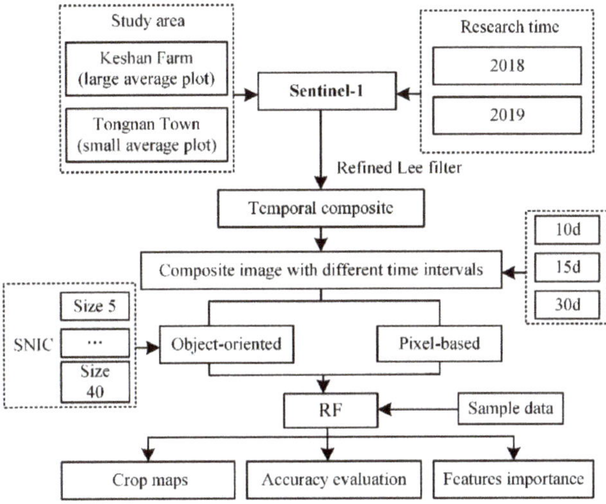


**Figure 1.** Flowchart.

### 2.1. Study Area

To verify the applicability of the method, two study areas with different agricultural production scales were selected in this study: Keshan farm (125°07′40″, 125°37′30″E, 48°11′15″, 48°24′07″N) and Tongnan town (124°54′15″, 125°12′44″E, 48°2′40″, 48°15′13″N) in central Heilongjiang Province, China, the areas are 3.35 $km^2$ and 3.15 $km^2$ respectively. As shown in Figure 2, the two study areas are adjacent to each other. Keshan farm is a state-owned farm. In Tongnan town, small-scale farmers are used as the mode of agricultural production and management. Due to the different agricultural production scales and management modes, the average plot sizes in the two areas are quite different. Corn, soybean and rice account for more than 95% of the planting area in Heilongjiang Province [37]. The two study areas are mainly planted with corn, soybean and rice, which are representative for Heilongjiang Province.

Keshan farm and Tongnan town are located in the northeastern Songnen Plain, with hilly terrain and fertile soil suitable for crop growth. Both study areas belong to the warm and cool type climate zone, which is characterized by a dry and windy spring, a high-temperature and rainy summer, rapid cooling in autumn, an early frost and a long winter that has snow and is cold and dry. The annual precipitation is approximately 502.5 mm and the precipitation in 6–8 months accounts for 68.3% of the annual amount. The frost-free period is only 120 days and only one growing season is guaranteed. The main crops are corn, soybean and rice and are generally sown in spring (April to May) and harvested in autumn (September to October). In fact, the crops have not grown at the end of April and the crops have basically been harvested at the beginning of October. Then we will take May to September as the study period. See Table 1 for details and more information on the major crop calendar in the study area.

**Table 1.** Main crop calendar of the study area.

| | Jan | Feb | Mar | Apr | May | Jun | Jul | Aug | Sep | Oct | Nov | Dec |
|---|---|---|---|---|---|---|---|---|---|---|---|---|
| corn | | | | | S | G | G | G | H | H | | |
| rice | | | | | S | G | G | G | H | H | | |
| soybeans | | | | | S | G | G | G | H | | | |

Note: S—Sowing; G—Growing; H—Harvesting.

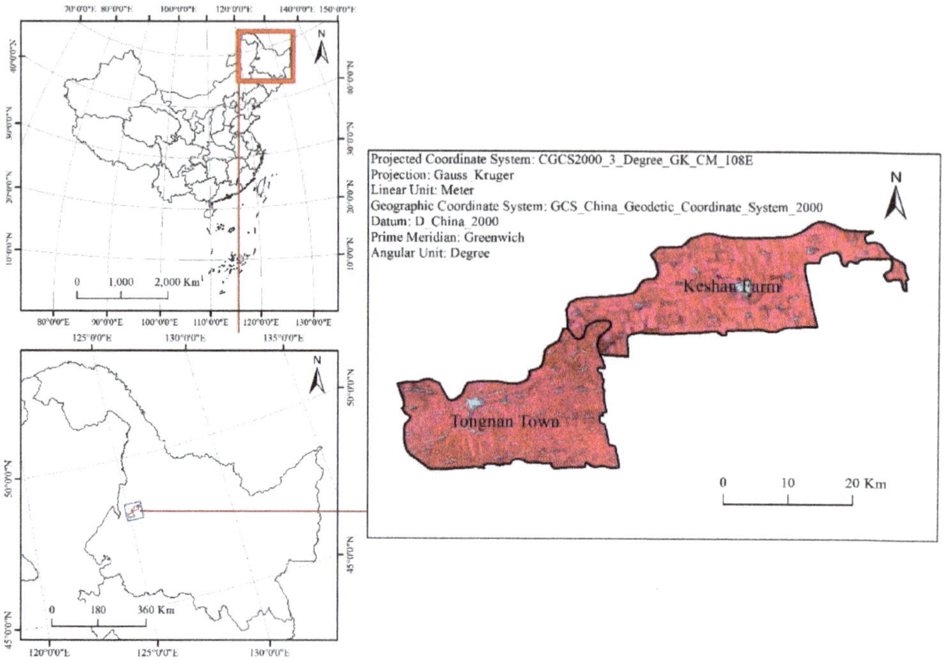


**Figure 2.** Overview of the study area.

## *2.2. Data Selection and Preprocessing*

### 2.2.1. Sentinel-1 SAR Image and Preprocessing

In this study, the Sentinel-1 SAR GRD dataset stored in the GEE cloud platform was used and it included all images covering the study area in 2018 and 2019 (May to September), including 27 images in 2018 and 22 images in 2019, it is very close to the temporal resolution of sentinel-1 which is usually reported as 6d [38]. The Sentinel-1 SAR GRD dataset was collected using the interference wide-band (IW) mapping mode, with a spatial resolution of 10 m, a width of 250 km and an average incidence angle of 30–45°. Each Sentinel-1 image stored by the GEE platform, which had been preprocessed using the European Space Agency's (ESA) Sentinel-1 Toolbox including orbit restitution, thermal noise removal, terrain correction and radiometric calibration [39–41].

In addition, all Sentinel-1 images to be used were filtered with refined Lee filter on the GEE platform to reduce speckle. The function of each speckle filter should be adjusted according to local image variations to smooth the values and thereby reduce speckle and lines and edges are enhanced to maintain the sharpness of the imagery. This refined Lee filter was chosen because it could better retain polarization information under the influence of speckle elimination [42,43].

To evaluate the impact of Sentinel-1 images with different time intervals on classification accuracy, the median value of observations in each temporal interval was obtained and image time series were built. We use day of year (DOY) to represent time series.

### 2.2.2. Reference Data

The sample plots of this study came from the land by insurance companies and the crop type was confirmed on the spot. In 2018, there were 483 sample plots in Keshan farm, including 200 corn plots, 268 soybean plots and 15 rice plots; in 2019, there were 486 sample plots, including 272 corn plots, 203 soybean plots and 11 rice plots. There were 511 sample plots in Tongnan town in 2018, including 204 corn plots, 305 soybean plots and 2 rice plots;

there were 542 sample plots in 2019, including 201 corn plots, 339 soybean plots and 2 rice plots. To avoid the time-out of GEE calculation, we used ArcGIS 10.3 to transform these sample plots into sample points. Each study area randomly selected 70% of the sample points as the training samples and 30% of the sample points as the verification samples. The plots in the study area are shown in Table 2.

**Table 2.** Plot characteristics of Keshan farm and Tongnan town.

| | Location | Number of Plots | Average Area ($m^2$) | Average Circumference (m) | Average Area Circumference Ratio |
|---|---|---|---|---|---|
| 2018 | Keshan Farm | 483 | 333,499.50 | 3126.49 | 101.77 |
| 2019 | Keshan Farm | 486 | 342,129.04 | 3137.81 | 103.94 |
| 2018 | Tongnan Town | 511 | 29,175.02 | 1256.92 | 21.15 |
| 2019 | Tongnan Town | 542 | 40,173.86 | 1313.92 | 26.85 |

### 2.3. Image Segmentation

The traditional pixel-based classification method may produce "salt and pepper" noise, especially for Sentinel-1 radar data. The object-based algorithm reduces this problem by considering the neighborhood information of a given pixel to divide the image into specific areas or objects according to certain parameters. In this study, the SNIC image segmentation algorithm in GEE was used for image segmentation [44]. Firstly, the centroid pixels on the regular grid in the image are initialized. Next, the distance of pixels in the five-dimensional space of color and spatial coordinates is used to determine the dependency of each pixel relative to the centroid. At last, the distance integrates the normalized spatial and color distances to produce efficient, compact and nearly uniform polygons [45]. The main parameters of the SNIC algorithm are "image", "size", "compactness", "connectivity", "neighborhood size" and "seeds". Among them, "image" is the image participating in segmentation. In this study, Sentinel-1 time series images with different time intervals composites in 2018 and 2019 (May to September) in two study areas were segmented. "Size" refers to the spacing of super-pixel seed positions based on pixels, that is, segmentation size. According to the situation of the study area, the "size" was set as "5", "10", "15", "20", "25", "30", "35" and "40" and the impact of different segmentation sizes on classification accuracy was evaluated. "Compactness" is compactness and the larger the value is, the closer the segmentation results are to the square; however, because the parcels in the study area were mostly rectangular, this study set the "compactness" value to 0. "Connectivity" stands for connectivity, which was set to 8 in this study. "Seeds" does not need to be set in this article, because the plot in this study area is closer to the rectangle. In the classification experiment using Sentinel-1 images, according to the experimental setting, each time interval has one original data control group and eight different size segmentation groups, for a total of three time interval composites, meaning each study area had 27 experimental groups every year.

### 2.4. Random Forest

Random forest is an improved self-classification and decision tree (CART), which is the most popular machine learning algorithm used for remote sensing classification [46,47]. The bootstrap sampling technique is employed to randomly extract a certain number of samples from the original dataset to generate a new training dataset. Each tree in the forest grows to the maximum extent without any trimming. The random sampling process avoids the occurrence of over-fitting. Random forest has many advantages over other algorithms [48,49]. First, random forest has been proven to be superior to other algorithms in classification accuracy. Second, random forest can process data with high dimensions without feature selection. The random forest algorithm uses bootstrap sampling technology to randomly select a certain number of samples from the original data set to generate a new training data set. Every tree in the forest can grow to its maximum without pruning. Random sampling process avoids overfitting [50].

It is very easy to use the random forest algorithm in the GEE cloud platform. In this study, we set the nTree value to 300, which can ensure accuracy and avoid overfitting [51]. Mtry was set to the default value, which was the square root of the input feature data.

Random forest classifiers usually use two methods to measure the importance of features relative to classification. MDA and MDG are, respectively, the average value of decreasing accuracy of a feature and the value of the decreasing Gini coefficient in random forest when other conditions are unchanged [46,47]. The higher the MDG value is, the more important this feature is. The higher the MDA value is, the more important the feature is. In this study, we used the MDA to evaluate the importance of different features. The principle of MDA value is to disrupt the eigenvalue order of each feature and to measure the influence of the order change on the accuracy of the model. This ingenious method uses out of pocket data to calculate importance. OOB (out-of-bag) data is a part of the training set but it is not used to train this special tree. OOB data is used to calculate the basic error and then the order of each feature is randomly scrambled. We used the R language package randomForest 4.6 to obtain the MDA value of the different features.

### 2.5. Accuracy Verification

In this study, the total accuracy (OA), kappa coefficient, producer accuracy (PA) and user accuracy (UA) were selected to evaluate the accuracy of crop classification [52]. The formulas are listed below:

$$\text{OA}\ (\%) = \frac{\sum_{i=1}^{n} p_{ii}}{N} \times 100 \tag{1}$$

$$\text{Kappa} = \frac{N\sum_{i=1}^{n} P_{ii} - \sum_{i=1}^{n}(P_{i+} \times P_{+i})}{N^2 - \sum_{i=1}^{n}(P_{i+} \times P_{+i})} \tag{2}$$

$$\text{UA}\ (\%) = \frac{P_{ii}}{P_{i+}} \times 100 \tag{3}$$

$$\text{PA}\ (\%) = \frac{P_{ii}}{P_{+i}} \times 100 \tag{4}$$

Here, $n$ is the total number of columns of the confusion matrix; that is, the total number of categories, $P_{ii}$ is the number of correct classifications of the upper crop-type sample in the $i$-th row and $i$-th column of the confusion matrix, $P_{i+}$ and $P_{+i}$ are the total number of crop-type samples in row $i$ and column $i$ and $N$ is the total number of samples used for verification.

## 3. Results

### 3.1. Sentinel-1 Time Series

Figure 3 shows the multitemporal images curves of the average backscatter coefficients of all samples of each crop type in the two study areas in 2018. In Figure 3, the X axis is the time series of images at different time intervals composites and the Y axis is the value of the backscattering coefficient. In the multitemporal images with a time interval of 30 d in the two study areas, the VH and VV bands in June, July, August and September easily identify rice, while the difference between corn and soybean is small. In the multitemporal images with a time interval of 15 d in the two study areas, the difference between rice and other crops is obvious in most periods. The difference between corn and soybean is more obvious in the first half of September and the second half of September in the VH band and more obvious in the second half of September in the VV band. In the multitemporal images with a time interval of 10 d in the two study areas, the difference between rice and other crops is still obvious. The difference between corn and soybean can be distinguished by the VH band in early September and late September or by the VV band in early August and late September.

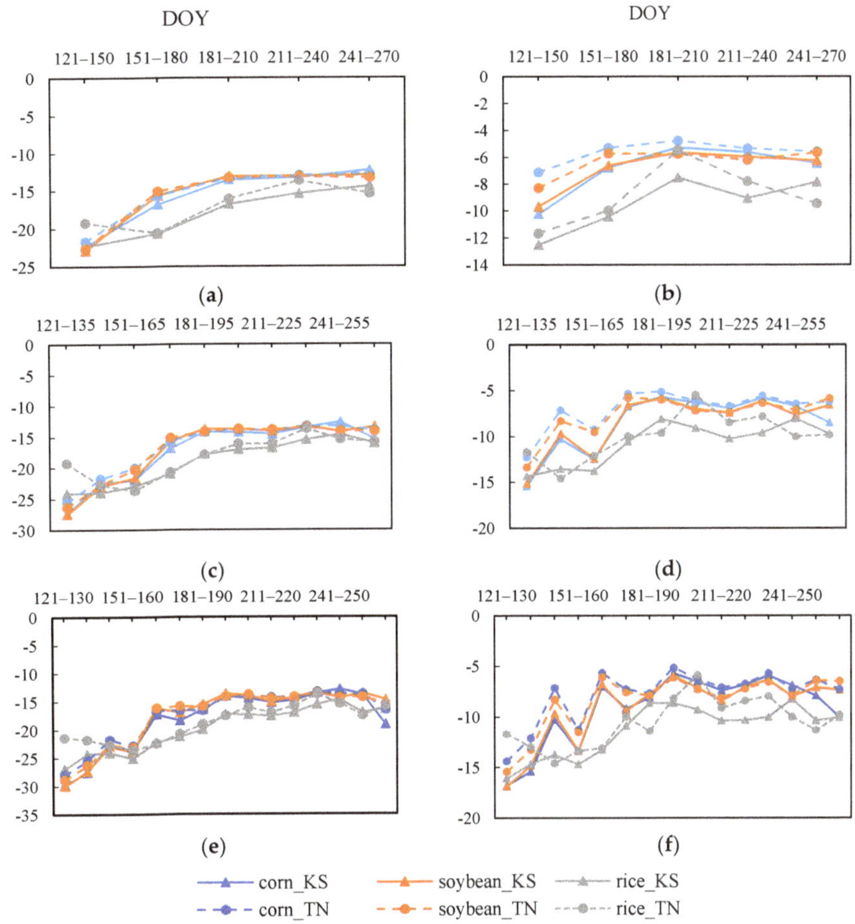


**Figure 3.** Variation characteristics of the average values of the polarization bands (VV and VH) of all samples of different crops in Keshan farm and Tongnan town with different time intervals ((**a**), VH_30d; (**b**), VV_30d; (**c**), VH_15d; (**d**), VV_15d; (**e**), VH_10d; (**f**), VV_10d).

### *3.2. Overall Accuracy Assessment*

Tables 3 and 4 list the OAs and kappa coefficients of the different classification schemes for Sentinel-1 time series images using different time intervals composite for Keshan farm and Tongnan town in 2018 and 2019. At Keshan farm in 2018, Sentinel-1 with a time interval of 10 d obtained the highest crop classification accuracy when the image segmentation size was 30 (Figure 4). At Keshan farm in 2019, Sentinel-1 with a time interval of 10 d obtained the highest crop classification accuracy when the image segmentation size was 25 (Figure 5). In Tongnan town in 2018, Sentinel-1 with a time interval of 10 d obtained the highest crop classification accuracy when the image segmentation size was 15 (Figure 6). In Tongnan town in 2019, Sentinel-1 with a time interval of 10 d obtained the highest crop classification accuracy when the image segmentation size was 5 (Figure 7).

**Table 3.** The overall accuracies obtained using different classification schemes in different study areas.

| Year | Study Area | Interval | Pixel | 5 | 10 | 15 | 20 | 25 | 30 | 35 | 40 |
|---|---|---|---|---|---|---|---|---|---|---|---|
| | | | Size | | | | | | | | |
| 2018 | Keshan Farm | 10 d | 84.99 | 88.10 | 93.20 | 93.77 | 94.62 | 92.92 | 95.47 | 94.33 | 94.05 |
| | | 15 d | 79.89 | 87.25 | 91.22 | 93.77 | 93.20 | 94.05 | 94.05 | 93.48 | 91.78 |
| | | 30 d | 69.41 | 77.62 | 84.42 | 87.82 | 88.10 | 88.95 | 86.69 | 88.10 | 86.69 |
| | Tongnan Town | 10 d | 72.04 | 76.08 | 76.88 | 76.88 | 74.19 | 72.04 | 73.12 | 75.54 | 69.62 |
| | | 15 d | 75.00 | 74.19 | 73.39 | 74.46 | 75.81 | 73.12 | 72.04 | 73.92 | 69.89 |
| | | 30 d | 71.51 | 75.81 | 71.51 | 72.31 | 71.77 | 67.20 | 68.01 | 64.78 | 68.55 |
| 2019 | Keshan Farm | 10 d | 80.66 | 90.33 | 92.75 | 94.86 | 93.96 | 94.86 | 94.26 | 94.56 | 93.66 |
| | | 15 d | 79.15 | 85.20 | 90.03 | 92.45 | 90.94 | 94.26 | 93.05 | 93.05 | 94.26 |
| | | 30 d | 76.44 | 83.08 | 85.20 | 88.82 | 91.54 | 91.24 | 92.45 | 92.15 | 92.75 |
| | Tongnan Town | 10 d | 79.58 | 80.90 | 78.51 | 77.45 | 73.47 | 75.60 | 72.94 | 72.68 | 72.15 |
| | | 15 d | 78.51 | 78.78 | 76.13 | 76.13 | 75.86 | 79.58 | 75.86 | 72.41 | 72.94 |
| | | 30 d | 68.70 | 74.80 | 70.56 | 70.56 | 71.88 | 71.35 | 70.29 | 72.68 | 70.29 |

**Table 4.** The kappa coefficients obtained using different classification schemes in different study areas.

| Year | | Interval | Pixel | 5 | 10 | 15 | 20 | 25 | 30 | 35 | 40 |
|---|---|---|---|---|---|---|---|---|---|---|---|
| | | | Size | | | | | | | | |
| 2018 | Keshan Farm | 10 d | 0.70 | 0.77 | 0.87 | 0.88 | 0.90 | 0.87 | 0.91 | 0.89 | 0.89 |
| | | 15 d | 0.60 | 0.75 | 0.83 | 0.88 | 0.87 | 0.89 | 0.89 | 0.88 | 0.84 |
| | | 30 d | 0.40 | 0.56 | 0.70 | 0.76 | 0.77 | 0.79 | 0.74 | 0.77 | 0.75 |
| | Tongnan Town | 10 d | 0.44 | 0.52 | 0.54 | 0.54 | 0.49 | 0.45 | 0.47 | 0.52 | 0.40 |
| | | 15 d | 0.50 | 0.49 | 0.47 | 0.50 | 0.52 | 0.47 | 0.44 | 0.48 | 0.41 |
| | | 30 d | 0.43 | 0.52 | 0.44 | 0.45 | 0.44 | 0.35 | 0.37 | 0.30 | 0.38 |
| 2019 | Keshan Farm | 10 d | 0.61 | 0.81 | 0.86 | 0.90 | 0.88 | 0.90 | 0.89 | 0.89 | 0.88 |
| | | 15 d | 0.57 | 0.70 | 0.80 | 0.85 | 0.82 | 0.89 | 0.86 | 0.86 | 0.89 |
| | | 30 d | 0.54 | 0.67 | 0.71 | 0.78 | 0.83 | 0.83 | 0.85 | 0.85 | 0.86 |
| | Tongnan Town | 10 d | 0.58 | 0.61 | 0.56 | 0.54 | 0.45 | 0.51 | 0.44 | 0.44 | 0.42 |
| | | 15 d | 0.57 | 0.57 | 0.51 | 0.52 | 0.51 | 0.58 | 0.51 | 0.45 | 0.45 |
| | | 30 d | 0.33 | 0.46 | 0.39 | 0.40 | 0.42 | 0.42 | 0.39 | 0.44 | 0.38 |

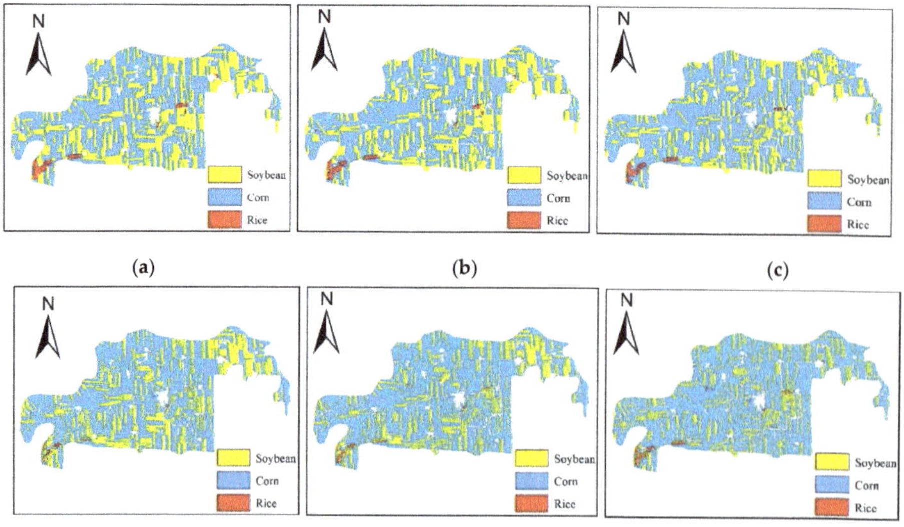


**Figure 4.** Crop classification results of highest crop classification accuracy with different time intervals in Keshan farm at 2018 ((**a**), 10d _size30; (**b**), 15d _size25; (**c**), 30d _size25; (**d**), 10d _Pixel; (**e**), 15d_Pixel; (**f**), 30d_Pixel).

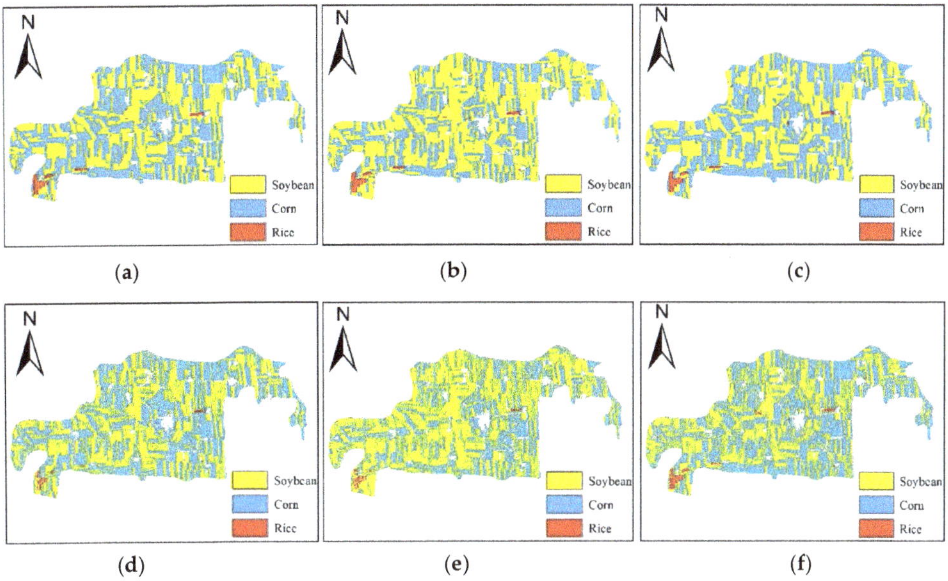


**Figure 5.** Crop classification results of highest crop classification accuracy with different time intervals in Keshan farm at 2019 ((**a**), 10d_size25; (**b**), 15d_size25; (**c**), 30d_size40; (**d**), 10d_Pixel; (**e**), 15d_ Pixel; (**f**), 30d_ Pixel).

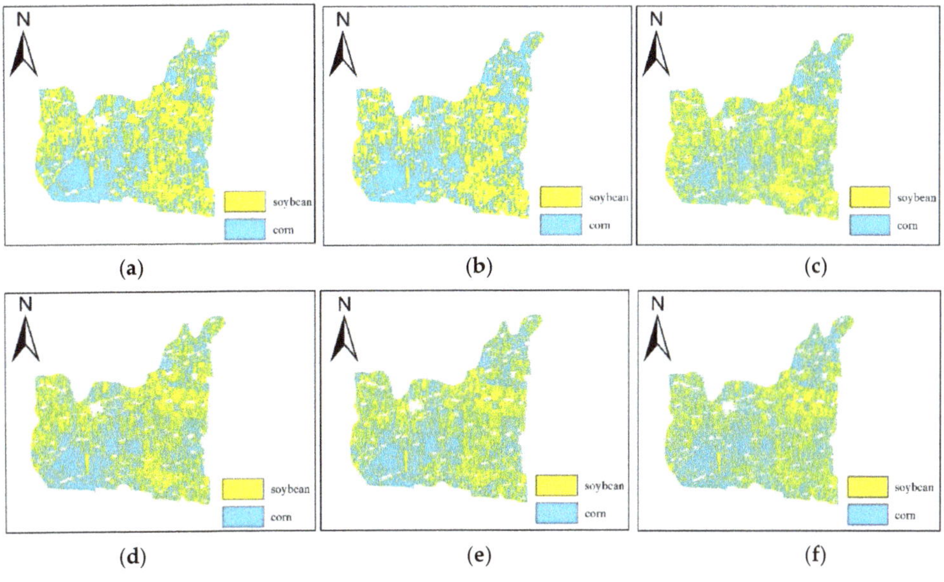


**Figure 6.** Crop classification results of highest crop classification accuracy with different time intervals in Tongnan town at 2018 ((**a**), 10d_size15; (**b**), 15d_size20; (**c**), 30d_size5; (**d**), 10d_Pixel; (**e**), 15d_ Pixel; (**f**), 30d_ Pixel).

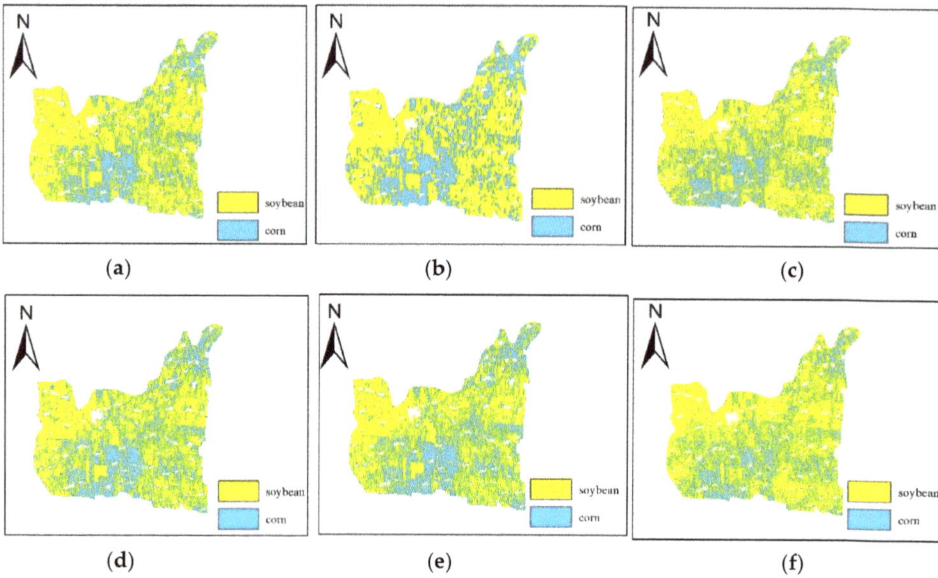


**Figure 7.** Crop classification results of highest crop classification accuracy with different time intervals in Tongnan town at 2019 ((**a**), 10d_size5; (**b**), 15d_size25; (**c**), 30d_size5; (**d**), 10d_Pixel; (**e**), 15d_ Pixel; (**f**), 30d_ Pixel).

In Keshan farm, where the average plot was large, the shorter time interval of Sentinel-1 composite was, the higher the accuracy of crop classification was; compared with the pixel-based classification method, the accuracy of using the object-oriented classification method was greatly improved. In Tongnan town, where the average plot was small, in most cases, the shorter time interval of Sentinel-1 composite was, the higher the accuracy of crop classification was; an exception was the pixel-based classification in 2018. when the time interval decreased from 15 d to 10 d, the accuracy of crop classification decreased; compared with the pixel-based classification accuracy, using the object-oriented classification method in Tongnan town did not result in an obvious improvement.

### 3.3. User Accuracy and Producer Accuracy

Figures 8 and 9 show the user accuracy (UA) and producer accuracy (PA) for images with different time intervals using pixel-based and object-oriented classification (optimal size) in Keshan farm and Tongnan town. For Keshan farm in 2018, Sentinel-1 with a time interval of 10 d combined with object-oriented classification achieved the highest UA and PA for corn and rice, Sentinel-1 with a time interval of 15 d combined with object-oriented classification achieved the highest PA for soybean (Figure 8). Compared with pixel-based classification, object-oriented classification always had a higher UA. In 2019, the PA and UA of different crops showed similar performance as that in 2018 (Figure 8) and the time interval between 10 d and 15 d was not much different between UA and PA.

In Tongnan town, there was no significant difference in the PA and UA between corn and soybean based on pixel classification compared with object-oriented classification and the PA and UA of corn and soybean changed irregularly with different time intervals (Figure 9).

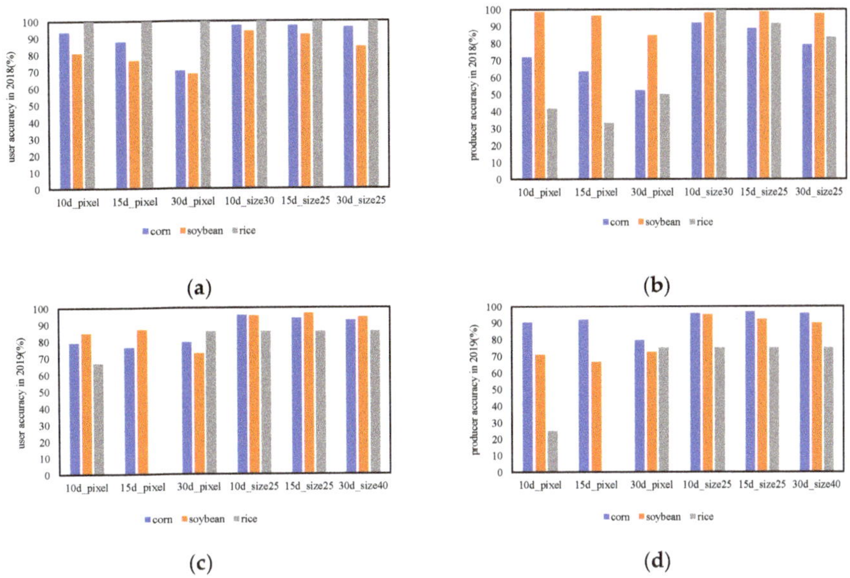


**Figure 8.** Producer accuracy and user accuracy of highest crop classification accuracy with different time intervals in Keshan farm at 2018 and 2019 ((**a**) user accuracy in 2018; (**b**) producer accuracy in 2018; (**c**) user accuracy in 2019; (**d**) producer accuracy in 2019).

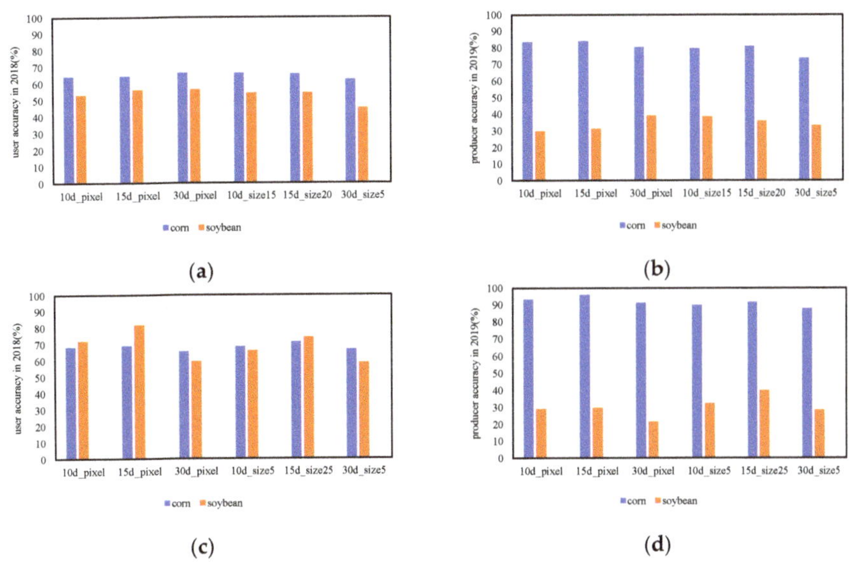


**Figure 9.** Producer accuracy and user accuracy of highest crop classification accuracy with different time intervals in Tongnan town at 2018 and 2019, there are only 2 rice sample sites in Tongnan Town, so no rice sample sites are assigned to the verification sample sites ((**a**) user accuracy in 2018; (**b**) producer accuracy in 2018; (**c**) user accuracy in 2019; (**d**) producer accuracy in 2019).

### 3.4. Features Importance Assessment

In addition, this study evaluated the importance features of Sentinel-1 images with different time intervals in Keshan farm and Tongnan town (object-oriented with optimal segmentation size). When using Sentinel-1 images with 10-d time interval for crop classification in Keshan farm and Tongnan town, the VH band in 2019 was more important than that in 2018, which may be due to the reduced importance of the water-sensitive VV band in crop classification due to the flood in 2019. This situation was not obvious when using Sentinel-1 images with a time interval of 15 d or 30 d, mainly because the effect of flood was eliminated by using the median value of a longer time interval (Figures 10 and 11).

The features with higher importance in Sentinel-1 images with different time intervals were mainly distributed in July, August and September, which was mainly due to the large differences in the structure of different crops in July, August and September. When corn enters the heading stage in July, the dry matter on the ground increases rapidly and the differences between corn and soybean increase. The features with higher importance in different years also have differences, which may be caused by the differences in crop phenology in different years.

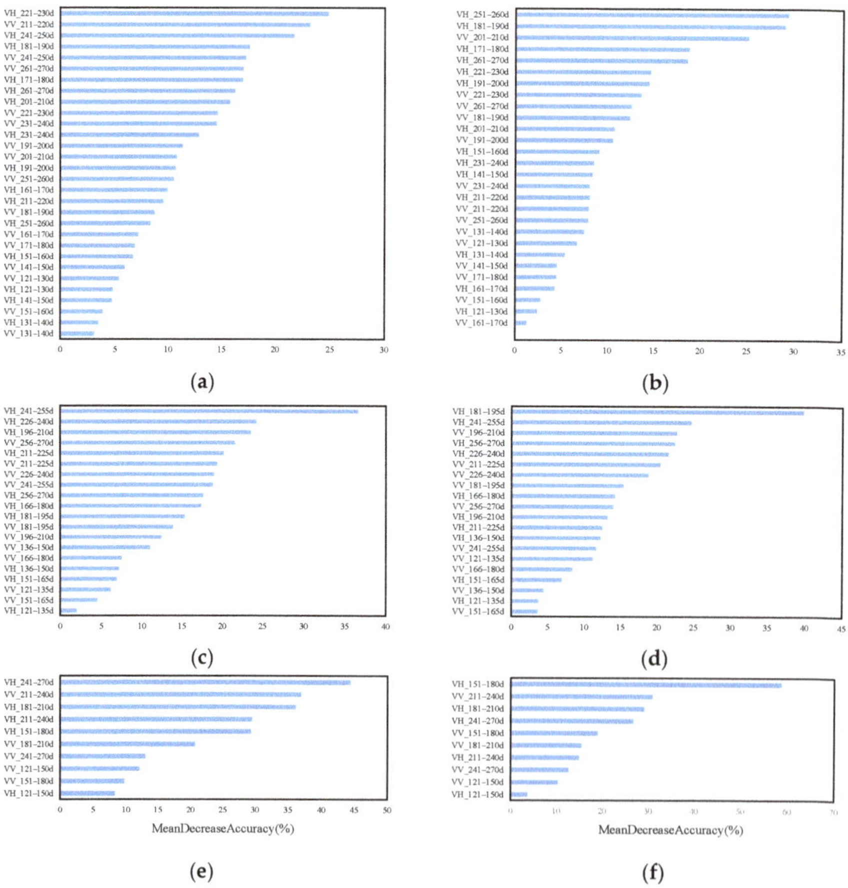


**Figure 10.** Importance assessment of crop classification features in 2018 and 2019 of Keshan farm with different time interval composite images ((**a**) MeanDecreaseAccuracy_2018_10d; (**b**) MeanDecreaseAccuracy_2019_10d; (**c**) MeanDecreaseAccuracy_2018_15d; (**d**) MeanDecreaseAccuracy_2019_15d; (**e**) MeanDecreaseAccuracy_2018_30d; (**f**) MeanDecreaseAccuracy_2019_30d).

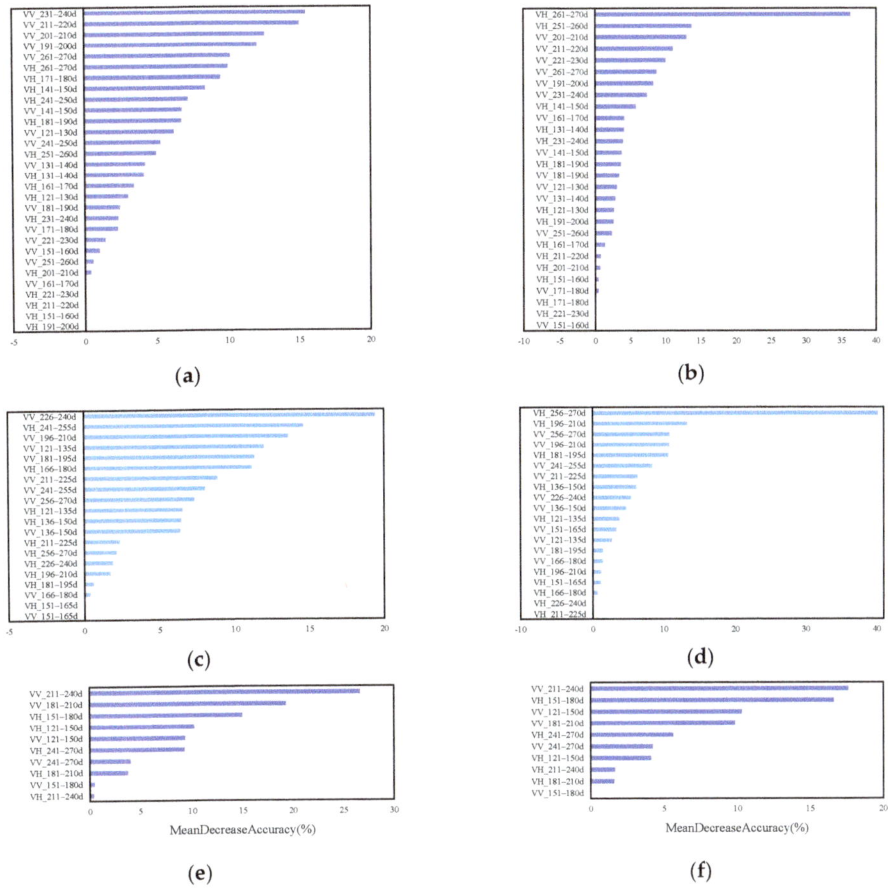


**Figure 11.** Importance assessment of crop classification features in 2018 and 2019 of Tongnan town with different time interval composite images ((**a**) MeanDecreaseAccuracy_2018_10d; (**b**) MeanDecreaseAccuracy_2019_10d; (**c**) MeanDecreaseAccuracy_2018_15d; (**d**) MeanDecreaseAccuracy_2019_15d; (**e**) MeanDecreaseAccuracy_2018_30d; (**f**) MeanDecreaseAccuracy_2019_30d).

### *3.5. Determination of the Optimal Segmentation Size*

The traditional method of determining the optimal segment size in object-oriented classification usually involves conducting repeated experiments. From Table 2, we can obtain the average area and average perimeter of sample plots in Keshan farm and Tongnan town in 2018 and 2019. We assume that the plots in the study area are evenly distributed rectangles and the average side length of the plots can be calculated from the perimeter and area. The long-side length and short-side length are shown in Table 5. We can obtain the following relationship: the best segmentation size ≈ average short side/image resolution. This relationship is better applied to the large plot of Keshan farm but is slightly less applicable to the small plot of Tongnan town.

**Table 5.** The relationship between the optimal segment size and the side length of the plot.

| | Number | Average Long Side | Average Short Side | Average Optimal Size | Average Short Side/Image Resolution |
|---|---|---|---|---|---|
| 2018 ks | 493 | 1308.03 | 254.96 | 26.67 | 25.50 |
| 2019 ks | 486 | 1306.04 | 261.95 | 30 | 26.20 |
| 2018 tn | 511 | 577.48 | 50.52 | 13.33 | 5.05 |
| 2019 tn | 542 | 587.64 | 68.36 | 11.67 | 6.84 |

## 4. Discussion

### 4.1. Advantages of Using Time Series Sentinel-1 Images Combined with Object-Oriented Classification Methods

In the past, research on crop classification using SAR data usually focused on obtaining more useful information from different polarizations to improve the accuracy of crop classification [53–55]. Therefore, RADARSAT-2 data are usually used as the data source because compared with single polarization and dual-polarization radar data, multi-polarization RADARSAT-2 data (HH, HV, VV, VH) have obvious advantages in crop classification [56].

Dense time series images can provide more detailed phenology information, which has been proven to improve the accuracy of crop classification [16] but RADARSAT-2 has a revisit period of 24 d, making it difficult to obtain dense time series data. The theoretical revisit period of Sentinel-1 data used in this study is 6 d and its temporal resolution can effectively meet the requirements of time series-based classification methods. The research in this study proves that using Sentinel-1 composite images with a shorter time interval can improve the accuracy of crop classification in most cases.

Many studies have proven that object-oriented classification is superior to pixel-based classification [57,58]. Previous studies have often used ready-made plot data directly so that crops within the parcel can be directly identified [26] but most developing countries do not have complete plot databases and there is no guarantee of the uniqueness of the crops grown on each plot. In this study, the image segmentation algorithm was used to segment the multitemporal images into "blocks" and then the pixels in the "blocks" were regarded as being the same class, thereby reducing the impact of speckle noise in SAR images on the classification accuracy. This study shows that the accuracy of crop classification can be improved to a very high level (OA > 90%) by using the method of object-oriented classification and time series Sentinel-1 in the case of a large plot size.

### 4.2. Advantages of Using GEE

The GEE cloud platform effectively promoted this research. GEE has been proved to be very suitable for high-speed data analysis with large spatial processing functions [34,35]. In this study, we used GEE to select Sentinel-1 images, used refined Lee filter to reduce speckles, composited the processed Sentinel-1 images according to different time intervals, used the SNIC algorithm for image segmentation and finally used random forest for crop classification. In this study, except for the band importance evaluation that was processed in R Studio, all other processes were performed in the GEE cloud platform. If these works are carried out offline, it may take several days to dozens of days. In the GEE cloud platform, we focus on experimental design and code writing, which greatly improved the efficiency of crop classification. In addition, GEE integrates many other machine learning algorithms and image segmentation algorithms, which provides the possibility of future algorithm improvements.

### 4.3. The Relationship between Image Resolution and Optimal Segmentation Size

Through the analysis of sample characteristics in Table 2, we see that the average plot area of Keshan farm is approximately 10 times that of Tongnan town and the average plot area perimeter ratio of Keshan farm is approximately 5 times that of Tongnan town, which shows that the plot of study area a is closer to the square, which is beneficial for image segmentation.

It can be seen from Figure 12 that when the segmentation size of Keshan farm is 25 or 30, the segmentation result is closest to the boundary of the plot. At this time, the accuracy of crop classification is the highest. If the segmentation size is too large, two adjacent plots are mixed together; in contrast, if the segmentation size is too small, the speckle noise cannot be removed well. Figure 13 shows that when the segmentation size of Tongnan town is 5, the segmentation result has mixed two adjacent plots together and the accuracy of object-oriented classification is not much better than that of pixel-based classification. Of course, this method of judging the optimal segmentation size is applicable only to areas with little difference in plot area.

### *4.4. Uncertainty of the Method*

The method proposed in this study has been proven to achieve good results in Keshan farm with a large plot area for two consecutive years. The highest OA of object-oriented classification was approximately 95%, which was 10% higher than that of pixel-based classification. However, in Tongnan town, which has a small plot area, the highest OA of the object-oriented classification method was only approximately 2% higher than that of the pixel-based classification method (Table 3). This difference is mainly because the resolution of the Sentinel-1 image is 10 m and the plot in Tongnan town is too small. If the segmentation size is small, the speckle noise of Sentinel-1 cannot be eliminated. If the segmentation size is large, the "block" after segmentation may contain adjacent plots, resulting in classification errors. We believe that this result does not mean that the object-oriented classification method does not work in the small area of the plot but rather that the object-oriented method needs to match the SAR image of the appropriate spatial resolution. When the resolution of the SAR image is too low to the short-side length of the plot, ideal accuracy cannot be obtained. In addition, since there are only three crops in this study, the effect of the model in the more complex area of crop types needs further study. Most of the previous studies highlight the advantages of object-oriented classification method [26,59,60]. Our research shows that object-oriented classification method has almost no advantages over pixel-based method in some areas, which provides a reference for future research.

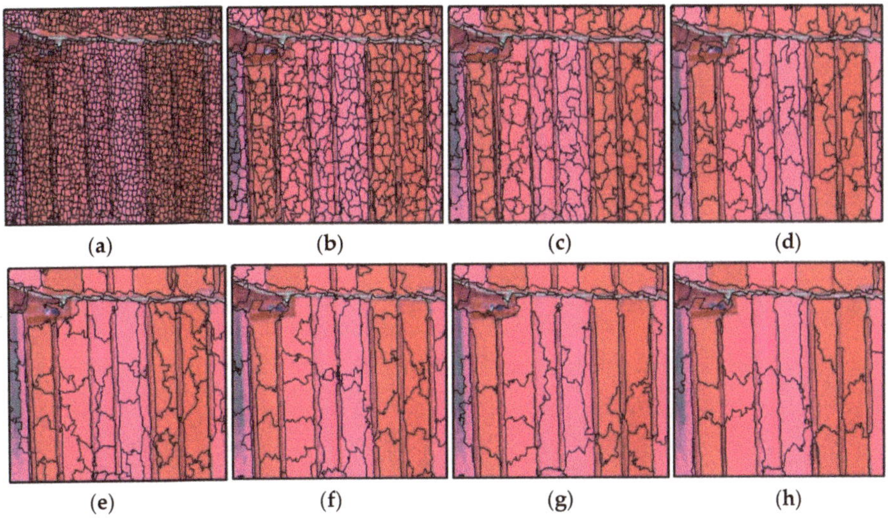


**Figure 12.** Segmentation effect of different segmentation sizes in Keshan farm ((**a**) size5; (**b**) size10; (**c**) size15; (**d**) size20; (**e**) size25; (**f**) size30; (**g**) size35; (**h**) size40).

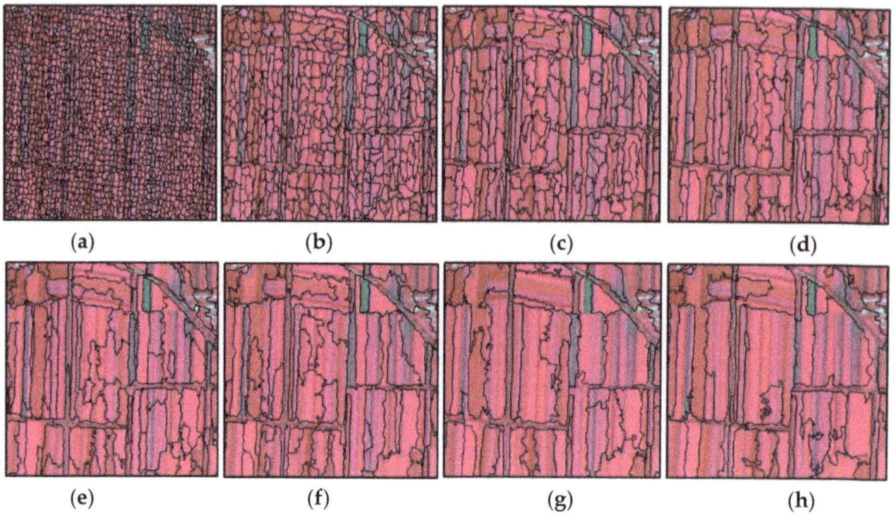

Figure 13. Segmentation effect of different segmentation sizes in Tongnan town. ((a) size5; (b) size10; (c) size15; (d) size20; (e) size25; (f) size30; (g) size35; (h) size40).

### 4.5. Future Research Directions

Crop classification maps are the basis of precise agricultural management [61]. It is of great significance for regional food security and sustainable development to produce annual crop distribution maps regularly [62]. Optical images have shown good performance in crop classification but in areas with hot and rainy seasons, there are many clouds covering the critical period of crop growth every year and cloud presence seriously affects the observation of optical images. At this time, it is very important to evaluate the potential use of SAR data for crop classification.

In this study, the method of using the object-oriented method combined with the multitemporal Sentinel-1 images method in GEE had an obvious effect in large plot areas but the effect in small plot areas was not obvious. The method to solve the poor effect of small plot area classification is to use SAR images with a higher spatial resolution combined with object-oriented methods. In addition, in this study, it was found that with the decrease in time interval composite from 15 d to 10 d, the accuracy of crop classification was not significantly improved; thus, we speculated that the object-oriented method combined with SAR images with higher spatial resolutions can achieve ideal accuracy without too high of a temporal resolution. In the next step, we will test whether the higher spatial resolution SAR data gaofen-3 (the highest spatial resolution is 1 m) combined with object-oriented method can significantly improve the accuracy of crop classification in small plot area. In addition, some studies have shown that band ratio or sentinel-1 radar vegetation index can better monitor agricultural land use [63] and other studies have proved that deep learning algorithm combined with SAR data can obtain higher crop classification accuracy [64], these are the directions that need further research.

## 5. Conclusions

The results emphasize the influence of the time interval and segmentation size of the composite image on the accuracy of crop classification when using sentinel-1 composite image combined with object-oriented classification. The composite image with shorter time interval can provide more information for crop classification, so as to improve the accuracy of crop classification. It was found that using the object-oriented classification method combined with Sentinel-1 data in GEE greatly improved the accuracy of crop classification in large plot area. The object-oriented classification method obviously improves the "salt and pepper phenomenon", which is common in SAR data classification results, because the method divides the image into "block" crop processing units to suppress noise. This approach also leads to the close relationship between object-oriented crop classification accuracy and the image segmentation effect. The closer the image segmentation result is to the real land distribution, the higher the accuracy of crop classification is. With the more popular use of GEE, more advanced image segmentation and more advanced machine learning combined with GEE will obtain more accurate crop classification results. In addition, this study evaluated the band importance of Sentinel-1 in two consecutive years crop classification. The results showed that the features with higher importance were mainly distributed in July, August and September and there were differences in the features with higher importance in different years. Our study also found that the optimal segmentation size of object-oriented classification was closely related to the short-side length and image resolution.

**Author Contributions:** Conceptualization, C.L. and H.L.; Methodology, C.L.; Software, C.L.; Validation, D.G. and B.Q.; Formal Analysis, C.L.; Investigation, C.L., Y.S. and L.L.; Resources, H.L.; Data Curation, Q.F.; Writing-Original Draft Preparation, C.L.; Writing-Review & Editing, C.L.; Visualization, C.L.; Supervision, H.L.; Project Administration, H.L.; Funding Acquisition, H.L. All authors have read and agreed to the published version of the manuscript.

**Funding:** This research was funded by the Special Foundation for Basic Research Program in wild China of CAS (XDA23070501) and the Natural Science Foundation of Heilongjiang Province of China (D2017001).

**Data Availability Statement:** The data presented in this study are available on request from the corresponding author.

**Acknowledgments:** We thank the Google Earth Engine platform for providing us with a free computing platform and the European Space Agency for providing us with free data. We also thank anonymous reviewers for their insightful advice.

**Conflicts of Interest:** The authors declare no conflict of interest.

## References

1. Carvalho, F.P. Agriculture, pesticides, food security and food safety. *Environ. Sci. Policy* **2006**, *9*, 685–692. [CrossRef]
2. McMichael, P. The power of food. *Agric. Hum. Values* **2000**, *17*, 21–33. [CrossRef]
3. Schmidhuber, J.; Tubiello, F.N. Global food security under climate change. *Proc. Natl. Acad. Sci. USA* **2007**, *104*, 19703–19708. [CrossRef]
4. Garnett, T.; Appleby, M.C.; Balmford, A.; Bateman, I.J.; Benton, T.G.; Bloomer, P.; Burlingame, B.; Dawkins, M.; Dolan, L.; Fraser, D. Sustainable intensification in agriculture: Premises and policies. *Science* **2013**, *341*, 33–34. [CrossRef] [PubMed]
5. Godfray, H.C.J.; Crute, I.R.; Haddad, L.; Lawrence, D.; Muir, J.F.; Nisbett, N.; Pretty, J.; Robinson, S.; Toulmin, C.; Whiteley, R. The future of the global food system. *Philos. Trans. R Soc. Lond. B Biol. Sci.* **2010**, *365*, 2769–2777. [CrossRef] [PubMed]
6. Opara, L.U. Traceability in agriculture and food supply chain: A review of basic concepts, technological implications, and future prospects. *J. Food Agric. Environ.* **2003**, *1*, 101–106.
7. Davis, K.F.; Rulli, M.C.; Seveso, A.; D'Odorico, P. Increased food production and reduced water use through optimized crop distribution. *Nat. Geosci.* **2017**, *10*, 919. [CrossRef]
8. Rockström, J.; Williams, J.; Daily, G.; Noble, A.; Matthews, N.; Gordon, L.; Wetterstrand, H.; DeClerck, F.; Shah, M.; Steduto, P. Sustainable intensification of agriculture for human prosperity and global sustainability. *AMBIO* **2017**, *46*, 4–17. [CrossRef]
9. Pervez, M.S.; Brown, J.F. Mapping irrigated lands at 250-m scale by merging MODIS data and national agricultural statistics. *Remote Sens.* **2010**, *2*, 2388–2412. [CrossRef]
10. Gumma, M.K.; Thenkabail, P.S.; Maunahan, A.; Islam, S.; Nelson, A. Mapping seasonal rice cropland extent and area in the high cropping intensity environment of Bangladesh using MODIS 500 m data for the year 2010. *ISPRS J. Photogramm. Remote Sens.* **2014**, *91*, 98–113. [CrossRef]
11. Joshi, N.; Baumann, M.; Ehammer, A.; Fensholt, R.; Grogan, K.; Hostert, P.; Jepsen, M.R.; Kuemmerle, T.; Meyfroidt, P.; Mitchard, E.T. A review of the application of optical and radar remote sensing data fusion to land use mapping and monitoring. *Remote Sens.* **2016**, *8*, 70. [CrossRef]
12. Huang, Y.; Chen, Z.-x.; Tao, Y.; Huang, X.-z.; Gu, X.-f. Agricultural remote sensing big data: Management and applications. *J. Integr. Agric.* **2018**, *17*, 1915–1931. [CrossRef]

13. Kussul, N.; Lemoine, G.; Gallego, F.J.; Skakun, S.V.; Lavreniuk, M.; Shelestov, A.Y. Parcel-based crop classification in ukraine using landsat-8 data and sentinel-1A data. *IEEE J. Sel. Top. Appl. Earth Obs. Remote Sens.* **2016**, *9*, 2500–2508. [CrossRef]
14. Kussul, N.; Lavreniuk, M.; Skakun, S.; Shelestov, A. Deep learning classification of land cover and crop types using remote sensing data. *IEEE Geosci. Remote Sens. Lett.* **2017**, *14*, 778–782. [CrossRef]
15. Belgiu, M.; Csillik, O. Sentinel-2 cropland mapping using pixel-based and object-based time-weighted dynamic time warping analysis. *Remote Sens. Environ.* **2018**, *204*, 509–523. [CrossRef]
16. Griffiths, P.; Nendel, C.; Hostert, P. Intra-annual reflectance composites from Sentinel-2 and Landsat for national-scale crop and land cover mapping. *Remote Sens. Environ.* **2019**, *220*, 135–151. [CrossRef]
17. Foga, S.; Scaramuzza, P.L.; Guo, S.; Zhu, Z.; Dilley Jr, R.D.; Beckmann, T.; Schmidt, G.L.; Dwyer, J.L.; Hughes, M.J.; Laue, B. Cloud detection algorithm comparison and validation for operational Landsat data products. *Remote Sens. Environ.* **2017**, *194*, 379–390. [CrossRef]
18. Zhu, Z.; Wang, S.; Woodcock, C.E. Improvement and expansion of the Fmask algorithm: Cloud, cloud shadow, and snow detection for Landsats 4–7, 8, and Sentinel 2 images. *Remote Sens. Environ.* **2015**, *159*, 269–277. [CrossRef]
19. Blaes, X.; Vanhalle, L.; Defourny, P. Efficiency of crop identification based on optical and SAR image time series. *Remote Sens. Environ.* **2005**, *96*, 352–365. [CrossRef]
20. Ndikumana, E.; Ho Tong Minh, D.; Baghdadi, N.; Courault, D.; Hossard, L. Deep recurrent neural network for agricultural classification using multitemporal SAR Sentinel-1 for Camargue, France. *Remote Sens.* **2018**, *10*, 1217. [CrossRef]
21. Xu, L.; Zhang, H.; Wang, C.; Zhang, B.; Liu, M. Crop Classification Based on Temporal Information Using Sentinel-1 SAR Time-Series Data. *Remote Sens.* **2019**, *11*, 53. [CrossRef]
22. Fu, W.; Ma, J.; Chen, P.; Chen, F. Remote Sensing Satellites for Digital Earth. In *Manual of Digital Earth*; Springer: Berlin, Germany, 2020; pp. 55–123.
23. Agrawal, S.; Raghavendra, S.; Kumar, S.; Pande, H. Geospatial Data for the Himalayan Region: Requirements, Availability, and Challenges. In *Remote Sensing of Northwest Himalayan Ecosystems*; Springer: Berlin, Germany, 2019; pp. 471–500.
24. Liu, C.-A.; Chen, Z.-X.; Yun, S.; Chen, J.-S.; Hasi, T.; Pan, H.-Z. Research advances of SAR remote sensing for agriculture applications: A review. *J. Integr. Agric.* **2019**, *18*, 506–525. [CrossRef]
25. Huang, X.; Wang, J.; Shang, J.; Liao, C.; Liu, J. Application of polarization signature to land cover scattering mechanism analysis and classification using multi-temporal C-band polarimetric RADARSAT-2 imagery. *Remote Sens. Environ.* **2017**, *193*, 11–28. [CrossRef]
26. Jiao, X.; Kovacs, J.M.; Shang, J.; McNairn, H.; Walters, D.; Ma, B.; Geng, X. Object-oriented crop mapping and monitoring using multi-temporal polarimetric RADARSAT-2 data. *Isprs J. Photogramm. Remote Sens.* **2014**, *96*, 38–46. [CrossRef]
27. Satalino, G.; Balenzano, A.; Mattia, F.; Davidson, M.W. C-band SAR data for mapping crops dominated by surface or volume scattering. *IEEE Geosci. Remote Sens. Lett.* **2013**, *11*, 384–388. [CrossRef]
28. Nagraj, G.M.; Karegowda, A.G. Crop Mapping using SAR Imagery: An Review. *Int. J. Adv. Res. Comput. Sci.* **2016**, *7*, 47–52.
29. Salehi, B.; Daneshfar, B.; Davidson, A.M. Accurate crop-type classification using multi-temporal optical and multi-polarization SAR data in an object-based image analysis framework. *Int. J. Remote Sens.* **2017**, *38*, 4130–4155. [CrossRef]
30. Tomppo, E.; Antropov, O.; Praks, J. Cropland Classification Using Sentinel-1 Time Series: Methodological Performance and Prediction Uncertainty Assessment. *Remote Sens.* **2019**, *11*, 2480. [CrossRef]
31. Gorelick, N.; Hancher, M.; Dixon, M.; Ilyushchenko, S.; Thau, D.; Moore, R. Google Earth Engine: Planetary-scale geospatial analysis for everyone. *Remote Sens. Environ.* **2017**, *202*, 18–27. [CrossRef]
32. Dong, J.; Xiao, X.; Menarguez, M.A.; Zhang, G.; Qin, Y.; Thau, D.; Biradar, C.; Moore III, B. Mapping paddy rice planting area in northeastern Asia with Landsat 8 images, phenology-based algorithm and Google Earth Engine. *Remote Sens. Environ.* **2016**, *185*, 142–154. [CrossRef]
33. Luo, C.; Liu, H.-j.; Fu, Q.; Guan, H.-x.; Ye, Q.; Zhang, X.-l.; Kong, F.-c. Mapping the fallowed area of paddy fields on Sanjiang Plain of Northeast China to assist water security assessments. *J. Integr. Agric.* **2020**, *19*, 1885–1896. [CrossRef]
34. Zurqani, H.A.; Post, C.J.; Mikhailova, E.A.; Cope, M.P.; Allen, J.S.; Lytle, B.A. Evaluating the integrity of forested riparian buffers over a large area using LiDAR data and Google Earth Engine. *Sci. Rep.* **2020**, *10*, 14096. [CrossRef] [PubMed]
35. Jia, M.; Wang, Z.; Mao, D.; Ren, C.; Wang, C.; Wang, Y. Rapid, robust, and automated mapping of tidal flats in China using time series Sentinel-2 images and Google Earth Engine. *Remote Sens. Environ.* **2021**, *255*, 112285. [CrossRef]
36. Malenovský, Z.; Rott, H.; Cihlar, J.; Schaepman, M.E.; García-Santos, G.; Fernandes, R.; Berger, M. Sentinels for science: Potential of Sentinel-1,-2, and-3 missions for scientific observations of ocean, cryosphere, and land. *Remote Sens. Environ.* **2012**, *120*, 91–101. [CrossRef]
37. You, N.; Dong, J. Examining earliest identifiable timing of crops using all available Sentinel 1/2 imagery and Google Earth Engine. *ISPRS J. Photogramm. Remote Sens.* **2020**, *161*, 109–123. [CrossRef]
38. Torres, R.; Snoeij, P.; Geudtner, D.; Bibby, D.; Davidson, M.; Attema, E.; Potin, P.; Rommen, B.; Floury, N.; Brown, M.; et al. GMES Sentinel-1 mission. *Remote Sens. Environ.* **2012**, *120*, 9–24. [CrossRef]
39. Sun, Z.; Luo, J.; Yang, J.; Yu, Q.; Zhang, L.; Xue, K.; Lu, L. Nation-Scale Mapping of Coastal Aquaculture Ponds with Sentinel-1 SAR Data Using Google Earth Engine. *Remote Sens.* **2020**, *12*, 3086. [CrossRef]
40. Hu, L.; Xu, N.; Liang, J.; Li, Z.; Chen, L.; Zhao, F. Advancing the Mapping of Mangrove Forests at National-Scale Using Sentinel-1 and Sentinel-2 Time-Series Data with Google Earth Engine: A Case Study in China. *Remote Sens.* **2020**, *12*, 3120. [CrossRef]

41. Gašparović, M.; Dobrinić, D. Comparative assessment of machine learning methods for urban vegetation mapping using multitemporal sentinel-1 imagery. *Remote Sens.* **2020**, *12*, 1952. [CrossRef]
42. Lee, J.-S.; Wen, J.-H.; Ainsworth, T.L.; Chen, K.-S.; Chen, A.J.J.I.T.o.G.; Sensing, R. Improved sigma filter for speckle filtering of SAR imagery. *IEEE Trans. Geosci. Remote Sens.* **2008**, *47*, 202–213.
43. Yommy, A.S.; Liu, R.; Wu, S. SAR image despeckling using refined Lee filter. In Proceedings of the 2015 7th International Conference on Intelligent Human-Machine Systems and Cybernetics, Hangzhou, China, 26–27 August 2015; pp. 260–265.
44. Achanta, R.; Susstrunk, S. Superpixels and polygons using simple non-iterative clustering. In Proceedings of the IEEE Conference on Computer Vision and Pattern Recognition, San Juan, PR, USA, 17–19 June 1997; pp. 4651–4660.
45. Mahdianpari, M.; Salehi, B.; Mohammadimanesh, F.; Homayouni, S.; Gill, E. The first wetland inventory map of newfoundland at a spatial resolution of 10 m using sentinel-1 and sentinel-2 data on the google earth engine cloud computing platform. *Remote Sens.* **2019**, *11*, 43. [CrossRef]
46. Liaw, A.; Wiener, M. Classification and regression by randomForest. *R News* **2002**, *2*, 18–22.
47. Pal, M. Random forest classifier for remote sensing classification. *Int. J. Remote Sens.* **2005**, *26*, 217–222. [CrossRef]
48. Belgiu, M.; Drăguţ, L. Random forest in remote sensing: A review of applications and future directions. *ISPRS J. Photogramm. Remote Sens.* **2016**, *114*, 24–31. [CrossRef]
49. Rodriguez-Galiano, V.F.; Ghimire, B.; Rogan, J.; Chica-Olmo, M.; Rigol-Sanchez, J.P. An assessment of the effectiveness of a random forest classifier for land-cover classification. *ISPRS J. Photogramm. Remote Sens.* **2012**, *67*, 93–104. [CrossRef]
50. Park, S.; Im, J.; Park, S.; Yoo, C.; Han, H.; Rhee, J. Classification and Mapping of Paddy Rice by Combining Landsat and SAR Time Series Data. *Remote Sens.* **2018**, *10*, 447. [CrossRef]
51. Tian, F.; Wu, B.; Zeng, H.; Zhang, X.; Xu, J. Efficient Identification of Corn Cultivation Area with Multitemporal Synthetic Aperture Radar and Optical Images in the Google Earth Engine Cloud Platform. *Remote Sens.* **2019**, *11*, 629. [CrossRef]
52. Story, M.; Congalton, R.G. Accuracy assessment: A user's perspective. *Photogramm. Eng. Remote Sens.* **1986**, *52*, 397–399.
53. Lee, J.-S.; Grunes, M.R.; Pottier, E. Quantitative comparison of classification capability: Fully polarimetric versus dual and single-polarization SAR. *IEEE Trans. Geosci. Remote Sens.* **2001**, *39*, 2343–2351.
54. Skriver, H. Crop classification by multitemporal C-and L-band single-and dual-polarization and fully polarimetric SAR. *IEEE Trans. Geosci. Remote Sens.* **2011**, *50*, 2138–2149. [CrossRef]
55. Wu, F.; Wang, C.; Zhang, H.; Zhang, B.; Tang, Y. Rice crop monitoring in South China with RADARSAT-2 quad-polarization SAR data. *IEEE Geosci. Remote Sens. Lett.* **2010**, *8*, 196–200. [CrossRef]
56. Huang, X.; Liao, C.; Xing, M.; Ziniti, B.; Wang, J.; Shang, J.; Liu, J.; Dong, T.; Xie, Q.; Torbick, N. A multi-temporal binary-tree classification using polarimetric RADARSAT-2 imagery. *Remote Sens. Environ.* **2019**, *235*, 111478. [CrossRef]
57. Qi, Z.; Yeh, A.G.-O.; Li, X.; Lin, Z. A novel algorithm for land use and land cover classification using RADARSAT-2 polarimetric SAR data. *Remote Sens. Environ.* **2012**, *118*, 21–39. [CrossRef]
58. Skakun, S.; Kussul, N.; Shelestov, A.Y.; Lavreniuk, M.; Kussul, O. Efficiency assessment of multitemporal C-band Radarsat-2 intensity and Landsat-8 surface reflectance satellite imagery for crop classification in Ukraine. *IEEE J. Sel. Top. Appl. Earth Obs. Remote Sens.* **2015**, *9*, 3712–3719. [CrossRef]
59. Tassi, A.; Vizzari, M. Object-Oriented LULC Classification in Google Earth Engine Combining SNIC, GLCM, and Machine Learning Algorithms. *Remote Sens.* **2020**, *12*, 3776. [CrossRef]
60. Shen, G.; Yang, X.; Jin, Y.; Luo, S.; Xu, B.; Zhou, Q. Land Use Changes in the Zoige Plateau Based on the Object-Oriented Method and Their Effects on Landscape Patterns. *Remote Sens.* **2020**, *12*, 14. [CrossRef]
61. Hao, P.-Y.; Tang, H.-J.; Chen, Z.-X.; Meng, Q.-Y.; Kang, Y.-P. Early-season crop type mapping using 30-m reference time series. *J. Integr. Agric.* **2020**, *19*, 1897–1911. [CrossRef]
62. Hao, P.-Y.; Tang, H.-J.; Chen, Z.-X.; Le, Y.; Wu, M.-Q. High resolution crop intensity mapping using harmonized Landsat-8 and Sentinel-2 data. *J. Integr. Agric.* **2019**, *18*, 2883–2897. [CrossRef]
63. Holtgrave, A.-K.; Röder, N.; Ackermann, A.; Erasmi, S.; Kleinschmit, B. Comparing Sentinel-1 and -2 Data and Indices for Agricultural Land Use Monitoring. *Remote Sens.* **2020**, *12*, 2919. [CrossRef]
64. Cué La Rosa, L.E.; Queiroz Feitosa, R.; Nigri Happ, P.; Del'Arco Sanches, I.; Ostwald Pedro da Costa, G.A. Combining Deep Learning and Prior Knowledge for Crop Mapping in Tropical Regions from Multitemporal SAR Image Sequences. *Remote Sens.* **2019**, *11*, 2029. [CrossRef]

*remote sensing*

MDPI

*Article*

# Agricultural Monitoring Using Polarimetric Decomposition Parameters of Sentinel-1 Data

**Katharina Harfenmeister [1,*], Sibylle Itzerott [1], Cornelia Weltzien [2,3] and Daniel Spengler [1]**

1 Helmholtz Centre Potsdam, GFZ German Research Centre for Geosciences, Telegrafenberg, 14473 Potsdam, Germany; itzerott@gfz-potsdam.de (S.I.); daniel.spengler@gfz-potsdam.de (D.S.)
2 Technische Universität Berlin, Chair of Agromechatronics, Straße des 17. Juni 144, 10623 Berlin, Germany; CWeltzien@atb-potsdam.de
3 Leibniz Institute for Agricultural Engineering and Bioeconomy (ATB), Max-Eyth-Allee 100, 14469 Potsdam, Germany
* Correspondence: katharina.harfenmeister@gfz-potsdam.de; Tel.: +49-331-288-28775

**Abstract:** The time series of synthetic aperture radar (SAR) data are commonly and successfully used to monitor the biophysical parameters of agricultural fields. Because, until now, mainly backscatter coefficients have been analysed, this study examines the potentials of entropy, anisotropy, and alpha angle derived from a dual-polarimetric decomposition of Sentinel-1 data to monitor crop development. The temporal profiles of these parameters are analysed for wheat and barley in the vegetation periods 2017 and 2018 for 13 fields in two test sites in Northeast Germany. The relation between polarimetric parameters and biophysical parameters observed in the field is investigated using linear and exponential regression models that are evaluated using the coefficient of determination ($R^2$) and the root mean square error (RMSE). The performance of single regression models is furthermore compared to those of multiple regression models, including backscatter coefficients in VV and VH polarisation as well as polarimetric decomposition parameters entropy and alpha. Characteristic temporal profiles of entropy, anisotropy, and alpha reflecting the main phenological changes in plants as well as the meteorological differences between the two years are observed for both crop types. The regression models perform best for data from the phenological growth stages tillering to booting. The highest $R^2$ values of the single regression models are reached for the plant height of wheat related to entropy and anisotropy with $R^2$ values of 0.64 and 0.61, respectively. The multiple regression models of VH, VV, entropy, and alpha outperform single regression models in most cases. $R^2$ values of multiple regression models of plant height (0.76), wet biomass (0.7), dry biomass (0.7), and vegetation water content (0.69) improve those of single regression models slightly by up to 0.05. Additionally, the RMSE values of the multiple regression models are around 10% lower compared to those of single regression models. The results indicate the capability of dual-polarimetric decomposition parameters in serving as meaningful input parameters for multiple regression models to improve the prediction of biophysical parameters. Additionally, their temporal profiles indicate phenological development dependent on meteorological conditions. Knowledge about biophysical parameter development and phenology is important for farmers to monitor crop growth variability during the vegetation period to adapt and to optimize field management.

**Keywords:** agriculture; crop monitoring; Sentinel-1; polarimetry; decomposition; field variability; crop parameters

**Citation:** Harfenmeister, K.; Itzerott, S.; Weltzien, C.; Spengler, D. Agricultural Monitoring Using Polarimetric Decomposition Parameters of Sentinel-1 Data. *Remote Sens.* **2021**, *13*, 575. https://doi.org/10.3390/rs13040575

Academic Editor: Olaniyi Ajadi
Received: 28 December 2020
Accepted: 4 February 2021
Published: 6 February 2021



## 1. Introduction

Agricultural production is of particular importance in ensuring food security in times of population growth, climate change, and land scarcity. The monitoring of agricultural fields using satellite data allows for statements about plant conditions in large areas to be made and enables site-specific management, also known as precision agriculture. Synthetic aperture radar (SAR) images are particularly suitable for agricultural monitoring. Due to their independence from cloud obstruction, they enable the observation of plant changes taking place in small time periods [1,2]. The advantages of SAR images taken by the Sentinel-1 twin satellites operated by the European Space Agency (ESA) are their high acquisition frequency with images taken under the same acquisition conditions every six days as well as their free availability.

The backscatter coefficient, the portion of the radar signal that is directly reflected towards to the radar antenna, is commonly used for agricultural applications such as crop type classification [3,4], often in combination with optical satellite data [5]. Furthermore, many studies monitored crop biomass [6,7], leaf area index (LAI) [8], or phenology [9,10] using SAR data from RADARSAT-2 or airborne SAR systems [11]. Both RADARSAT-2 and Sentinel-1 acquire images using a C-band wavelength of around 5 cm, whereas some airborne systems also carry SAR systems operating at L-band wavelengths from 15 to 30 cm. In contrast to RADARSAT-2, which acquires images with equal acquisition conditions every 24 days, Sentinel-1 data are available with a higher temporal resolution every six days. Therefore, several studies use dense backscatter time series of Sentinel-1 to characterize and monitor crops, mainly wheat, barley, rapeseed, maize, and soybean. Time series in combination with regression analysis estimating biophysical parameters are used for this purpose [12,13]. Furthermore, the time series are often compared with vegetation indices derived from optical satellite data in different study areas in France [14], the Netherlands [15], or Germany [16] to evaluate the synergistic use of optical and SAR data. Additionally, phenological growth stages such as stem elongation and harvest are successfully extracted from Sentinel-1 time series [17,18].

Polarimetric decomposition, which uses the phase information of a radar signal, is commonly used and often essential for the extraction of biophysical parameters and allows for the identification of different scattering mechanisms of the surface. In an agricultural context, it is mainly used for soil moisture retrieval under vegetation cover using airborne L-band SAR systems [19–21] or crop mapping and identification [22,23]. Furthermore, polarimetric decomposition is used for the extraction of phenological stages [24,25] and for the estimation of crop parameters such as height, LAI, and biomass [7,26]. Polarimetric decomposition was originally designed for full-polarimetric data such as those of RADARSAT-2, but approaches for dual-polarimetric data such as Sentinel-1 exist as well, although with some limitations [27,28]. While it is not possible to extract single scattering mechanisms using only dual-polarimetric data with VV (vertical-vertical) and VH (vertical-horizontal) polarisation [27], polarimetric parameters such as entropy, anisotropy, and alpha can still be useful as additional data sources and are used in several studies, e.g., for the mapping of burnt areas [29] or of flooded vegetation [30], for ship detection [31], as well as for land cover [32] and vegetable classification [33]. In an agricultural context, polarimetric decomposition parameters are used to detect crop lodging [34]. Mercier et al. [35] used the polarimetric parameters span and Shannon entropy of Sentinel-1 data to the predict phenological stages of wheat and rapeseed. Furthermore, dual-polarimetric data with different polarisation combinations such as HH and VV, e.g., from TerraSAR-X, are used for the monitoring of reed belts [36], grasslands [37], or rice phenology [38].

However, polarimetric decomposition of dual-polarised Sentinel-1 images is rarely used in an agricultural context. This study investigates the contribution of polarimetric decomposition parameters of Sentinel-1 data in monitoring the six biophysical crop parameters wet biomass, dry biomass, leaf area index (LAI), plant height, absolute vegetation water content (VWC), and relative VWC of wheat and barley. It particularly focuses on differences between the relatively wet year 2017, with precipitation amounts higher than those of the reference period from 1981 to 2010 and of the extremely dry year 2018, when precipitation amounting to only half as high as in the reference period was measured.

This study builds on the work of Harfenmeister et al. [13], who analysed temporal profiles of Sentinel-1 backscatter data and performed a regression analysis between backscatter and biophysical crop parameters. The temporal profile analysis and the regression analysis are performed as well in the present study using the polarimetric decomposition parameters entropy, anisotropy, and alpha angle. Furthermore, a multiple regression with backscatter parameters VV and VH and polarimetric decomposition parameters entropy and alpha is performed to evaluate their potential to predict crop parameters.

## 2. Test Sites and Meteorological Conditions

Ground reference data is necessary to relate satellite data with actual conditions on the Earth's surface. For this purpose, extensive field work took place on 13 agricultural fields in two test sites in Northeast Germany in 2017 and 2018 (Figure 1). Both test sites are formed by glacial and periglacial processes [39] and are characterized by extensive agricultural areas with field sizes often larger than 100 ha. Field heterogeneity is mostly due to differences in the soil and morphology as well as heterogeneously distributed rainfall events, among others.

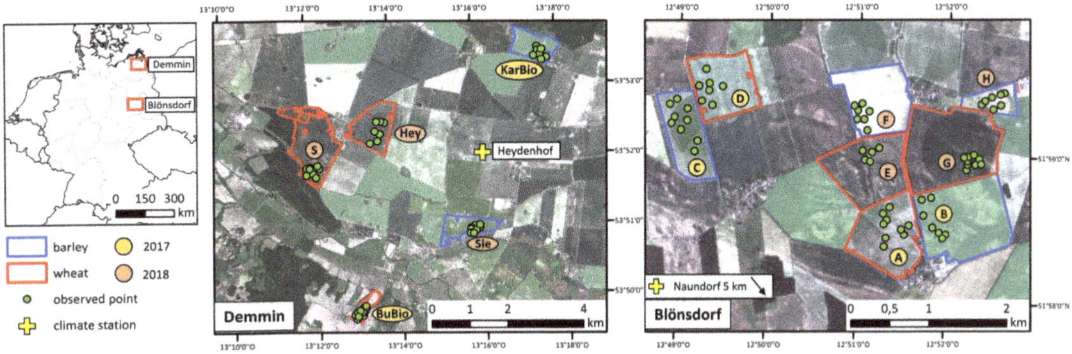


**Figure 1.** Location of the two test sites DEMMIN (durable environmental multidisciplinary monitoring information network) and Blönsdorf in Germany (**left**) and observed wheat and barley fields with observation points in 2017 and 2018 in Demmin (**center**) and Blönsdorf (**right**): background colors of the field names indicate the year of observation.

The test site *DEMMIN* (durable environmental multidisciplinary monitoring information network) in the federal state Mecklenburg-West Pomerania is part of the Joint Experiment of Crop Assessment and Monitoring (JECAM) [40] and of the long-term monitoring project TERENO (Terrestrial Environmental Observatories) [41,42]. The area is extensively agriculturally used; furthermore, numerous lakes, wetlands, grassland, and pine and deciduous forests shape the landscape.

The second test site, *Blönsdorf*, in the federal state Brandenburg is located in a glacial hill chain called "Fläming Heath" with fertile sand-less soils. In contrast to DEMMIN, the Blönsdorf region has more forests and less water bodies.

In both study areas, the observed fields are managed conventionally with common applications of fertilizers and pesticides, whereas precision farming is not widely used yet. Only field G was irrigated in 2018. The average size of the observed fields is 80 ha, whereas the smallest field has a size of 20 ha and the largest field has a size of 180 ha. Next to wheat and barley, which are cultivated on 27% (wheat) and on 11% (barley) of the agricultural land in Mecklenburg-West Pomerania and Brandenburg (averaged for 2017 and 2018), the other main cultivated crops in both study areas are corn (18%), rapeseed (17%), and rye (11%) [43–45].

With 9.3 °C, the mean annual air temperature of Blönsdorf is slightly higher compared to DEMMIN, with 8.8 °C, whereas the total yearly precipitation amount is around 600 mm higher in DEMMIN compared to around 540 mm in Blönsdorf in the reference period 1981–2010 reported by weather stations operated by the German Weather Service in Naundorf b. Seyda, which is close to Blönsdorf (station code: 13,146), and in Demmin (station code: 939) [46]. In 2017, precipitation in the observed time period from March until the end of July added up to 456 mm in DEMMIN and to 300 mm in Blönsdorf. Compared to the reference period 1981–2010 with precipitation of 259 mm (DEMMIN) and 238 mm (Blönsdorf) between March and July, precipitation in 2017 was higher in both study areas. The remarkably lower precipitation sums in 2018 with 155 mm in DEMMIN and 118 mm in Blönsdorf emphasize the extremely dry conditions in this year. Particularly, the spring and summer months from April to June, which are the main growing months of wheat and barley, suffered from scarce precipitation events.

## 3. Data

### 3.1. SAR data

The analyses were carried out using SAR data from Sentinel-1, a constellation of two identical satellites provided by the European Copernicus program of the ESA. Each Sentinel-1 satellite carries a C-band SAR at 5.405 GHz and acquires SAR data in two polarisations: VH and VV (Table 1). The images were downloaded as single look complex (SLC) data in interferometric wide swath (IW) mode. Sentinel-1 data are available free of charge.

**Table 1.** Selected features of the Sentinel-1 data.

| | Sentinel-1 |
|---|---|
| Antenna | C-Band (5.405 GHz) |
| Mode | Interferometric Wide Swath (IW) |
| Swath | 250 km |
| Range Spacing | 2.33 m |
| Azimuth Spacing | 13.89 m |
| Polarization | VV, VH |
| Incidence Angles | 36°–46° |
| Revisit Frequency | 6 days |
| Pass-Orbit Settings | ASC 44, ASC 146<br>DESC 95, DESC 168 |

The two Sentinel-1 satellites passed the test sites in ascending and in descending pass directions each, which resulted in four different acquisition settings with different foot prints, orbits, and incidence angles: ASC 44, ASC 146, DESC 95, and DESC 168 (named after the pass direction and relative orbit). Consequently, they reached a very high revisit frequency of one or two days. Images with equal acquisition settings were available every six days. In total, 400 Sentinel-1 images from all acquisition settings (about 100 images per acquisition setting) were used for the analyses (Figure 2).

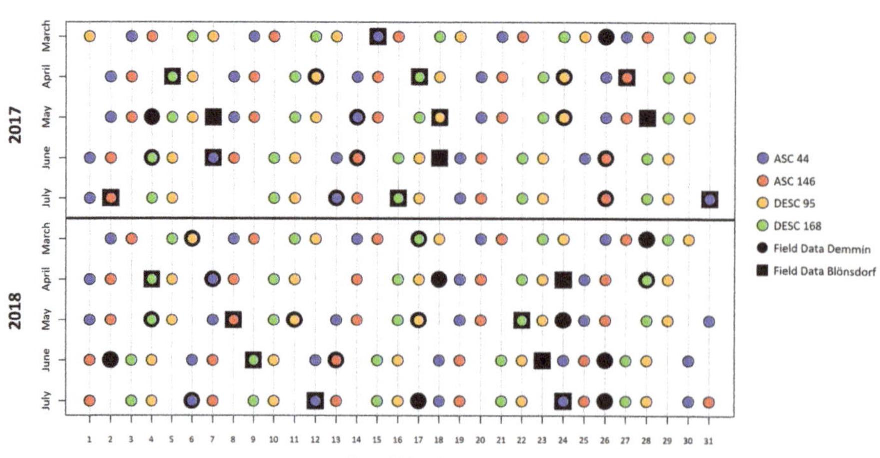


**Figure 2.** Sentinel-1 acquisitions and field measurement dates 2017 and 2018: the data gaps resulted from unavailable images at the time of analysis.

### 3.2. Field Data

In total, 48 observation points on three wheat and two barley fields in DEMMIN and 60 points on four wheat and four barley fields in Blönsdorf were monitored every 10–14 days in 2017 and 2018 from March until the end of July or until harvest (Figures 1 and 2). The observation points were distributed within the fields using a stratified sampling design in order to represent the field heterogeneity, which was investigated beforehand using soil maps of the German Soil Survey Description [47], aerial images from drones taken during several previous field observations, optical satellite images of the Sentinel-2 satellites operated by ESA, as well as digital elevation models of both test sites with a 10 m × 10 m resolution. The actual observation points were each placed into a homogeneous area to avoid edge effects and mixed pixels.

At each field measurement date (Figure 2), eight biophysical crop parameters were measured in a homogeneous 1 m × 1 m surrounding of each observation point (Table 2). Three crop parameters (dry biomass, absolute, and relative vegetation water content (VWC)) were later obtained in the laboratory (Table 2). Six of the observed crop parameters were selected to perform the analyses: plant height, LAI, wet and dry biomass, VWC, and relative VWC. The crop parameters phenology, number of leaves, row distance, crop coverage, and chlorophyll content were obtained for each point as well. Since field measurements and satellite acquisitions did not necessarily take place on the same day, all crop parameters were linearly interpolated over time to obtain daily values.

The temporal development of the six crop parameters as well as their characteristic appearance in different BBCH (*Biologische Bundesanstalt, Bundessortenamt, und CHemische Industrie* [48]) stages are shown in Figures 3 and 4. The SAR data are sensitive to changes of the plant structure and their water content. During the first BBCH stages from tillering to booting (development of the flag leave) from March to May, the plant appearance changed remarkably, whereas it was rather constant in the following growth stages. This is expressed by a strong increase of LAI and plant height as well as biomass and VWC for both crop types (Figure 4). During the last BBCH stages from the ripening stage in June, the water content of the plants changed notably, which was also captured by SAR data. The main influencing factor on the SAR data was the change in plant structure by water content over the course of the phenological development. Therefore, the analysis was also executed separately to account for the different BBCH stages.

**Figure 3.** Characteristic appearance of wheat and barley in different BBCH (*Biologische Bundesanstalt, Bundessortenamt, und CHemische Industrie*) stages.

**Table 2.** Crop parameters obtained during field campaigns 2017 and 2018.

| Crop Parameter (Field) | Unit | Method |
|---|---|---|
| Phenology | | Phenological growth stage on the BBCH scale [48] |
| Plant Height | cm | Average height of ca. 10 plants, measured by the folding rule |
| Number of Leaves | | Average number of green leaves of ca. 10 plants |
| Row Distance | cm | Row distance of crops, stable during vegetation period |
| Crop Coverage | % | Estimated percentage of ground surface covered by crops |
| Chlorophyll Content | $\mu g/cm^2$ | Average chlorophyll content of 10 leaves of several plants, measured by SPAD device (2017) and DUALEX (2018) |
| Leaf Area Index (LAI) | $m^2/m^2$ | Green leaf area per unit ground surface area, average of ca. 10 measurements with SunScan LAImeter |
| Wet Biomass | $g/m^2$ | Weight of fresh plants, destructive sampling of plants in five 1 m rows |
| **Crop Parameter (Lab)** | | |
| Dry Biomass | $g/m^2$ | Weight of oven-dried plants (at 105 °C until constant weight) |
| Vegetation Water Content (VWC) | $g/m^2$ | Weight difference between wet and dry biomass |
| Relative Vegetation Water Content | % | Weight difference between wet and dry biomass divided by wet biomass |

### 3.3. Meteorological Data

SAR data are sensitive to moisture; therefore, daily precipitation data were plotted together with the temporal development of the polarimetric decomposition parameters. The weather station "Heydenhof" is located close to the surveyed fields and is part of the TERENO meteorological network [41**?** ]. It is on average 3.5 km away from the observed fields, whereas the closest field (Sie) is 2 km and the farthest field (BuBio) is 5.5 km away from the station (Figure 1). Precipitation data were available every 15 min and were summed up daily.

The weather station "Naundorf b. Seyda" is located around 6 km southwest of the observed Blönsdorf fields. The closest fields (A and B) are 5 km and the farthest field (D) is 7.5 km away from the station. It is operated by the German Weather Service and provides daily precipitation data [46].

Both weather stations also provide daily air temperature data, which were used to calculate growing degree days and the accumulation of average daily temperatures (Equation (1)). Growing degree days are based on the assumption that plant development takes place from a certain temperature and are used to predict plant growth depending on current meteorological conditions [50]. The base temperature $T_{base}$ below which no development takes place was set at 10 °C. Cumulative growing degree days were additionally plotted together with temporal profiles of polarimetric decomposition parameters as an indicator for plant development and to show temperature differences between both test sites and both years.

$$GDD = [\frac{T_{max} + T_{min}}{2}] - T_{base} \quad (1)$$

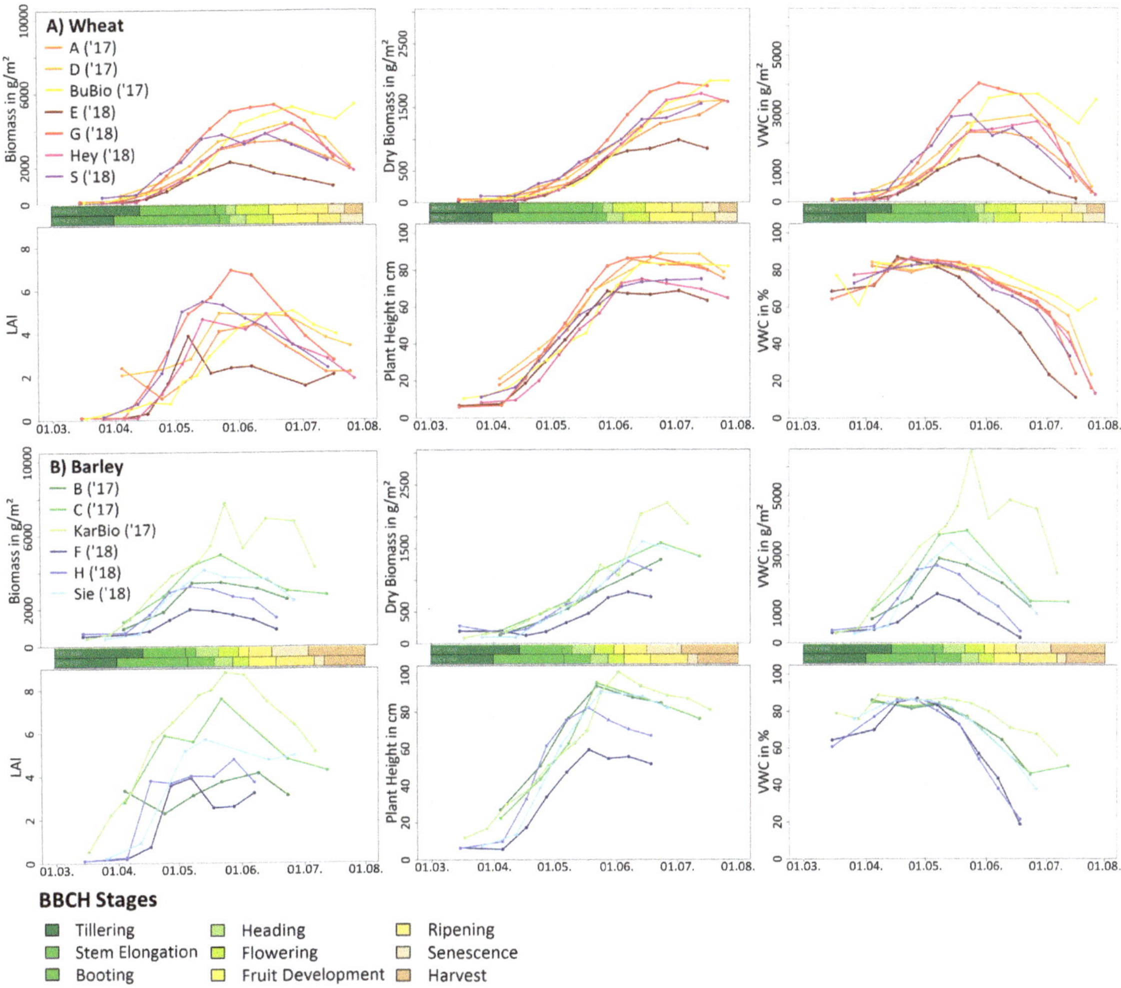


**Figure 4.** Temporal development of crop parameters for all wheat fields (**A**) and barley fields (**B**): the greenish and brownish bars between the plots represent the BBCH stages in 2017 (lower bar) and 2018 (upper bar).

## 4. Methods

### *4.1. SAR Data Processing and Polarimetric Decomposition*

Polarimetric decomposition was originally designed for full-polarimetric data and aimed to differentiate the three scattering mechanisms surface scattering, volume scattering, and dihedral scattering. The concept was based on the complex scattering matrix [S], for which the elements were the complex scattering coefficients (Equation (2)). The point scatterers were completely represented by a scattering matrix, whereas distributed scatterers, which were randomly distributed in a resolution cell, could not be described by a single scattering matrix [51]. By vectorisation of the scattering matrix, the covariance and coherency matrices were derived also to characterize distributed scatterers [52].

$$[S_{full}] = \begin{bmatrix} S_{HH} & S_{HV} \\ S_{VH} & S_{VV} \end{bmatrix} \quad [S_{dual}] = \begin{bmatrix} 0 & S_{VH} \\ S_{VH} & S_{VV} \end{bmatrix} \tag{2}$$

There are different decomposition approaches to extract the scattering mechanisms: eigenvector and eigenvalue-based decomposition and model-based decomposition [52]. In this study, the eigenvector and eigenvalue-based H-A-$\alpha$ decomposition by Cloude and Pottier [53] with its modification for dual-polarimetric data was used [27,28]. This approach used the eigenvalues and the eigenvector of the coherency matrix [T] or covariance matrix [C] (Equations (3) and (4)) to calculate the three parameters entropy (H), anisotropy (A), and alpha angle ($\alpha$) [51,52].

$$[T] = \frac{1}{2}\begin{bmatrix} \langle|S_{HH}+S_{VV}|^2\rangle & \langle(S_{HH}+S_{VV})(S_{HH}-SVV)^*\rangle & 2\langle(S_{HH}+S_{VV})S_{XX}^*\rangle \\ \langle(S_{HH}-S_{VV})(S_{HH}-S_{VV})^*\rangle & \langle|S_{HH}-S_{VV}|^2\rangle & 2\langle(S_{HH}-S_{VV})S_{XX}^*\rangle \\ 2\langle S_{XX}(S_{HH}+S_{VV})^*\rangle & 2\langle S_{XX}(S_{HH}-S_{VV})^*\rangle & 4\langle|S_{XX}|^2\rangle \end{bmatrix} \quad (3)$$

$$[C] = \begin{bmatrix} \langle S_{VV}S_{VV}^*\rangle & \langle S_{VV}S_{VH}^*\rangle \\ \langle S_{VH}S_{VV}^*\rangle & \langle S_{VH}S_{VH}^*\rangle \end{bmatrix} \quad (4)$$

Entropy is a measure of the randomness of the scattering process and ranges from 0 to 1. The lower the value, the purer and more polarised the surface, whereas a high value indicates a random scattering process with a completely depolarised wave [52]. High entropy indicates a mixture of possible point scatterer types and that the identification of a dominant scattering mechanism is reduced [51]. It is calculated using the logarithmic sum of the eigenvalues of the coherency matrix [T] (Equations (5) and (6)). $P_i$ refers to the probability of each eigenvalue contribution, whereas $n$ assumes 2 in the dual-polarimetric case and 3 for full-polarimetric data, depending on the size of [T].

$$H = -\sum_{i=1}^{n} P_i \, log_n \, P_i \quad (5)$$

$$P_i = \frac{\lambda_i}{\sum_{j=1}^{n} \lambda_j} \quad (6)$$

For full-polarimetric data, the anisotropy indicates the presence of a second scattering mechanism and is therefore complementary to entropy. It is particularly useful to improve the separation of different scattering mechanism when entropy is high [51]. It is calculated using the normalized difference of the second and third eigenvalues (Equation (7)). In the dual-polarimetric case, it is calculated using the first and second eigenvalues; therefore, its interpretation as an indicator of a second scattering mechanism is controversial.

$$A_{full} = \frac{\lambda_2 - \lambda_3}{\lambda_2 + \lambda_3}; \qquad A_{dual} = \frac{\lambda_1 - \lambda_2}{\lambda_1 + \lambda_2} \quad (7)$$

The alpha angle describes the dominant scattering mechanism. It is defined as the mean of the scattering angles of the eigenvector $v$ (Equations (8) and (9)). Values close to 0° indicate surface scattering, values close to 45° refer to volume scattering, and values close to 90° indicate dihedral scattering.

$$\alpha = \sum_{i=1}^{n} P_i \, \alpha_i \quad (8)$$

$$\alpha_i = cos^{-1}\left(|v_{1i}|\right) \quad (9)$$

All Sentinel-1 data processing steps including the polarimetric decomposition were performed using the Sentinel Application Platform (SNAP) [54]. Images from both satellites were calibrated to a complex output first, whereas Sentinel-1 images had to be split and the appropriate orbit file had to be applied beforehand. After calibration, Sentinel-1 images had to be debursted. The polarimetric speckle filter "Refined Lee" [55] with a window size of 5 × 5 was applied for all images to reduce speckle and to enhance the image quality. Afterwards, all images were decomposed using the H-$\alpha$ dual polarimetric decomposition algorithm with a window size of 5 × 5. Range-Doppler terrain correction using the digital elevation model of SRTM (Shuttle Radar Topography Mission) with a resolution of 1 arc-second was performed subsequently. Coregistration with ground control points (GCPs), which were uniformly spaced in a master image automatically selected by the SNAP software, was applied as a last step to georeference the images. Images from different pass directions (ASC and DESC) were coregistrated separately.

The resulting parameters were entropy, anisotropy, and alpha angle with a spatial resolution of 10 m × 10 m. The resulting alpha angle had to be inverted by subtracting it from 90° to obtain the appropriate value range, as reported in previous studies [29]. The resulting decomposition parameters from the SNAP toolbox were compared to the results generated by the PolSARpro software [56], and no considerable differences were observed.

The three decomposition parameters entropy, anisotropy, and alpha were extracted from each SAR image at each sampling point and additionally averaged for each field in the appropriate year. The point extraction was performed using a buffer of 7.5 m around the centroid of the exact measurement locations, documented via GPS devices with an accuracy of 3–5 m at each measurement date (Figure 5). In consequence, all point extraction areas covered an area of 176.7 $m^2$ and included 4–9 pixels. The point value was determined using the weighted mean of all pixels inside the extraction area. This approach accounts for possible inaccuracies of the GPS measurements during the field observations by taking into account the region of measurements without weighting outliers too strongly. For calculating the average field value, the outer field boundary (20 m) was excluded to avoid influences by the surrounding area and the headland.

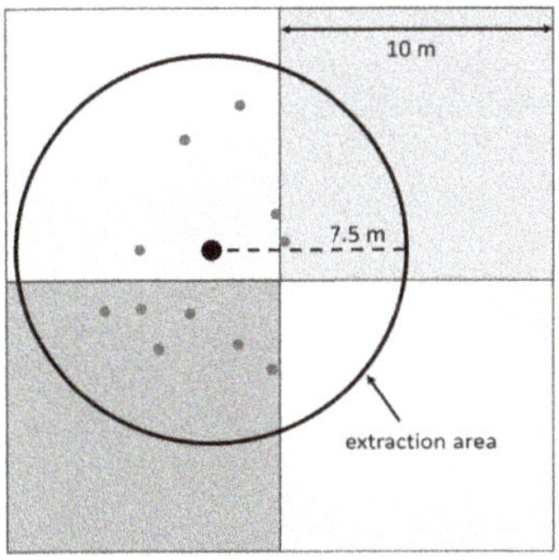


**Figure 5.** Exemplary extraction area of 176.7 $m^2$ with the four relevant 10 m × 10 m pixels of synthetic aperture radar (SAR) data depicted in the background: the center of the circle represents the centroid of the observation points of all dates (grey dots).

### 4.2. Temporal Variation Analysis

A main goal of the study was to investigate the temporal development of the three decomposition parameters in the course of the vegetation period from the beginning of March until the end of July. The plant appearance in different phenological stages changed; therefore, how these changes influence entropy, anisotropy, and alpha was figured out. Additionally, differences between both years 2017 and 2018 were investigated because they remarkably differed in their meteorological conditions.

The parameter values at each satellite acquisition are represented by line diagrams. In every plot, dark and light blue bars represent daily precipitation sums for each test site, whereas reddish lines illustrate cumulative growing degree days. The differences between satellite missions, acquisition settings (pass and orbit), crop types, years, and fields are described and interpreted. Except for the comparison of satellite missions and acquisition settings, only images with equal pass and orbit were used to ensure comparability: ASC 146 images for wheat and ASC 44 images for barley.

### 4.3. Regression Analysis

The second analysis of the study aimed to detect relationships between decomposition parameters and biophysical crop parameters measured in the field. For every decomposition and crop parameter combination, linear and exponential regression models were calculated, whereas the polarimetric decomposition parameters served as predictor variables and the crop parameters served as response variables. The quality of the regression models was evaluated using the coefficient of determination $R^2$ and the root mean square error (RMSE). $R^2$ indicates the proportion of variation in the response variable that is explained by the regression model. It assumes values between 0 and 1, whereas 1 indicates a perfect fit. $R^2$ is calculated as the quotient of the variances of the fitted values $\hat{y}_i$ and observed values $y_i$ of the response variable with the mean value $\bar{y}_i$ (Equation (10)) [57]:

$$R^2 = \frac{\Sigma(\hat{y}_i - \bar{y}_i)^2}{\Sigma(y_i - \bar{y}_i)^2} \tag{10}$$

The RMSE is a measure of the error of the regression model and is calculated as the square root of the averaged squared differences between the fitted values $\hat{y}_i$ and observed values $y_i$ of $n$ observations (Equation (11)) [57]:

$$RMSE = \sqrt{\sum_{i=1}^{n} \frac{(\hat{y}_i - y_1)^2}{n}} \tag{11}$$

Additionally, a stepwise multiple regression was performed using the two backscatter parameters VH and VV as well as the two decomposition parameters entropy and alpha. The stepwise multiple regression was performed using the R-package "caret" [58]. It searched for the best one-variable model, two-variable model, three-variable model, and four-variable model using a 20-fold cross validation and minimizing the RMSE. Anisotropy was neglected due to multicollinearity with entropy and its ambiguous interpretation. Multicollinearity was tested using the variance inflation factor (VIF) [59]. It was calculated using multiple regression models of the predictor variables, with each variable alternately serving as a response variable explained by the remaining variables. $R^2$ values of the resulting regression models were used for the calculation of the VIF, whereas values higher than 10 indicated multicollinearity (Equation (12)).

$$VIF = \frac{1}{1 - R^2} \tag{12}$$

The VH/VV ratio was not considered as a predictor variable for the multiple regression because it is already a combination of both VH and VV backscatter. The number and names of the variables building the best model as well as its $R^2$ and RMSE were analysed. Furthermore, the multiple regression results were compared with those of single regression models of backscatter parameters VH, VV, and VH/VV as well as of the polarimetric decomposition parameters entropy, anisotropy, and alpha.

## 5. Results

### 5.1. Temporal Behavior of Polarimetric Parameters

The temporal development of the three parameters entropy, anisotropy, and alpha of the four acquisition settings was plotted for wheat (Figure 6) and barley (Figure 7) for both years 2017 and 2018. The values from all fields were averaged to observe general trends in the course of the year.

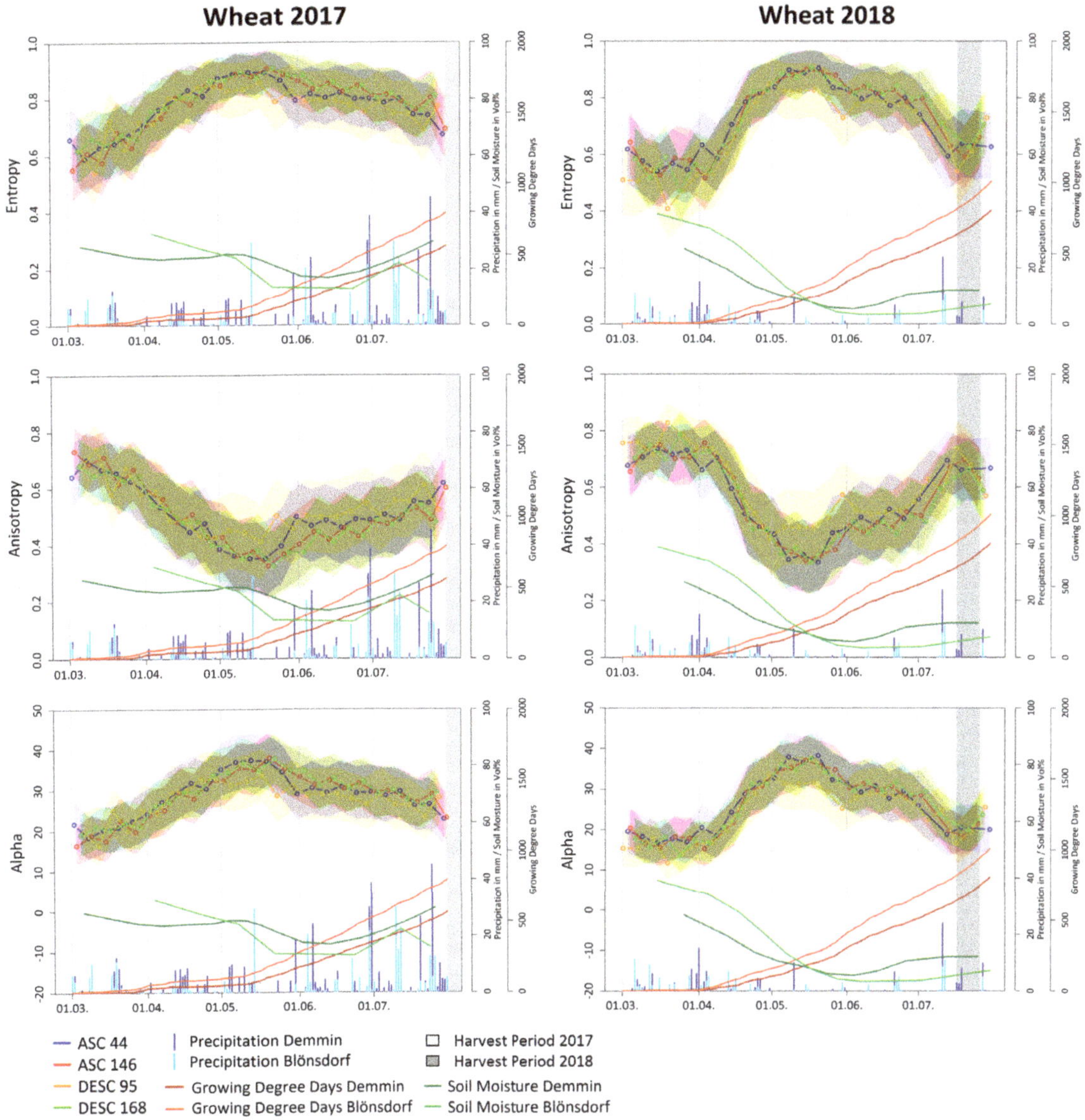


**Figure 6.** Temporal development of decomposition parameters entropy, anisotropy, and alpha for all wheat fields in 2017 (**left**) and 2018 (**right**): polygons around the lines indicate ±one standard deviation. Blue bars represent the daily precipitation sums, red lines show the cumulative growing degree days, and green lines represent soil moisture at both test sites.

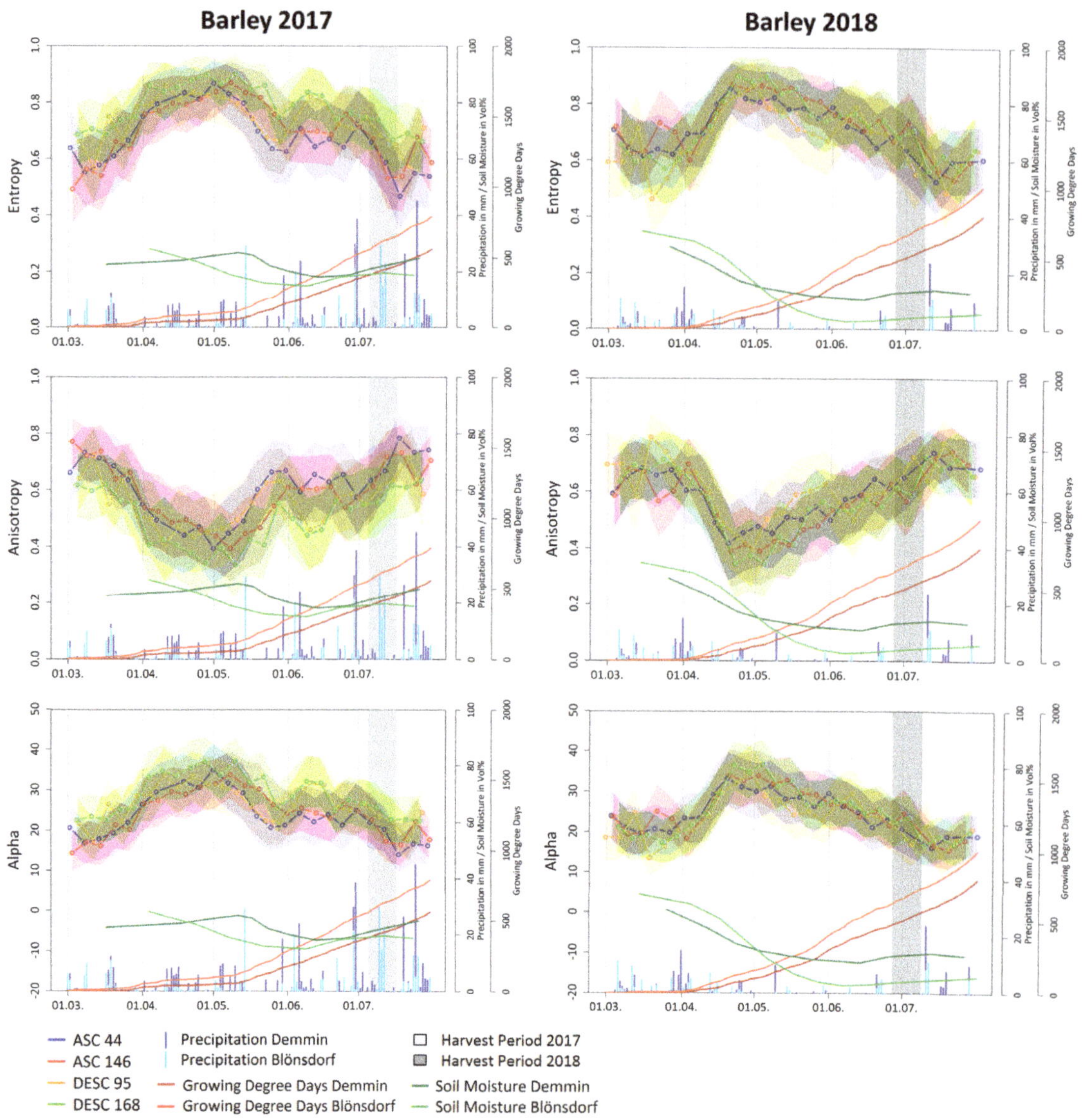


**Figure 7.** Temporal development of decomposition parameters entropy, anisotropy, and alpha for all barley fields in 2017 (**left**) and 2018 (**right**): polygons around the lines indicate ±one standard deviation. Blue bars represent the daily precipitation sums, red lines show the cumulative growing degree days, and green lines represent soil moisture at both test sites.

Entropy and alpha of wheat and barley start with values around 0.6 and around 20° in March and increase during April up to their maximum values around 0.85–0.9 and around 35° in May. This increase can be explained by a decreasing soil attribution to the signal due to plant growth in the early vegetation stages. In March, when plants undergo the tillering stage (BBCH 20–29), plants are still short, with 10–15 cm height, and have only a few and narrow leaves (Figures 3 and 4). Consequently, the soil fraction is still high, leading to lower entropy and alpha values. Low alpha values indicate a high portion of surface scattering, whereas lowest entropy values in March indicate less depolarization of the radar signal due to a more homogeneous surface compared to later phenological stages. At this time of the year, the entropy and alpha values of barley are slightly higher than those of wheat because of a higher number of tillers and consequently a higher crop coverage of barley plants. Furthermore, the entropy and alpha values of wheat are on average higher in March 2017 than in March 2018. Plants in 2017 are up to 10 cm larger and further developed than at the same time in 2018, which is caused by a later seeding in autumn 2017 due to a high soil moisture leading to a missing trafficability as well as to a cold winter 2018. Consequently, plant development in 2018 did not start before April; therefore, the increase in entropy and alpha of both crop types did not take place until April. The high fluctuation in the values in March 2018 with outliers of the descending pass directions (wheat) and ASC 146 (barley) might be explained by several snow events.

Next to precipitation and snow events, the row orientation of fields leads to varying soil contribution to the signal depending on the viewing direction that differs between pass directions. Furthermore, soil moisture and surface roughness particularly influence the SAR signal in the early growing stages. Soil moisture is particularly high in early phenological stages in both years (Figures 6 and 7). Additionally, the acquisition time might be an influencing factor because images in descending pass direction are acquired in the early morning, when plants might still be moist with dew.

The cumulative growing degree days also confirm the growing differences between 2017 and 2018 (Figures 6 and 7). Temperatures in March 2018 are still below 10 °C; consequently, plant development only starts in April. Afterwards, the growing degree days continuously rise. At the end of the vegetation period end of July, the accumulated growing degree days were higher in 2018, when they added up to 804 °C in DEMMIN and 1007 °C in Blönsdorf. In contrast, the cumulative growing degree days in 2017 added up to 562 °C in DEMMIN and 792 °C in Blönsdorf. In 2017, the growing degree days began to rise already in March but stayed constant from April to mid-May and only increased again from this time on. This explains the higher entropy and alpha values in March 2017 compared to March 2018 and the strong increase in both parameters in April 2018.

In April and May, the phenological stages stem elongation (BBCH 30–39) and booting (BBCH 41–49) dominated (Figure 4). Plants developed stems, longer and thicker leaves, as well as flag leaves and grew in height, leading to a crop coverage up to 100%, maximum LAI values up to 8, and high biomass values over 3000 $g/m^2$. With evolving plants, the portion of surface scattering of the soil decreased due to an increasing volume scattering by plants, which explains the increasing alpha values. Furthermore, the randomness in the signal represented by entropy increased due to the ongoing depolarisation of the wave by vegetation, which consists of multiple point scatterers instead of a single equivalent point scatterer [34,51]. At the course of the booting stage, often connected with the development of the flag leave, entropy and alpha values reached maximum values of around 0.85–0.9 and around 35°.

After reaching their maximum, the entropy and alpha values started to decrease in May. The entropy and alpha values of wheat decreased until around 0.75 and around 25° until the end of July 2017 and until around 0.6 and around 20° in 2018. The decrease in barley values already started at the beginning of May in both years. The entropy values decreased to around 0.73 (DESC 168) and 0.62 (other pass directions), and the alpha values decreased down to around 25° until the end of May 2017. In 2018, entropy and alpha decreased down to around 0.65 and around 25° until harvest at the end of June. This decrease in entropy and alpha can be associated with the heading stage of the plants (BBCH 51–59). At this time, the plants evolved heads and awns (only barley) and often reached their maximum height of up to 100 cm and a very high biomass up to 5000 $g/m^2$. Furthermore, vegetation water content (VWC), which was stable around 80% until the heading stage, now started to decrease by around 20% per month. From this time on, the radar signal was mainly influenced by the vegetation and reflected plant structure and water content. In 2018, the heading stage reached a few weeks earlier than in 2017, which explains the earlier decline of entropy and alpha in 2018. Furthermore, barley reached the heading stage earlier than wheat in both years. The decrease in entropy and alpha values during and after heading reflects the ongoing drying of the plants and the consequently higher soil contribution to the signal. Surface scattering of the soil gained importance again, and the portion of volume scattering from vegetation decreased, which was indicated by decreasing alpha values. Depolarisation of the wave by vegetation was furthermore extenuated due to the decreasing biomass, which results in decreasing entropy values.

In June 2017, the entropy curves of barley from different acquisition settings behaved differently. All curves except for ASC 146 showed a short-time increase to around 0.82 (descending pass directions) and around 0.7 (ASC 44) at the beginning of June. Until harvest, entropy values decreased (descending pass directions) or increased (ascending pass directions) again until they met at a level of around 0.72. Alpha values of barley from ascending orbits fluctuated constantly between 20° and 28°, whereas alpha values from descending orbits increased again up to 35° before declining down to around 28°. Since the row orientation only marginally affected the signal at this time of the year [60], the vegetation structure might be the cause for differing courses. In June, the heads of the barley plants bent over to a horizontal position. Due to wind, relief, and other factors, the heads of a field often point in the same direction. Therefore, depending on the viewing direction of the sensor and the varying lengths of the signal path through the vegetation because of different incidence angles, the signal more or less depolarised. In 2018, this was not observable, maybe because of the more random positions of the bent heads. Furthermore, there were noticeable plant density differences between barley fields, which also explains the variable development in June 2017.

The strong decrease in entropy and alpha of wheat shortly before harvest in July 2018 indicates the fast drying of the plants in the course of the ripening due to the very dry conditions in this year, also expressed by extremely low soil moisture. Consequently, the soil contribution to the signal increased again and the surface scattering of the soil increased at the expense of the volume scattering of the vegetation. During the ripening stage (BBCH 83–89), LAI and biomass shrunk again down to around 3 and 3000 $g/m^2$ due to a significant decline in the water content below 50% and the resulting lower crop coverage. This was particularly visible in 2018 because of the extremely dry and hot weather conditions leading to a premature ripening of the plants. Barley plants were more dense than wheat plants; therefore, the decline in entropy and alpha values was only visible during and after harvest. The entropy and alpha values of wheat tended to rise again during harvest 2018. Those of barley decreased down to around 0.65 and around 22° (descending pass orbits), and around 0.5 and around 18° (ascending pass orbits) during harvest 2017 and only slightly rose again afterwards. During harvest period 2018, entropy and alpha showed a steep decline to values down to around 0.55 and around 20°, where they stayed after harvest as well. Coverage of the fields after harvest differed as well and included vegetation remains, soil tilling, or replanting, leading to no consistent temporal entropy and alpha trends after harvest. Furthermore, a short-term increase in entropy and alpha values at the end of July 2017 was observable for both wheat and barley and was due to a large precipitation event. Because barley fields were already harvested at this time, the higher soil moisture particularly influenced the SAR signal in this case.

Temporal profiles of anisotropy behaved complementary to those of entropy. In both years, anisotropy of wheat started with values around 0.7 at the beginning of March 2017 and decreased until it reached values around 0.35 mid-May. Afterwards, anisotropy increased again up to around 0.6 at the end of July. Anisotropy of barley also started with values around 0.62 in March and decreased down to around 0.4 in May before they rose again until around 0.62 in June and up to 0.75 during harvest. Anisotropy can be used as a source of discrimination in cases in which entropy is high (>0.7) [51]. When entropy is lower than 0.7, anisotropy is highly affected by noise. For quad-polarimetric data, a high entropy and low anisotropy indicates random scattering. For the existing dual-polarimetric data, this would be the case for April for wheat, and for April and May and partly for

June for barley. This corresponds to volume scattering of the vegetation. At the beginning and at the end of the vegetation period, anisotropy was higher than or equally as high as entropy, which indicates a change in the scattering mechanism. At these times, the soil contribution was higher and the presence of a second scattering mechanism next to the volume scattering of the vegetation was the surface scattering of the soil.

Remarkable for all parameters and both years are their high standard deviations of around 0.1 for entropy and anisotropy and around 6.6° for alpha, which is quite high considering the overall value ranges. However, comparing the coefficients of variation of polarimetric parameters with those of backscatter parameters, the relative variation around the mean of polarimetric parameters fluctuates around 0.1 and 0.35 and is thus only slightly higher than the coefficients of variation of VH and VV backscatter, which move between 0.05 and 0.1 (Figure 8). Only the VH/VV ratio shows temporary higher coefficients of variation and a distinct temporal behavior, similarly to anisotropy.

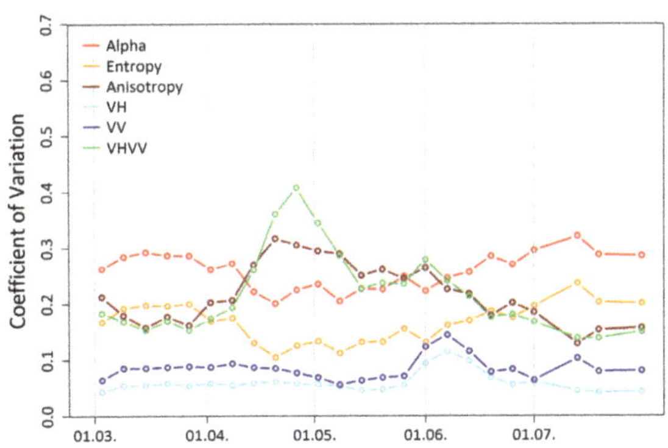


**Figure 8.** Temporal development of the coefficients of variation for all polarimetric and backscatter parameters: exemplarily shown are the mean values for barley fields 2018.

## 5.2. Between- and Within-Field Variability

Differences not only between crop types and years but also between and within individual fields exist. These differences are mainly caused by factors such as geographical location, relief, and soil type but can also be caused by small-scale meteorological variability such as unevenly distributed rainfall events or by differences in field management such as seeding date and density, row orientation, cultivated species, time, and amount of fertilization or irrigation. To find out whether temporal profiles of polarimetric decomposition parameters are able to reflect between- and within-field variability, temporal profiles of mean values of entropy, anisotropy, and alpha were plotted for each single field (Figure 9). Only images from equal pass direction and orbit number were used to avoid variability by acquisition settings. ASC 44 images were used for wheat, whereas ASC 146 images were used for barley. Ascending images were used due to their advantageous acquisition time at late afternoon, and a lower orbit number provided a longer signal path through vegetation. Because one barley field is located outside the ASC 44 footprint, ASC 146 was used instead.

The characteristic temporal behavior of entropy, anisotropy, and alpha becomes apparent for single wheat fields as well. However, some differences between fields are visible. In March, entropy values of three fields (A, BuBio and S) are around 0.2 higher than those of the other fields (Figure 9). Plants on the three mentioned field are already further developed with a higher crop coverage, which strengthens the assumption that entropy and alpha increase with a decreasing soil fraction. On the first days of April 2018, fields in DEMMIN were covered with snow, which is recognizable as peaks in the profiles of fields S and Hey.

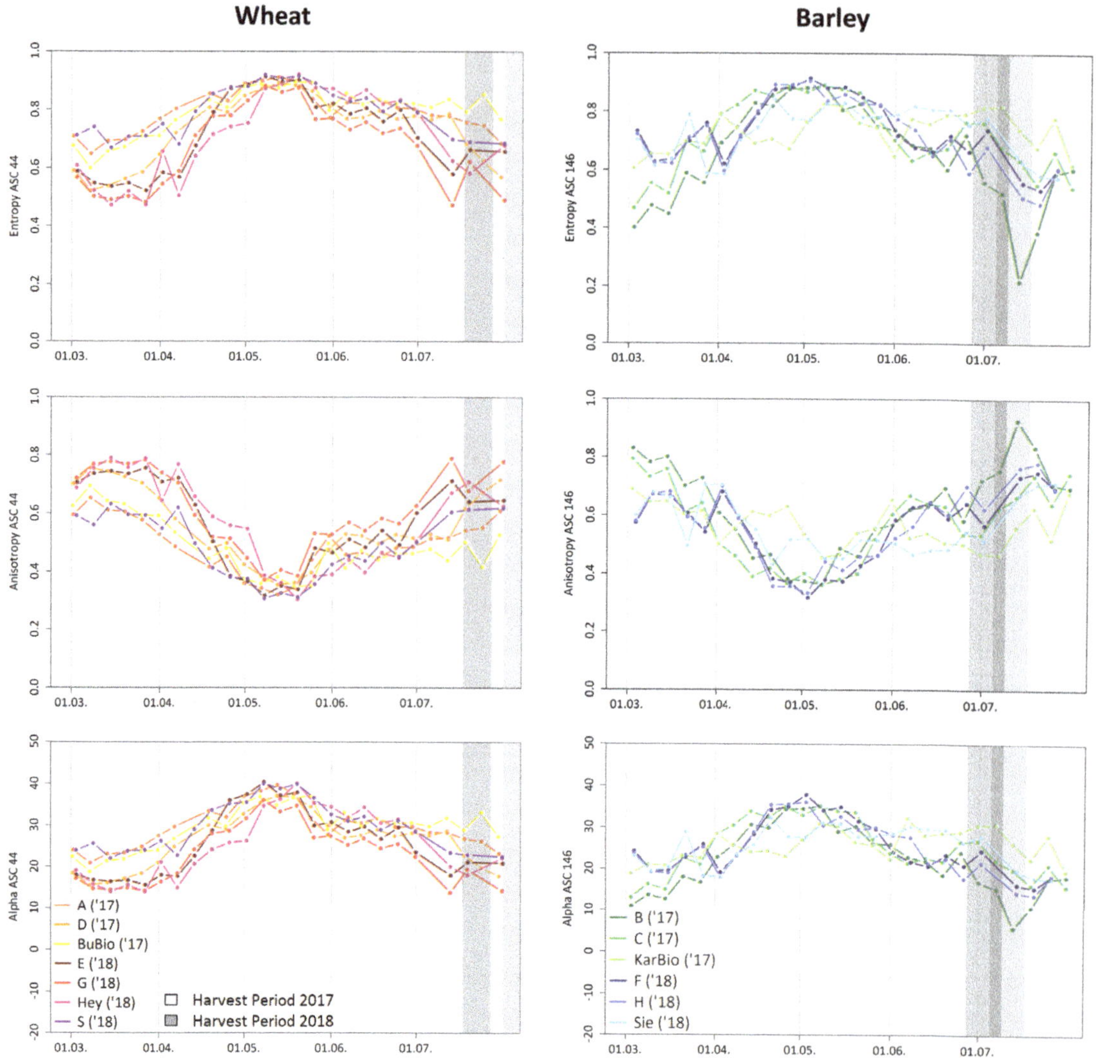


**Figure 9.** Temporal development of decomposition parameters entropy, anisotropy, and alpha for all observed wheat fields (**left**) and all observed barley fields (**right**).

In the course of April, all fields show the characteristic increase in entropy whereas the absolute height of the entropy values is still variable and fluctuates between 0.5 and 0.8. The closer the profiles are to the maximum of around 0.9 at the beginning of May, the more the values approximate each other. It is assumed that entropy saturates at this value level. The profiles of alpha are similarly approximated at the maximum value around 40°, but differences in March and April are not as pronounced as for entropy. After the heading stage in May, entropy and alpha of all fields slightly decrease and have values between 0.7 and 0.85 in June. It is noticeable that field Hey, which had the lowest values until May, now has very high values in June and thus the highest value range. This behavior is not observable for biophysical parameters (Figure 4).

In July shortly before harvest, greater differences between fields emerged. The entropy values of the fields in 2018, which were harvested at the end of July, decreases more, whereas the entropy values of the fields in 2017 were rather constant, except for field D. This might be related to the dryness of the plants as well as their head position. Whereas the head of fields G and E mostly hung down, the heads of all other fields were still upright.

The temporal profiles of barley fields show more variability between fields and the curves are not as uniform as those of wheat fields. At the beginning of the vegetation period in March, barley fields from 2017, particularly fields B and C, had up to 0.3 lower entropy and up to 10° lower alpha values than the barley fields in 2018 (Figure 9). This is due to a lower crop coverage of barley in March 2017. The higher soil fraction is reflected in the signal: the lower entropy indicates a more polarised surface, whereas the lower alpha value indicates a higher portion of surface scattering from the soil. In March and April, the entropy and alpha values of fields B and C increased steeply and overtook the fields of 2018 at the beginning of April. This is also visible in most biophysical parameters, where the fields of 2017 had in general higher biomass, height, and LAI values in April compared to the fields of 2018 (Figure 4).

At the end of April, most fields except for fields Sie and KarBio met at the presumed saturation point of entropy around 0.9 and around 38° for alpha. It is noticeable that particularly field KarBio had very high biomass values, similar to field C at this time, but around 0.2 lower entropy and around 10° lower alpha values. Comparing the biophysical parameters as well as images of field KarBio and field C, there are no recognizable differences, and the variability in polarimetric decomposition parameters cannot be explained yet. A possible reason is the orientation of the leaves in relation to the viewing direction of the sensor. In June before harvest, fields KarBio and Sie again behaved differently compared to the other fields. The entropy and alpha values stayed constant around 0.8 and 28°, whereas the values of all other fields decreased until harvest. The specific behavior of fields KarBio and Sie was not explainable with the observed biophysical parameters (Figure 4).

To investigate whether within-field differences can be detected using time series of dual-polarimetric parameters, fields with a high field heterogeneity were selected and the temporal behavior of the observed points were compared with corresponding biophysical parameters. The temporal profiles of single points turned out to be much more irregular than averaged values of crops and fields. For this reason, the data were smoothed using the LOESS algorithm (locally estimated scatterplot smoothing) to find clearer trends between individual points of a field. A one-degree polynomial regression and different sizes of the neighborhood defining the smoothness of the fitting were tested. However, no distinct relationship between polarimetric parameters and the observed biophysical parameters could be detected.

### 5.3. Regression Analysis

To quantify the relation between the polarimetric decomposition parameters and the observed biophysical parameters described in Table 2, a regression analysis was performed. The quality of the linear and exponential models was evaluated using $R^2$ and root mean square error (RMSE). Additionally, a multiple regression algorithm using the four parameters VH, VV, entropy, and alpha was tested and compared to the one-variable models.

As a first step, data of all fields of a crop type for the whole vegetation period were used for the regression analysis but $R^2$ values of most of the combinations were poor to moderate (Table 3). Exponential regression of entropy achieved the highest $R^2$ values in combination with wet biomass (0.41), VWC (0.46), and LAI (0.46). THe $R^2$ values of wheat are in general higher than these of barley, where no relationship was found for any combination.

Because temporal profiles of wheat and barley show diverging trends for specific development steps, the data of the whole vegetation period were divided into three groups based on their BBCH stages, and the regression analysis was performed again for each group. The first group ranged from BBCH stages 21–49 (tillering to booting), the second group contained BBCH stages 51–77 (heading to fruit development), and the third group represented BBCH stages 83–99 (ripening to harvest). The first BBCH group achieves remarkably higher $R^2$ values compared to those of the whole vegetation period as well as compared to the second and third BBCH group (Table 4). Particularly, wheat reaches high $R^2$ values for all three polarimetric decomposition parameters for linear and exponential regression and for most biophysical parameters except for relative VWC. The exponential regression of entropy and plant height achieves the highest $R^2$ values (0.64), but anisotropy (0.61) and alpha (0.6) also reached high $R^2$ values for plant height. Wet and dry biomass, VWC, and LAI achieved $R^2$ values higher than 0.5 for the exponential regression of all three parameters. Barley reaches the highest $R^2$ values for the exponential regression of alpha and plant height (0.29), but most of the regression models of barley still performed poorly.

**Table 3.** $R^2$ values of the regression between crop and dual-polarimetric parameters per crop type for all fields: values in parentheses are not significant ($p > 0.01$). The darker the shade of green, the higher the $R^2$ value.

| ASC146 | | Entropy | | Anisotropy | | Alpha | |
|---|---|---|---|---|---|---|---|
| | | Wheat | Barley | Wheat | Barley | Wheat | Barley |
| Wet Biomass | lin | 0.22 | 0.05 | 0.19 | 0.04 | 0.17 | 0.08 |
| | exp | 0.41 | 0.08 | 0.35 | 0.07 | 0.34 | 0.11 |
| Dry Biomass | lin | 0.08 | (0) | 0.06 | (0) | 0.05 | 0.01 |
| | exp | 0.29 | 0.02 | 0.24 | 0.02 | 0.23 | 0.04 |
| VWC | lin | 0.26 | 0.06 | 0.24 | 0.05 | 0.22 | 0.08 |
| | exp | 0.46 | 0.12 | 0.4 | 0.11 | 0.39 | 0.16 |
| VWC in % | lin | 0.04 | 0.07 | 0.05 | 0.07 | 0.05 | 0.06 |
| | exp | 0.04 | 0.07 | 0.05 | 0.06 | 0.05 | 0.06 |
| LAI | lin | 0.3 | 0.04 | 0.28 | 0.03 | 0.26 | 0.07 |
| | exp | 0.46 | 0.09 | 0.4 | 0.08 | 0.39 | 0.12 |
| Plant Height | lin | 0.26 | 0.04 | 0.22 | 0.03 | 0.2 | 0.06 |
| | exp | 0.4 | 0.08 | 0.33 | 0.07 | 0.33 | 0.1 |

**Table 4.** $R^2$ values of the regression between crop and dual-polarimetric parameters per crop type for all fields in BBCH stages 21–49. The darker the shade of green, the higher the $R^2$ value.

| ASC146 BBCH 21–49 | | Entropy | | Anisotropy | | Alpha | |
|---|---|---|---|---|---|---|---|
| | | Wheat | Barley | Wheat | Barley | Wheat | Barley |
| Wet Biomass | lin | 0.4 | 0.09 | 0.43 | 0.09 | 0.42 | 0.13 |
| | exp | 0.56 | 0.12 | 0.54 | 0.12 | 0.54 | 0.16 |
| Dry Biomass | lin | 0.37 | 0.09 | 0.4 | 0.09 | 0.4 | 0.13 |
| | exp | 0.52 | 0.1 | 0.52 | 0.1 | 0.52 | 0.13 |
| VWC | lin | 0.4 | 0.09 | 0.43 | 0.09 | 0.42 | 0.13 |
| | exp | 0.56 | 0.16 | 0.54 | 0.16 | 0.53 | 0.2 |
| VWC in % | lin | 0.24 | 0.12 | 0.2 | 0.11 | 0.2 | 0.13 |
| | exp | 0.24 | 0.11 | 0.2 | 0.1 | 0.2 | 0.12 |
| LAI | lin | 0.42 | 0.1 | 0.44 | 0.09 | 0.43 | 0.14 |
| | exp | 0.57 | 0.18 | 0.53 | 0.17 | 0.53 | 0.21 |
| Plant Height | lin | 0.54 | 0.21 | 0.55 | 0.22 | 0.55 | 0.25 |
| | exp | 0.64 | 0.26 | 0.61 | 0.25 | 0.6 | 0.29 |

The relationship between entropy and all biophysical parameters in BBCH stages 21–49 are shown for wheat in Figure 10. Except for VWC, biophysical parameters increase exponentially with increasing entropy; therefore, the exponential regression models result in higher $R^2$ values compared to linear models. For high entropy values (>0.7), the related biophysical parameter values are widely spread and a discrimination is not possible anymore. The good performance of plant height might be due to the comparatively low number of very small values. The RMSE values of all combinations are relatively high compared to the value range.

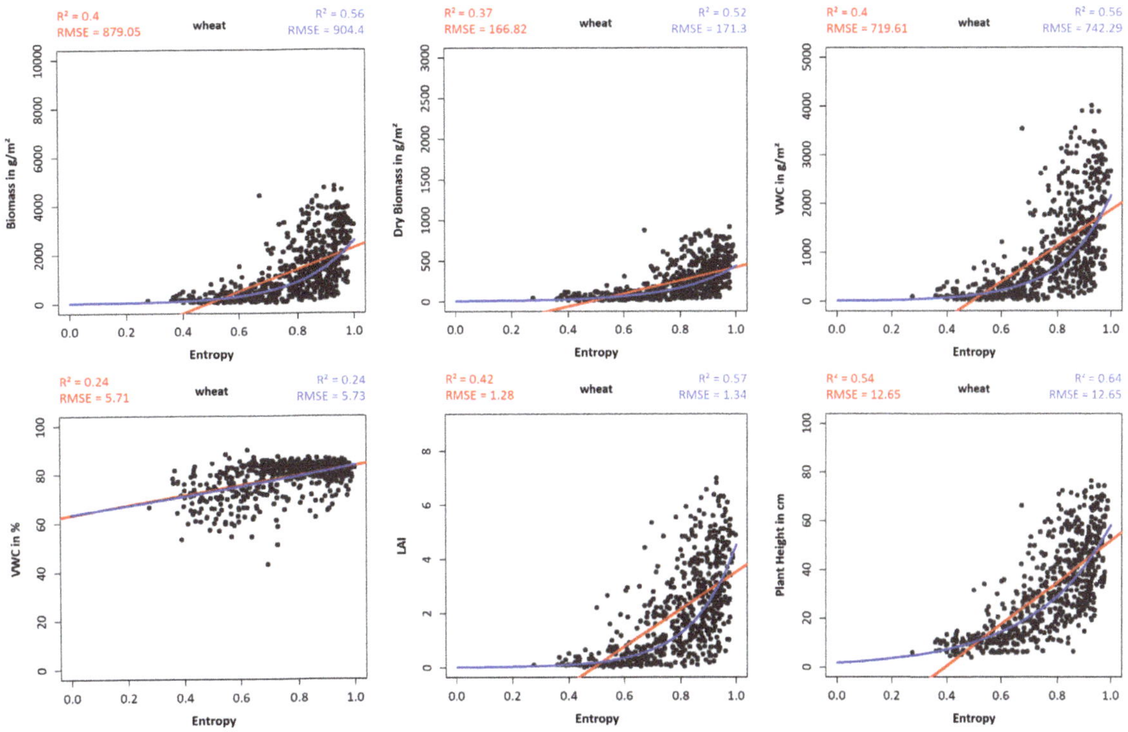


**Figure 10.** Scatter plots of entropy and biophysical parameters of wheat for BBCH stages 21–49: the regression lines, $R^2$, and root mean square error (RMSE) are shown in red (linear regression) and blue (exponential regression).

Figure 11 shows the scatter plots for plant height related to entropy, anisotropy, and alpha. Similar to entropy, the alpha values increase with increasing plant height, whereas anisotropy values decrease with increasing plant height. The $R^2$ values are similar for both linear and exponential regression. The advantage of anisotropy in explaining variability in cases where entropy is high is visible as plant heights between 20 cm and 80 cm are spread over a larger range of anisotropy (around 0.1–0.7) compared to entropy (around 0.6–1). The RMSE values of all three dual-polarimetric parameters are similar (12.46–12.65), with slightly higher values of exponential models of anisotropy and alpha (13.61 and 13.89).

The regression analysis with data from BBCH stages 21–49 was additionally performed for each individual field to detect fields with consistently good regression results and fields with rather poor results (Figure 12). Most of the wheat fields reach equally high $R^2$ values except for field A, which consistently has very poor regression results with $R^2$ values lower than 0.1. Individual wheat fields reach very high $R^2$ values of 0.7 and up to 0.76, for example fields G, S, E and Hey. These high $R^2$ values are reached for exponential regression of all three polarimetric decomposition parameters combined with biomass, VWC, LAI, and plant height, whereas LAI and plant height perform best.

Only two barley fields (F and H) reach $R^2$ values higher than 0.5, with a maximum value of 0.56 for the exponential regression of alpha and plant height for field H.

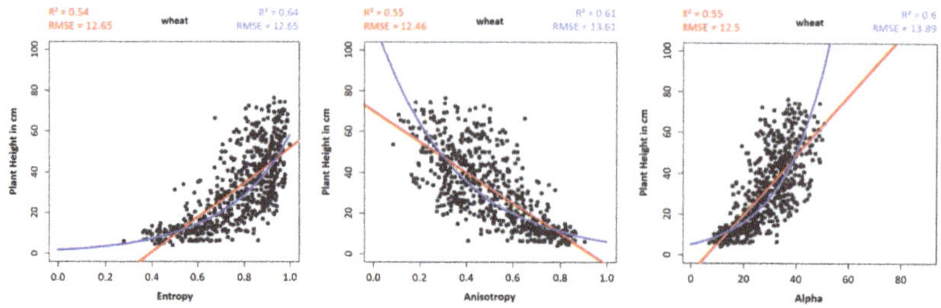


**Figure 11.** Scatter plots of entropy, anisotropy, and alpha related to the plant height of wheat for BBCH stages 21–49: the regression lines, $R^2$, and RMSE are shown in red (linear regression) and blue (exponential regression).

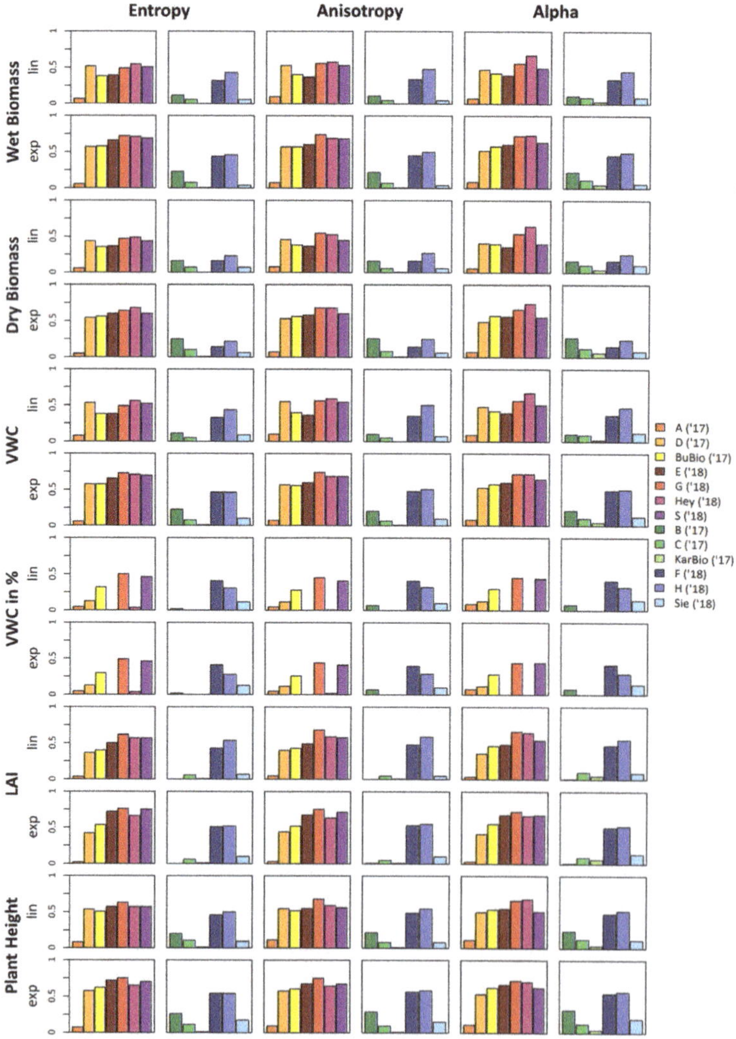


**Figure 12.** Bar plots of $R^2$ values of regression between crop and dual-polarimetric parameters per crop type for single fields in BBCH stages 21–49.

As a last step, a multiple regression including the backscatter parameters VV and VH and the polarimetric decomposition parameters entropy and alpha was performed for BBCH stages 21–49. Anisotropy was neglected because of multicollinearity with entropy. Using a stepwise approach enabled us to define the best performing model variables. All one-variable models of backscatter and polarimetric decomposition parameters are shown to compare their performance with that of the multiple model (Tables 5 and 6).

**Table 5.** $R^2$ values of the regression between crop parameters and backscatter and dual-polarimetric parameters for all wheat fields in BBCH stages 21–49: values in parentheses are not significant ($p > 0.01$). The parameters of values with asterisk contribute to the multiple regression model. The darker the shade of green, the higher the $R^2$ value.

| ASC146 Wheat BBCH 21–49 | | VH | VV | VH/VV | Entropy | Anisotropy | Alpha | Multiple |
|---|---|---|---|---|---|---|---|---|
| Wet Biomass | lin | 0.22 * | 0.69 * | 0.46 | 0.4 | 0.43 | 0.42 * | 0.7 |
| | exp | 0.06 * | 0.6 * | 0.61 | 0.56 * | 0.54 | 0.54 * | 0.68 |
| Dry Biomass | lin | 0.23 | 0.69 * | 0.44 | 0.37 | 0.4 | 0.4 * | 0.69 |
| | exp | 0.09 * | 0.65 * | 0.6 | 0.52 * | 0.52 | 0.52 | 0.7 |
| VWC | lin | 0.22 * | 0.68 * | 0.45 | 0.4 | 0.42 | 0.42 * | 0.69 |
| | exp | 0.05 * | 0.57 * | 0.6 | 0.56 * | 0.54 | 0.53 * | 0.67 |
| VWC in % | lin | (0.01) * | 0.09 * | 0.21 | 0.24 * | 0.2 | 0.2 * | 0.26 |
| | exp | (0.01) * | 0.09 * | 0.2 | 0.23 * | 0.2 | 0.2 * | 0.25 |
| LAI | lin | 0.19 | 0.65 * | 0.46 | 0.42 | 0.44 | 0.43 * | 0.67 |
| | exp | 0.03 * | 0.52 * | 0.58 | 0.57 * | 0.53 | 0.53 * | 0.64 |
| Plant Height | lin | 0.12 * | 0.71 * | 0.61 | 0.54 | 0.55 | 0.55 * | 0.75 |
| | exp | 0.05 * | 0.65 * | 0.69 | 0.64 * | 0.61 | 0.6 * | 0.76 |

**Table 6.** $R^2$ values of the regression between crop parameters and backscatter and dual-polarimetric parameters for all barley fields in BBCH stages 21–49: the parameters of values with asterisk contribute to the multiple regression model. The darker the shade of green, the higher the $R^2$ value.

| ASC146 Barley BBCH 21–49 | | VH | VV | VH/VV | Entropy | Anisotropy | Alpha | Multiple |
|---|---|---|---|---|---|---|---|---|
| Wet Biomass | lin | 0.36 * | 0.38 * | 0.11 | 0.09 * | 0.09 | 0.12 * | 0.44 |
| | exp | 0.2 * | 0.3 * | 0.14 | 0.12 | 0.12 | 0.16 * | 0.32 |
| Dry Biomass | lin | 0.32 * | 0.35 * | 0.11 | 0.09 * | 0.09 | 0.13 * | 0.41 |
| | exp | 0.14 * | 0.22 * | 0.1 | 0.1 * | 0.1 | 0.13 * | 0.24 |
| VWC | lin | 0.38 * | 0.41 * | 0.12 | 0.09 * | 0.09 | 0.13 * | 0.47 |
| | exp | 0.28 * | 0.42 * | 0.2 | 0.16 * | 0.16 | 0.2 * | 0.44 |
| VWC in % | lin | 0.2 | 0.33 * | 0.17 | 0.12 | 0.11 | 0.13 | 0.33 |
| | exp | 0.19 | 0.32 * | 0.17 | 0.11 | 0.1 | 0.12 | 0.32 |
| LAI | lin | 0.34 * | 0.4 * | 0.14 | 0.1 * | 0.09 | 0.14 * | 0.45 |
| | exp | 0.27 * | 0.46 * | 0.24 | 0.18 | 0.17 | 0.21 | 0.46 |
| Plant Height | lin | 0.43 * | 0.62 * | 0.28 | 0.21 * | 0.22 | 0.25 * | 0.64 |
| | exp | 0.39 * | 0.64 * | 0.33 | 0.26 | 0.26 | 0.29 * | 0.65 |

The regression results of the multiple regression of wheat and all biophysical parameters reaches $R^2$ higher than 0.64 and up to 0.76 except for relative VWC. Both linear and exponential regression perform similarly well. The best results achieve plant height with $R^2$ values of 0.75 (linear) and 0.76 (exponential). The backscatter parameter VV reaches highest $R^2$ values for the one-variable regression model and consequently contributes to all multiple models as well. Particularly, exponential multiple regression models benefit from the inclusion of further parameters, and strongest effects are observed for plant height. The $R^2$ values are up to 0.07 higher compared to the best one-variable exponential model, and linear models slightly improve their $R^2$ values but are not higher than around 0.04. Considering both exponential and linear models, $R^2$ values of multiple regression improve by up to 0.05 for plant height. Most exponential multiple models include all four variables, whereas linear models mostly consist of two or three variables. RMSE values of multiple regression models are lower than those of one-variable models, for example, 9.2 cm compared to higher than 12 cm for plant height, 0.98 compared to higher than 1.2 for LAI, or 625 g/m$^2$ compared to higher than 800 g/m$^2$ for wet biomass.

For barley, VV backscatter provides the best regression results for the one-variable models as well, but $R^2$ are lower than these for wheat (Table 6). Multiple regression models of plant height reach the highest $R^2$ values but do not exceed 0.65, whereas multiple regression models of biomass, VWC, and LAI reach $R^2$ values in the range of 0.41 to 0.47. Multiple regression models improve the $R^2$ values only slightly, the difference in $R^2$ values of the one-variable model of VV backscatter, and the multiple model is up to 0.06 for VWC. In contrast to wheat, linear models yield better (particularly wet and dry biomass) or equal $R^2$ values than exponential models. Models predicting relative VWC reach higher $R^2$ values for barley (0.33) than for wheat (0.26).

Exponential multiple regression models predicting plant height for wheat and barley are exemplarily applied for the Blönsdorf fields observed in 2018 using Equation (13) for wheat and Equation (14) for barley (Figure 13). The resulting maps of four dates during BBCH stages 21–49 represent the plant development in the course of the vegetation period starting with plant heights around 10 cm in March. With the beginning of the stem elongation in April, plant growth increased remarkably. At the end of April, fields E, F, and G had plant heights around 40 cm whereas field H already reached plant heights around 60 cm, which was also observed during the field measurements (Figure 4). In May, plant heights between 40 and 60 cm were reached for wheat, which is consistent with observed heights in the field. For field F, the model overestimates plant height and suggests heights from 60 to 80 cm, whereas heights only up to 60 cm were measured in the field. Field heterogeneity can only be observed at the end of April for field H.

$$\log(height_{wheat}) = 0.077 * VH - 0.157 * VV - 0.005 * alpha + 1.343 * entropy + 1.668 \quad (13)$$

$$\log(height_{barley}) = -0.067 * VH - 0.172 * VV + 0.01 * alpha - 0.62 \quad (14)$$

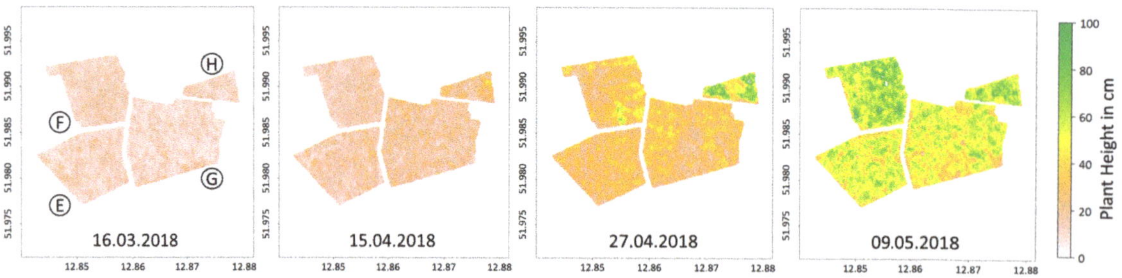


**Figure 13.** Plant heights of four dates during BBCH stages 21–49 in 2018 estimated by multiple exponential regression for the Blönsdorf fields E and G (wheat) and F and H (barley).

## 6. Discussion

Analysis of the temporal behavior as well as of regression models using polarimetric decomposition parameters entropy, anisotropy, and alpha is both helpful in enhancing the understanding of these parameters regarding their ability to enable agricultural monitoring.

The temporal development of polarimetric decomposition parameters in the course of the vegetation period is mainly caused by structural changes of the vegetation as well as by varying soil and vegetation contributions to the signal. The temporal profiles show recurrent trends in both years for all acquisition settings but are also sensitive to specific meteorological conditions such as intense rainfall events in 2017 or snowfall in March and April 2018 as well as summer drought, particularly in 2018. Compared to the temporal behavior of backscatter parameters, entropy, anisotropy, and alpha show similar reactions to phenological changes [12–14]. Both reflect the growing vegetation at the beginning of the vegetation period with decreasing or increasing values and find their maximum or minimum during the booting stage. The heading stage marks a turning point of the course, and from that time on, the parameter values increase or decrease again. An advantage of backscatter profiles is their ability to reflect the time of the bending of the heads of barley [13]. Entropy, anisotropy, and alpha profiles are not that sensitive to this structural change and the time of the bending of the heads is not observable. Similar to this study, Mercier et al. [35] also observed a highly varying polarisation during stem elongation due to heterogeneous plant structures, which was expressed by an increasing Shannon entropy.

Although there are tremendous differences between fields regarding biophysical parameters such biomass, VWC, and LAI (Figure 4), these differences are not definitely distinguishable using temporal profiles of entropy, anisotropy, and alpha. Whereas at least few differences between fields could be observed, e.g., at the beginning of the vegetation period or during ripening, within-field variability was not detectable. Furthermore, some observed between- and within-field differences are more likely to explain variability in the plant structure, leaf, or head orientation, which is not recorded by the existing field measurements. Within-field variability could not be detected using backscatter data as well, whereas between-field variability was more pronounced in backscatter data particularly for barley [13].

As a next step, the comparison of breakpoints and local extrema of a smoothed temporal profile with phenological changes would be reasonable to estimate the beginning of phenological stages such as booting, ripening. or harvest [17,18]. This information would be particularly interesting for farmers adopting their field management such as irrigation or fertilization as well as harvesting according to current phenological stages.

Additionally, the performance of regression models differs remarkably between phenological stages, which was also found by previous studies for backscatter [7,13,16]. The best results are reached in the early BBCH stages from tillering to booting (21–49) because changes in the plant appearance indicated by increasing biomass, LAI, and plant height are most evident at this time. In later BBCH stages, the biophysical parameters change slower and are more subtle. These changes, such as moisture changes in the grains or structural changes of the plants in general, are not observable by the applied measurements. Because relative VWC does not remarkably change during BBCH stages 21–49 (Figure 4), the regression models predicting this parameter perform rather poorly.

Multiple models combining backscatter parameters and polarimetric decomposition parameters improve $R^2$ values by up to 0.05 for plant height compared to those of single regression models. Multiple regression achieves highest $R^2$ values for plant height because this parameter is easy to measure and measurement errors are unlikely. The main reason for the good performance might be the value range. All height values are evenly distributed over the value range and do not increase as fast and steep as for example biomass values. Furthermore, many small values do not significantly influence the regression line as is the case for biomass.

Because barley plants are already higher and more dense compared to wheat plants at the beginning of the vegetation period, resulting regression models reach not as high $R^2$ values compared to wheat. The missing of small values prevent the fitting of an exponential regression line, and the value range is smaller. This also explains the better performance of fields F and H because these fields are comparatively smaller and not as dense as other barley fields in the same period.

The polarimetric decomposition parameters of full-polarimetric SAR data such as RADARSAT-2 reach generally higher or comparably high $R^2$ values of single parameters compared to dual-polarimetric decomposition parameters of our study. Betbeder et al. [26] used a different decomposition algorithm and calculated Shannon entropy for Sentinel-1 data, which similar to entropy derived from H-A-$\alpha$ decomposition describes the randomness of the scattering. They reached $R^2$ values up to 0.7 for Shannon entropy and rapeseed height. Mercier et al. [35] calculated the Shannon entropy for Sentinel-1 data and found that it is an important parameter for predicting the phenological stages of wheat and rapeseed. Wiseman et al. [7] used a H-A-$\alpha$ decomposition for RADARSAT-2 data and linearly correlated entropy, anisotropy, alpha, and further SAR parameters with dry biomass of spring wheat. They reached $R^2$ values of 0.215 for entropy and 0.418 for alpha. These values were comparable to our wheat results (0.4 for dry biomass and alpha) or lower (0.37 for dry biomass and entropy). The reasons for the comparatively lower performance of RADARSAT-2 data in this case might be that data from the whole vegetation period was used and might be furthermore due to the lower temporal resolution of RADARSAT-2. Canisius et al. [25] also used RADARSAT-2 data to correlate the biophysical parameters of spring wheat with the polarimetric decomposition parameters. They found a high correlation between wheat height and alpha angle, with an $R^2$ value of 0.66, which is slightly higher than our $R^2$ value of 0.6 for alpha and plant height of wheat. The $R^2$ value of Canisius et al. [25] was even higher for a smoothed time series and reached a value of 0.88.

Similar to temporal profiles, it is not possible to detect the field variability of biophysical parameters using regression models of backscatter and polarimetric decomposition parameters of single dates. The RMSE values are too high for a meaningful estimation of biophysical parameters and are sometimes even higher than the expected variability within a field on a single date, for example, around 9 cm for plant height. Varying meteorological conditions between years make it difficult to establish generally valid regression models as well. Therefore, the identification of phenological stages, e.g., based on temporal profiles, is meaningful.

Irregular patterns in images of polarimetric parameters make it difficult to make statements about individual points and complicates the fitting of a regression model as well as the detection of between- and within-field variability. These patterns are due to the radar speckle, the "salt-and-pepper-effect" of the images, which is individual for every image. The extraction of values of a small area includes only a few pixels; therefore, the exact extraction location has a high influence on the extracted value because neighboring pixels might be extremely different. This is a general characteristic of radar images and is furthermore fostered by speckle filtering and other averaging processes such as multilooking. These processing steps affect the inherent scattering mechanisms of each pixel and a higher amount of averaging results, for instance, in increasing entropy values and decreasing anisotropy values [51]. The high spatial variability of decomposition parameters is also mentioned in other studies [29,36]. To obtain more meaningful information for individual points, a larger extraction area or a smoothed image could improve the results, but there is again the trade-off between smoothing and information loss. Further error sources might be the different quality of field measurements, geometric inaccuracies of the SAR data caused by image processing, and differing local incidence angles, among others.

In the future, the current results of dual-polarimetric decomposition of Sentinel-1 data can be compared to the decomposition parameters of full polarimetric SAR data such as RADARSAT-2. Additionally, a smoothing of the time series could lead to better regression results, as found by Canisius et al. [25]. Furthermore, the multiple regression analysis could be performed using additional parameters such as the Radar Vegetation Index (RVI) [61], Shannon entropy [35], or field statistics of SAR parameters, as was done by Holtgrave et al. [16]. Additionally, it would be advantageous if existing field measurements are completed by additional parameters with the consideration of water content in different parts of the plant to catch also small water content differences, particularly during the ripening in the grains. Furthermore, the documentation of structural changes such as the emergence of flag leaves or heads as well as the bending of the heads must be documented in detail to explain scattering changes.

## 7. Conclusions

This study confirms that the polarimetric decomposition parameters of the dual-polarimetric Sentinel-1 data are useful for the monitoring of biophysical parameters on agricultural fields with some limitations. Temporal profiles of entropy, anisotropy, and alpha are sensitive to changes in the plant appearance in the course of the phenological development of both wheat and barley, whereas the variable contribution of soil and vegetation as well as the changing water content of the plants mainly influence the SAR signal. Furthermore, differences between the two years 2017, which was rather wet, and 2018, a very dry year, are clearly evident in the temporal profiles. However, temporal profiles of individual fields do not satisfactorily reflect differences between and within fields caused by varying biophysical parameters.

Regression models of polarimetric decomposition parameters related to biophysical parameters show particularly high $R^2$ values for wheat in BBCH stages 21–49 (tillering to booting). Plant height could be predicted using single exponential regression models of entropy and anisotropy with $R^2$ of 0.64 and 0.61, respectively. Multiple regression models including entropy and alpha as well as backscatter coefficients VH and VV resulted in even higher $R^2$ values for plant height (0.76), wet biomass (0.7), dry biomass (0.7), and vegetation water content (0.69). Therefore, polarimetric decomposition parameters are beneficial as additional input parameters for multiple regression models to improve the prediction of biophysical parameters. The $R^2$ values of multiple regression models improved by up to 0.05 by taking the dual-polarimetric parameters into account compared to using only backscatter parameters. Furthermore, the RMSE values are around 10% and up to 20% lower compared to those of single regression models but are still too high for a meaningful prediction of biophysical parameters on single dates.

The presented results serve as a basis for further research in agricultural monitoring and show the potential of polarimetric decomposition parameters of Sentinel-1 data as an additional source of information next to backscatter coefficients.

**Author Contributions:** Conceptualization, K.H., S.I., D.S., and C.W.; methodology, K.H.; formal analysis, K.H.; investigation, K.H.; data curation, K.H.; writing—original draft preparation, K.H.; writing—review and editing, K.H., S.I., D.S., and C.W.; visualization, K.H.; supervision, S.I., D.S., and C.W.; project administration, D.S.; funding acquisition, D.S. All authors have read and agreed to the published version of the manuscript.

**Funding:** This research is funded by the Federal Ministry of Food and Agriculture (BMEL) based on a decision of the Parliament of the Federal Republic of Germany via the Federal Office for Agriculture and Food (BLE) under the innovation support programme, grant number 2815710715.

**Data Availability Statement:** Publicly available datasets were analyzed in this study. Sentinel-1 data can be found here: https://scihub.copernicus.eu/. Field data and resulting data sets presented in this study are available on request from the corresponding author.

**Acknowledgments:** The authors would like to thank numerous interns and students for their great support in the fieldwork.

**Conflicts of Interest:** The authors declare no conflict of interest. The funders had no role in the design of the study; in the collection, analyses, or interpretation of data; in the writing of the manuscript; or in the decision to publish the results.

## References

1. Mc Nairn, H.; Brisco, B. The application of C-band polarimetric SAR for agriculture: A review. *Can. J. Remote. Sens.* **2004**, *30*, 525–542. [CrossRef]
2. Steele-Dunne, S.C.; McNairn, H.; Monsivais-Huertero, A.; Mahdian, M.; Homayouni, S.; Fazel, M.A.; Mohammadimanesh, F.; Baghdadi, N.; Cresson, R.; El Hajj, M.; et al. Radar Remote Sensing of Agricultural Canopies: A Review. *Remote Sens. Environ.* **2016**, *187*, 1607–1621. [CrossRef]
3. Satalino, G.; Balenzano, A.; Mattia, F.; Davidson, M.W. C-band SAR data for mapping crops dominated by surface or volume scattering. *IEEE Geosci. Remote Sens. Lett.* **2014**, *11*, 384–388. [CrossRef]
4. Skriver, H.; Mattia, F.; Satalino, G.; Balenzano, A.; Pauwels, V.R.N.; Verhoest, N.E.C.; Davidson, M. Crop Classification Using Short-Revisit Multitemporal SAR Data. *IEEE J. Sel. Top. Appl. Earth Obs. Remote Sens.* **2011**, *4*, 423–431. [CrossRef]
5. Bargiel, D. A new method for crop classification combining time series of radar images and crop phenology information. *Remote Sens. Environ.* **2017**, *198*, 369–383. [CrossRef]

6. Macelloni, G.; Paloscia, S.; Pampaloni, P.; Marliani, F.; Gai, M. The relationship between the backscattering coefficient and the biomass of narrow and broad leaf crops. *IEEE Trans. Geosci. Remote Sens.* **2001**, *39*, 873–884. [CrossRef]
7. Wiseman, G.; McNairn, H.; Homayouni, S.; Shang, J. RADARSAT-2 Polarimetric SAR response to crop biomass for agricultural production monitoring. *IEEE J. Sel. Top. Appl. Earth Obs. Remote Sens.* **2014**, *7*, 4461–4471. [CrossRef]
8. Inoue, Y.; Sakaiya, E.; Wang, C. Capability of C-band backscattering coefficients from high-resolution satellite SAR sensors to assess biophysical variables in paddy rice. *Remote Sens. Environ.* **2014**, *140*, 257–266. [CrossRef]
9. Vicente-Guijalba, F.; Martinez-Marin, T.; Lopez-Sanchez, J.M. Dynamical approach for real-time monitoring of agricultural crops. *IEEE Trans. Geosci. Remote Sens.* **2015**, *53*, 3278–3293. [CrossRef]
10. Susan Moran, M.; Alonso, L.; Moreno, J.F.; Pilar Cendrero Mateo, M.; Fernando de la Cruz, D.; Montoro, A.; Alonso, L.; Moreno, J.F.; Cendrero Mateo, M.P. A RADARSAT-2 Quad-Polarized Time Series for Monitoring Crop and Soil Conditions in Barrax, Spain Biological Resource Management for Sustainable Agricultural Systems to encourage international research on sustainable use of natural resources in agriculture. *Ieee Trans. Geosci. Remote Sens.* **2011**, *1*, 1–14. [CrossRef]
11. Skriver, H.; Svendsen, M.T.; Thomsen, A.G. Multitemporal C- and L-band polarimetric signatures of crops. *IEEE Trans. Geosci. Remote Sens.* **1999**, *37*, 2413–2429. [CrossRef]
12. Vreugdenhil, M.; Wagner, W.; Bauer-Marschallinger, B.; Pfeil, I.; Teubner, I.; Rüdiger, C.; Strauss, P. Sensitivity of Sentinel-1 backscatter to vegetation dynamics: An Austrian case study. *Remote Sens.* **2018**, *10*, 1396. [CrossRef]
13. Harfenmeister, K.; Spengler, D.; Weltzien, C. Analyzing Temporal and Spatial Characteristics of Crop Parameters Using Sentinel-1 Backscatter Data. *Remote Sens.* **2019**, *11*, 1569. [CrossRef]
14. Veloso, A.; Mermoz, S.; Bouvet, A.; Le Toan, T.; Planells, M.; Dejoux, J.F.; Ceschia, E. Understanding the temporal behavior of crops using Sentinel-1 and Sentinel-2-like data for agricultural applications. *Remote Sens. Environ.* **2017**, *199*, 415–426. [CrossRef]
15. Khabbazan, S.; Vermunt, P.; Steele-Dunne, S.; Ratering Arntz, L.; Marinetti, C.; van der Valk, D.; Iannini, L.; Molijn, R.; Westerdijk, K.; van der Sande, C. Crop Monitoring Using Sentinel-1 Data: A Case Study from The Netherlands. *Remote Sens.* **2019**, *11*, 1887. [CrossRef]
16. Holtgrave, A.K.; Röder, N.; Ackermann, A.; Erasmi, S.; Kleinschmit, B. Comparing Sentinel-1 and -2 data and indices for agricultural land use monitoring. *Remote Sens.* **2020**, *12*, 2919. [CrossRef]
17. Schlund, M.; Erasmi, S. Sentinel-1 time series data for monitoring the phenology of winter wheat. *Remote Sens. Environ.* **2020**, *246*, 111814. [CrossRef]
18. Nasrallah, A.; Baghdadi, N.; El Hajj, M.; Darwish, T.; Belhouchette, H.; Faour, G.; Darwich, S.; Mhawej, M. Sentinel-1 data for winter wheat phenology monitoring and mapping. *Remote Sens.* **2019**, *11*, 2228. [CrossRef]
19. Hajnsek, I.; Jagdhuber, T.; Schön, H.; Papathanassiou, K.P. Potential of estimating soil moisture under vegetation cover by means of PolSAR. *IEEE Trans. Geosci. Remote Sens.* **2009**, *47*, 442–454. [CrossRef]
20. Jagdhuber, T.; Hajnsek, I.; Bronstert, A.; Papathanassiou, K.P. Soil moisture estimation under low vegetation cover using a multi-angular polarimetric decomposition. *IEEE Trans. Geosci. Remote Sens.* **2013**, *51*, 2201–2215. [CrossRef]
21. Wang, H.; Magagi, R.; Goita, K. Comparison of different polarimetric decompositions for soil moisture retrieval over vegetation covered agricultural area. *Remote Sens. Environ.* **2017**, *199*, 120–136. [CrossRef]
22. McNairn, H.; Shang, J.; Jiao, X.; Champagne, C. The contribution of ALOS PALSAR multipolarization and polarimetric data to crop classification. *IEEE Trans. Geosci. Remote Sens.* **2009**, *47*, 3981–3992. [CrossRef]
23. Jiao, X.; Kovacs, J.M.; Shang, J.; McNairn, H.; Walters, D.; Ma, B.; Geng, X. Object-oriented crop mapping and monitoring using multi-temporal polarimetric RADARSAT-2 data. *ISPRS J. Photogramm. Remote Sens.* **2014**, *96*, 38–46. [CrossRef]
24. Mascolo, L.; Lopez-Sanchez, J.M.; Vicente-Guijalba, F.; Nunziata, F.; Migliaccio, M.; Mazzarella, G. A Complete Procedure for Crop Phenology Estimation with PolSAR Data Based on the Complex Wishart Classifier. *IEEE Trans. Geosci. Remote Sens.* **2016**, *54*, 6505–6515. [CrossRef]
25. Canisius, F.; Shang, J.; Liu, J.; Huang, X.; Ma, B.; Jiao, X.; Geng, X.; Kovacs, J.M.; Walters, D. Tracking crop phenological development using multi-temporal polarimetric Radarsat-2 data. *Remote Sens. Environ.* **2018**, *210*, 508–518. [CrossRef]
26. Betbeder, J.; Fieuzal, R.; Philippets, Y.; Ferro-Famil, L.; Baup, F. Contribution of multitemporal polarimetric synthetic aperture radar data for monitoring winter wheat and rapeseed crops. *J. Appl. Remote Sens.* **2016**, *10*, 026020. [CrossRef]
27. Ji, K.; Wu, Y. Scattering Mechanism Extraction by a Modified Cloude-Pottier Decomposition for Dual Polarization SAR. *Remote Sens.* **2015**, *7*, 7447–7470. [CrossRef]
28. Cloude, S.R. The dual polarisation entropy/alpha decomposition: A PALSAR case study. *Sci. Appl. Sar Polarim. Polarim. Interferom.* **2007**, *644*, 2.
29. Engelbrecht, J.; Theron, A.; Vhengani, L.; Kemp, J. A simple normalized difference approach to burnt area mapping using multi-polarisation C-Band SAR. *Remote Sens.* **2017**, *9*, 764. [CrossRef]
30. Plank, S.; Jüssi, M.; Martinis, S.; Twele, A. Mapping of flooded vegetation by means of polarimetric sentinel-1 and ALOS-2/PALSAR-2 imagery. *Int. J. Remote. Sens.* **2017**, *38*, 3831–3850. [CrossRef]
31. Velotto, D.; Bentes, C.; Tings, B.; Lehner, S. First Comparison of Sentinel-1 and TerraSAR-X Data in the Framework of Maritime Targets Detection: South Italy Case. *IEEE J. Ocean. Eng.* **2016**, *41*, 993–1006. [CrossRef]
32. Banqué, X.; Lopez-Sanchez, J.M.; Monells, D.; Ballester, D.; Duro, J. Polarimetry-Based Land Cover Classification with Sentinel-1 Data. *Polinsar* **2015**, *729*, 1–5.

33. Li, M.; Bijker, W. Vegetable classification in Indonesia using Dynamic Time Warping of Sentinel-1A dual polarization SAR time series. *Int. J. Appl. Earth Obs. Geoinf.* **2019**, *78*, 268–280. [CrossRef]
34. Chauhan, S.; Darvishzadeh, R.; Boschetti, M.; Nelson, A. Discriminant analysis for lodging severity classification in wheat using RADARSAT-2 and Sentinel-1 data. *ISPRS J. Photogramm. Remote. Sens.* **2020**, *164*, 138–151. [CrossRef]
35. Mercier, A.; Betbeder, J.; Baudry, J.; Le Roux, V.; Spicher, F.; Lacoux, J.; Roger, D.; Hubert-Moy, L. Evaluation of Sentinel-1 & 2 time series for predicting wheat and rapeseed phenological stages. *ISPRS J. Photogramm. Remote. Sens.* **2020**, *163*, 231–256. [CrossRef]
36. Heine, I.; Jagdhuber, T.; Itzerott, S. Classification and monitoring of reed belts using dual-polarimetric TerraSAR-X time series. *Remote Sens.* **2016**, *8*, 552. [CrossRef]
37. Voormansik, K.; Jagdhuber, T.; Zalite, K.; Noorma, M.; Hajnsek, I. Observations of Cutting Practices in Agricultural Grasslands Using Polarimetric SAR. *IEEE J. Sel. Top. Appl. Earth Obs. Remote Sens.* **2016**, *9*, 1382–1396. [CrossRef]
38. Lopez-Sanchez, J.M.; Cloude, S.R.; Ballester-Berman, J.D. Rice phenology monitoring by means of SAR polarimetry at X-band. *IEEE Trans. Geosci. Remote Sens.* **2012**, *50*, 2695–2709. [CrossRef]
39. Ratzke, U.; Mohr, H.J. *Böden in Mecklenburg-Vorpommern: Abriss ihrer Entstehung, Verbreitung und Nutzung*; Beiträge zum Bodenschutz in Mecklenburg-Vorpommern, Landesamt für Umwelt, Naturschutz und Geologie Mecklenburg-Vorpommern: Güstrow, Germany, 2003.
40. Spengler, D.; Förster, M.; Borg, E. Editorial. *PFG-J. Photogramm. Remote. Sens. Geoinf. Sci.* **2018**, *86*, 49–51. [CrossRef]
41. Zacharias, S.; Bogena, H.; Samaniego, L.; Mauder, M.; Fuß, R.; Pütz, T.; Frenzel, M.; Schwank, M.; Baessler, C.; Butterbach-Bahl, K.; et al. A Network of Terrestrial Environmental Observatories in Germany. *Vadose Zone J.* **2011**, *10*, 955–973. [CrossRef]
42. Heinrich, I.; Balanzategui, D.; Bens, O.; Blasch, G.; Blume, T.; Böttcher, F.; Borg, E.; Brademann, B.; Brauer, A.; Conrad, C.; et al. Interdisciplinary Geo-ecological Research across Time Scales in the Northeast German Lowland Observatory (TERENO-NE). *Vadose Zone J.* **2018**, *17*, 1–25. [CrossRef]
43. Amt für Statistik Berlin-Brandenburg. *Statistischer Bericht C I 1–j/17. Bodennutzung der Landwirtschaftlichen Betriebe im Land Brandenburg 2017*; Amt für Statistik Berlin-Brandenburg: Potsdam, Germany, 2017.
44. Amt für Statistik Berlin-Brandenburg. *Statistischer Bericht C I 1–j/18. Bodennutzung der Landwirtschaftlichen Betriebe im Land Brandenburg 2018*; Amt für Statistik Berlin-Brandenburg: Potsdam, Germany, 2018.
45. Ministerium für Landwirtschaft und Umwelt Mecklenburg-Vorpommern. *Statistisches Datenblatt 2020*; Ministerium für Landwirtschaft und Umwelt Mecklenburg-Vorpommern: Schwerin, Germany, 2020.
46. DWD Climate Data Center (CDC). *Historical Daily Precipitation Observations for Germany, Version v007*; Deutscher Wetterdienst: Offenbach, Germany, 2020.
47. Eckelmann, W.; Sponagel, H.; Grottenthaler, W.; Hartmann, K.J.; Hartwich, R.; Janetzko, P.; Joisten, H.; Kühn, D.; Sabel, K.J.; Traidl, R. *Bodenkundliche Kartieranleitung. KA5*; Schweizerbart'sche Verlagsbuchhandlung: Stuttgart, Germany, 2005.
48. Meier, U.; Bleiholder, H.; Buhr, L.; Feller, C.; Hack, H.; Heß, M.; Lancashire, P.; Schnock, U.; Stauß, R.; Van den Boom, T.; et al. The BBCH system to coding the phenological growth stages of plants-history and publications. *J. Für Kult.* **2009**, *61*, 41–52.
49. Itzerott, S.; Hohmann, C.; Stender, V.; Maass, H.; Borg, E.; Renke, F.; Jahncke, D.; Berg, M.; Conrad, C.; Spengler, D. TERENO (Northeast), Climate station Heydenhof, Germany. V. 2.0 GFZ Data Services. 2018. Available online: https://dataservices.gfz-potsdam.de/tereno-new/showshort.php?id=escidoc:3311065 (accessed on 28 December 2020).
50. Russelle, M.P.; Wilhelm, W.W.; Olson, R.A.; Power, J.F. Growth Analysis Based on Degree Days. *Crop Sci.* **1984**, *24*, 28–32. [CrossRef]
51. Lee, J.S.; Pottier, E. *Polarimetric Radar Imaging: From Basics to Applications*; CRC Press: Boca Raton, FL, USA, 2009. [CrossRef]
52. Moreira, A.; Prats-Iraola, P.; Younis, M.; Krieger, G.; Hajnsek, I.; Papathanassiou, K.P. A tutorial on synthetic aperture radar. *IEEE Geosci. Remote. Sens. Mag.* **2013**, *1*, 6–43. [CrossRef]
53. Cloude, S.R.; Pottier, E. An entropy based classification scheme for land applications of polarimetric SAR. *IEEE Trans. Geosci. Remote. Sens.* **1997**, *35*, 68–78. [CrossRef]
54. ESA. SNAP—ESA Sentinel Application Platform v6.0.6, 2018. Available online: https://step.esa.int/main/snap-6-0-released/ (accessed on 28 December 2020).
55. Lee, J.S.; Grunes, M.R.; de Grandi, G. Polarimetric SAR speckle filtering and its implication for classification. *IEEE Trans. Geosci. Remote Sens.* **1999**, *37*, 2363–2373. [CrossRef]
56. Pottier, E.; Ferro-Famil, L. PolSARPro V5.0: An ESA educational toolbox used for self-education in the field of POLSAR and POL-INSAR data analysis. In Proceedings of the International Geoscience and Remote Sensing Symposium (IGARSS), Munich, Germany, 22–27 July 2012. [CrossRef]
57. Bahrenberg, G.; Giese, E.; Mevenkamp, N.; Nipper, J. *Statistische Methoden in der Geographie-Band 2: Multivariate Statistik*; Schweizerbart'sche Verlagsbuchhandlung: Stuttgart, Germany, 2008.
58. Kuhn, M. Caret: Classification and Regression Training; R Package Version 6.0-86. 2020. Available online: https://cran.r-project.org/web/packages/caret/index.html (accessed on 28 December 2020).
59. Robinson, C.; Schumacker, R.E. Interaction effects: Centering, variance inflation factor, and interpretation issues. *Mult. Linear Regres. Viewpoints* **2009**, *35*, 6–11.

60. Joerg, H.; Pardini, M.; Hajnsek, I.; Papathanassiou, K.P. Sensitivity of SAR Tomography to the Phenological Cycle of Agricultural Crops at X-, C-, and L-band. *IEEE J. Sel. Top. Appl. Earth Obs. Remote Sens.* **2018**, *11*, 3014–3029. [CrossRef]
61. Mandal, D.; Kumar, V.; Ratha, D.; Dey, S.; Bhattacharya, A.; Lopez-Sanchez, J.M.; McNairn, H.; Rao, Y.S. Dual polarimetric radar vegetation index for crop growth monitoring using sentinel-1 SAR data. *Remote Sens. Environ.* **2020**, *247*, 111954. [CrossRef]

remote sensing

MDPI

*Article*

# Sentinel-1 Time Series for Crop Identification in the Framework of the Future CAP Monitoring

**Emilie Beriaux [1,*,†], Alban Jago [1,†], Cozmin Lucau-Danila [1], Viviane Planchon [1] and Pierre Defourny [2]**

[1] Agriculture, Territory and Technologies Integration Unit, Production in Agriculture Department, Walloon Agricultural Research Centre, rue de Liroux 9, 5030 Gembloux, Belgium; a.jago@cra.wallonie.be (A.J.); c.lucau-danila@cra.wallonie.be (C.L.-D.); v.planchon@cra.wallonie.be (V.P.)

[2] Earth and Life Institute, Université Catholique de Louvain, Croix du Sud 2, 1348 Louvain-la-Neuve, Belgium; pierre.defourny@uclouvain.be

* Correspondence: e.beriaux@cra.wallonie.be; Tel.: +32-(0)-485476756

† These authors contributed equally to this work and should be considered as co-first authors.

**Abstract:** In this upcoming Common Agricultural Policy (CAP) reform, the use of satellite imagery is taking an increasing role for improving the Integrated Administration and Control System (IACS). Considering the operational aspect of the CAP monitoring process, the use of Sentinel-1 SAR (Synthetic Aperture Radar) images is highly relevant, especially in regions with a frequent cloud cover, such as Belgium. Indeed, SAR imagery does not depend on sunlight and is barely affected by the presence of clouds. Moreover, the SAR signal is particularly sensitive to the geometry and the water content of the target. Crop identification is often a pre-requisite to monitor agriculture at parcel level (ploughing, harvest, grassland mowing, intercropping, etc.) The main goal of this study is to assess the performances and constraints of a SAR-based crop classification in an operational large-scale application. The Random Forest object-oriented classification model is built on Sentinel-1 time series from January to August 2020 only. It can identify crops in the Walloon Region (south part of Belgium) with high performance: 93.4% of well-classified area, representing 88.4% of the parcels. Among the 48 crop groups, the six most represented ones get a F1-score higher or equal to 84%. Additionally, this research documents how the classification performance is affected by different parameters: the SAR orbit, the size of the training dataset, the use of different internal buffers on parcel polygons before signal extraction, the set of explanatory variables, and the period of the time series. In an operational context, this allows to choose the right balance between classification accuracy and model complexity. A key result is that using a training dataset containing only 3.2% of the total number of parcels allows to correctly classify 91.7% of the agricultural area. The impact of rain and snow is also discussed. Finally, this research analyses how the classification accuracy depends on some characteristics of the parcels like their shape or size. This allows to assess the relevance of the classification depending on those characteristics, as well as to identify a subset of parcels for which the global accuracy is higher.

**Keywords:** Sentinel-1; SAR; multitemporal analysis; crop identification; parcel-based classification; remote sensing; Common Agricultural Policy

**Citation:** Beriaux, E.; Jago, A.; Lucau-Danila, C.; Planchon, V.; Defourny, P. Sentinel-1 Time Series for Crop Identification in the Framework of the Future CAP Monitoring. *Remote Sens.* **2021**, *13*, 2785. https://doi.org/10.3390/rs13142785

Academic Editors: Olaniyi Ajadi and David McAlpin

Received: 31 May 2021
Accepted: 12 July 2021
Published: 15 July 2021



## 1. Introduction

As part of the upcoming reform of the Common Agricultural Policy (CAP) to be implemented from 1 January 2023, the European Commission (EC) has adopted new rules that will allow to increase the use of modern technologies during verifications related to the area-based CAP payments. The legal and technical framework to make use of the advantages of Earth Observation data in the context of CAP controls was provided in 2018 by the EC. In particular, this covers data coming from the EU's Copernicus Sentinel satellites.

The actual On-the-Spot-Checks (OTSC) control system is based on a yearly verification done by each EU Member State, who must carry out controls on at least 5% of the farms applying for subsidies. The OTSC are fulfilled either by visiting farms, by interpreting

Remote Sensing images or combining both methods (RS imagery and rapid field visits on farms). This system will be replaced by a new monitoring approach (Area Monitoring System—AMS), which is a procedure of regular observation, tracking, and assessment of all eligibility criteria, commitments, and other obligations which can be monitored in a continuous way (for instance by using mainly Copernicus Sentinel satellites images).

This allows to move from an a posteriori control to a continuous assessment of a parcel eligibility. The goal is to warn the farmers in due time about a possible detected discrepancy, in order to avoid penalties. For example, in the case of an activity that must be completed before a given date (such as grass mowing), if it has not been performed approaching the deadline, the farmer could be informed that he should ensure compliance before the deadline to avoid a sanction. This will guarantee that farmers are able to carry out their environmental and other obligations in due time and avoid penalties for non-compliance with CAP rules. Reducing the number of farm visits will also significantly decrease the time spent by farmers with inspectors in the field. This new approach also aims at encouraging farmers to benefit from digital technologies, such as crop monitoring at parcel level, to improve agronomic performances while reducing fertilizer costs and environmental impacts.

In this context of agriculture monitoring by remote sensing, accurate and up-to-date crop type maps are a prerequisite for agriculture practices analyses, such as ploughing or harvest detection. Moreover, identification of land cover types provides basic information for generation of other thematic maps and establishes a baseline for monitoring activities.

High temporal and high spatial resolution remote sensing imagery provides a major asset to monitor agriculture and identify crop types. Satellite image crop classifications are mostly obtained from optical imagery [1–7]. Indeed, surface reflectances are often key explanatory variables for land cover classification [8,9] as well as indices derived from reflectance data, such as the NDVI, NDWI, and Brightness [8,10–13]. Temporal metrics derived from time series indices have also been used in several studies [8,11,14–16]. However, clouds and cloud shadows remain a major challenge for optical time series, as they lead to gaps and missing data. Furthermore, the reliance of operational applications on cloud-free optical images is a critical issue when precise timeliness is strictly required. The cropland mapping methods applied to time series images have proven to perform better than single-date mapping methods [17]. Indeed, as the phenological status of different crop types evolves differently with time, having information over time allows to better discriminate the crops. Therefore, the use of optical data over large and cloudy areas involves working with composite images (including sometimes spatiotemporal context for residual missing pixels in composite images [18,19]), linearly temporally gap-filled images [20], or pixel-wise weighted least-squares smoothing of the values over time [21,22].

The advantages of SAR images allow meeting the rigid data requirements of operational crop monitoring in the CAP policy context. Unlike optical data, SAR images are not affected by the presence of clouds or haze. The SAR sensors can obtain data during both day and night [23]. As a result, temporal series of high spatial resolution can be regularly and reliably recorded throughout a growing season, whereas optical acquisition is never guaranteed. Agricultural crop mapping performs better with regular information during the whole growing season. Another reason to use microwaves is that they can penetrate more deeply into vegetation than optical wavelengths. SAR signal is very sensitive to plant water content. Polarized microwaves respond differently to shapes and orientations of scattering elements of the plant canopy. Such interactions lead to differences both in the backscattered power in those different polarizations and in the degree of penetration through the canopy. Moreover, the Sentinel-1 sensors present a high temporal resolution and a dual-polarization mode.

Before the launch of Sentinel-1, a number of research works have been carried out to use satellite radar images for crop classification, with different bands of acquisition—L-band [24,25], C-band [24,26,27] (and different polarizations), VV, HH, and HV [27]. The C-band Sentinel-1 SAR data have been analyzed temporally to recognize which agricultural crops grow in fields [28–34]. In [32], the authors classified 14 crop types in Denmark from SAR Sentinel-1 data time series with an average pixel-base accuracy of 86%.

Studies combining both optical and SAR sensors showed that the synergistic use of radar and optical data for crop classification led to richer information increasing classification accuracies compared to optical-only classification [22,35–38]. In [39], the authors used a deep learning-based architecture for crop classification using Landsat-8 and Sentinel-1 data and obtained accuracies of more than 85% for five major crops.

While other initiatives already used Sentinel-1 data to classify crops, the main innovation of this research is to highlight which choices lead to the crop map with the highest quality, while discussing the constraints associated to them (e.g., using a longer period implies that the crop map is only available later in the year). This allows to choose the right balance between classification accuracy and the operational interest of keeping the model simple. More specifically, this study documents the effects on the classification performance of different choices that can be made for five parameters: the SAR orbit(related to Ascending or Descending modes), the set of explanatory variables, the quality and size of the training sample, the use of different internal buffers on parcel polygons before SAR signal extraction, and the selected period of the time series.

In order to compare the different models, a statistical analysis is conducted to assess whether the performance differences are significant or solely due to the random fluctuations in the results.

The presence of rain or snow during SAR acquisitions, which is a major constraint of using a SAR dataset, is also discussed. Finally, this research analyses how the classification accuracy depends on some characteristics of the parcels such as their size, shape, and classification confidence level. This allows to assess whether the classification of a parcel is relevant depending on that characteristic (e.g., is the classification relevant for small parcels). Furthermore, considering a threshold on those characteristics allows to identify a subset of the parcels for which the global classification accuracy is higher.

## 2. Materials and Methods

The study is carried out in the context of the preparation by the Walloon region of their implementation of the Area Monitoring System. The cloud frequency in Belgium is quite significant, and the availability of cloud free image time series over the growing season can be a challenge.

This section first presents the study area and the data used for this research. The data used include the Land Parcel Identification System (LPIS) and the Sentinel-1 images. Then, the method is developed.

### 2.1. Study Area

The Walloon Region (WR) is located in the south part of Belgium (in grey on Figure 1) and covers 16,901 square kilometers.

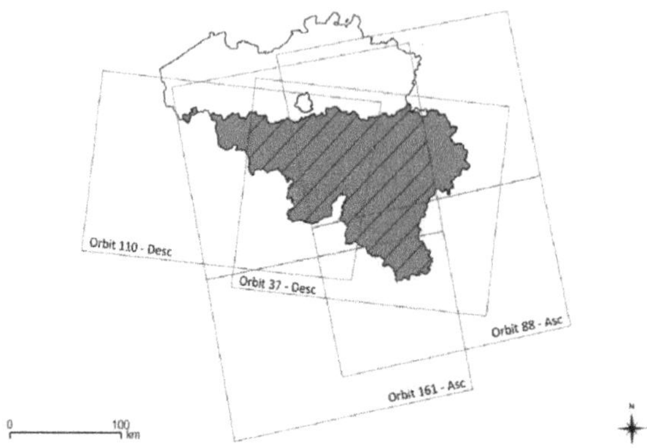


**Figure 1.** Walloon Region (South part of Belgium) and the Sentinel-1 orbits covering the biggest parts of the Walloon Region.

In this region, agricultural land represents about half of the total area. As shown on Figure 1, the WR is covered by several Sentinel-1 orbits (see below for details), out of which only two will be used in the present study: 37—Descending and 161—Ascending.

Because those two orbits will be compared and used together, only the part of the WR that is covered by both orbits is considered in the study. It corresponds to the hatched part on Figure 1, which is our study area.

## 2.2. Data

### 2.2.1. Land Parcel Identification System (LPIS)

The Land Parcel Identification System (LPIS) was designed as the main instrument for the implementation of the CAP's first pillar, whereby direct payments are made to the farmer once the land and area eligible for payments have been identified and quantified.

The Walloon LPIS is maintained and updated each year using aerial orthoimages with a spatial resolution of 25 cm. In the context of the CAP, applicants use the LPIS and available RS image to indicate the location of the parcels and the total area under cultivation of a particular crop. In the WR, the farmers do the applications online. The LPIS includes a shapefile with polygons representing the delineation of the parcels as declared by the farmers each year. The declared crop must correspond to the crop type present on field on the 31st of May of each delineated plot. In this study, the 2020 LPIS is used.

At the WR level, the farmers 2020 declarations contain 285,776 parcels, which cover a total area of approximately 764,000 hectares. Only 269,278 parcels, referred as "agricultural field" used for food or feed production and covering 742,173 ha, are considered. Among the discarded parcels, one finds for instance fields that are managed for horticulture or environmental concern only, such as ornamental crops or 6 m wide grassed headlands along watercourses. For this study we considered the part of the WR covered by both orbits 37 and 161, which contains 211,875 agricultural fields and covers 617,077 hectares.

The crop types of the same family (e.g., grain maize and fodder maize) are clustered into different crop groups. These groups were defined by the Walloon authorities in the context of the greening of the CAP in 2013. The selected parcels located in the study area represent 48 crop groups. These 48 crop groups exclude the crop groups with less than 7 occurrences (as consequence 18 agricultural fields with exceedingly rare crop types were not included).

The most represented agricultural groups in our study area are grassland, winter wheat, maize, potato, and sugar beet, as detailed on Table 1.

**Table 1.** Most represented agricultural groups, with their relative appearance in terms of number of fields and of area.

| Agricultural Group | % of Number of Fields | % of Area |
|---|---|---|
| Grassland | 55.7% | 46.8% |
| Winter wheat | 11.31% | 17.1% |
| Maize | 9.9% | 8.5% |
| Potato | 3.8% | 5.9% |
| Sugar beet | 3.6% | 5.4% |

#### 2.2.2. Sentinel-1 Images-Backscattering Coefficient Time Series (8 Months from January until August 2020)

The Sentinel-1 mission is composed of a constellation of two identical satellites performing C-band Synthetic Aperture Radar (SAR) imaging at 5.6 GHz (5.4 cm wavelength), with an effective revisit time of 12 days (6 days considering both satellites).

Interferometric Wide Swath (IW) Mode is the main operational mode over land and features a 5-by-20-metre spatial resolution and a 250 km swath. The level 1 products made up of Single Look Complex (SLC) and Ground Range-Detected (GRD) outputs in single (HH or VV) or double (HH + HV or VV + VH) polarization.

Four different Sentinel-1 orbits (two descending and two ascending) are covering the majority of the WR area. The hours of acquisition are around 5:30 p.m. and 6:00 a.m. (UTC) for Ascending and Descending orbits, respectively. In order to quantify the impact of different local incidence angle and acquisition time for this study, we considered only data coming from one Ascending orbit (161) and one Descending orbit (37).

GRD products (resolution of 20 × 22 m) are pre-processed to obtain calibrated, geocoded backscattering coefficients sigma nought ($\sigma 0$) using the European Space Agency's SNAP Sentinel-1 toolbox software. The GRD SAR data pre-processing chain include geometric and radiometric calibration, with correction for the local incidence angle, using the Shuttle Radar Topography Mission (STRM 90) digital elevation model (DEM). The images are interpolated at a 6.5 × 6.5 m pixel size during the projection in the UTM-31N coordinate system. The 6.5 m pixel size was considered in order to include in our data set a maximum of small fields, with tiny shapes.

Time series of GRD Sentinel-1 data acquired during the considered period, from January until August 2020, are used to extract explanatory variables for the crop classification of this study. The considered period begins in January as crops are in place in more than half of the parcels, i.e., in grasslands and winter crops, at this time of the year. It ends in August when the winter crops are already harvested but not the spring crops. The reason to end the period no later than August is that, in the operational context of the CAP, a crop map is needed by the Paying Agencies in early autumn to pay the famers. For each orbit, using both Sentinel-1 satellites, one acquisition is available every 6 days. The first available acquisition for orbit 161 is on 12 January 2020 (the 6 January acquisition is missing because of a technical problem on a Sentinel satellite). Since the study aims at comparing orbits 37 and 161, the first acquisition of orbit 37 is taken to be the closest one to the first orbit 161 acquisition, which is on 10 January. For the same reason of keeping similar acquisition dates for the two orbits, the orbit 161 acquisition on 2 September is also used (however, we still refer to the period as extending from January to August). In total, this gives 40 acquisition dates per orbit.

### 2.3. Methodology

#### 2.3.1. Methodology Overview

This research describes the effect on the classification performance of 5 parameters: the set of explanatory variables, the SAR orbit (related to local incidence angle and acquisition time), the period of the year covered by the time series, the use of different internal buffers on parcel polygons before SAR signal extraction, and the training set characteristics (size of the set and specific parcels kept in the set). To this aim, several scenarios are defined, each corresponding to a different set of choices made for each of those parameters. For all the scenarios, the classification model is chosen to be a random forest. The idea is then to compare the accuracy of the classification model of each scenario using cross-validation. In order to take into account the random fluctuations of the performances due to the inherent randomness of the model calibration and the calibration/validation sampling, 10 random forest models are built for each scenario providing 10 different results. A statistical analysis is conducted to test the significance of the score differences.

In the following, the parameters defining the scenarios are first detailed. Then, the classification model is presented, together with the training and validation procedure. Finally, the statistical analysis is explained.

#### 2.3.2. Parameters and Scenarios

Here follow the possible choices made for the 5 parameters, which define the different scenarios.

Parameter 1: the SAR input dataset

The explanatory variables used to build the classification model are computed from data coming either from the orbit 37 only, or from the orbit 161 only, or from both orbits 37 and 161. This impacts the number of images used during a considered period and the acquisition time of the images.

Comparing the results obtained from both orbits and from a single one allows to know if a higher temporal resolution (by doubling the number of acquisitions for a considered period) improves the quality of the crop type identification.

Images are acquired at different times depending on the orbit: early in the morning or late in the afternoon. The interest of testing each orbit separately is also to assess whether the acquisition time is important. For instance, images acquired early in the morning could be affected by dew.

Parameter 2: the explanatory variables

The starting idea is that the different crop types can be distinguished using temporal profiles of spectral features extracted from the SAR signal, these features are the explanatory variables of the model to be calibrated. Object-based features are preferred over pixel-based features because of the speckle effect, which needs to be filtered out or averaged out by working at object level. In the context of this study, this is possible since the parcel boundaries are known, and this is also true in each member states in the context of the CAP.

In this study, we have used the per-field mean backscattering coefficient ($\sigma 0$), which is extracted for each parcel and each available date. The mean is computed as the average over the pixels whose center falls inside the parcel polygons after internal buffer application. Notice that all computations are done in linear units. This is done for both VV and VH polarizations. For each parcel, this gives two time series per orbit, which are used in all scenarios.

In addition to VV and VH, the effect of adding a third time series per orbit, called VRAT, is considered. It is defined as the ratio VH/VV. Indeed, in [40], the authors recommend using VH/VV to separate maize, soybean, and sunflower, which could be applied to other crop types. This quantity is also considered as a variable sensitive to crop growth in several other studies [41–43] and thus seems to be possibly a variable of interest for our purpose of crops classification.

Finally, the effect of adding some static variables is studied. For each orbit, the per-field mean of the local incidence angle is computed, together with its standard deviation. This is motivated by the fact that all types of cultural practices—that differ depending on the crop type—cannot be carried on in steep parcels and inhomogeneous reliefs. Hence, the local incidence angle and its intra-field variation could be correlated to the crop type cultivated in a parcel. Since the incidence angle is an almost constant value in time, we take it at a unique date only, which is on 3 May 2020 for orbit 37 and on 29 April 2020 for orbit 161. Some temporal statistics are also added: the mean, the minimum, the maximum, the range, the standard deviation, and the variance are computed for each of the VV, VH, and VRAT time series. For each orbit, there are thus 6 static variables for each time series and 2 for the local incidence angle.

The results obtained with and without these additional variables are compared in order to assess whether they are worth the additional model complexity and data preparation.

Parameter 3: the buffer size

The noise in the time series caused by heterogeneous pixels on the borders of parcels might be reduced by introducing an internal buffer before computing the per-field mean of the backscattering coefficient, thus keeping only clean pixels to represent the parcel. In some cases, it might also mitigate the negative effect of the imperfect geometric accuracy of the image. As was suggested in [44] in the context of a Sentinel-2 analysis, a buffer size of −5 m is chosen in the present study. Another possible source of noise is the fact that the edges of the fields are sometimes managed differently from the central part of the fields (different machines passage orientation or inputs interdiction for example), which causes heterogeneity in the parcel signal. From that point of view, a bigger buffer size of −15 m might ensure a higher probability of having homogeneous pixels for a given parcel. However, a bigger buffer size implies that less pixels are used to average the parcel signal, which is less efficient to filter out the speckle. Therefore, the three following buffer sizes are tested in this study: 0 m (no buffer), −5 m and −15 m. In the −5 m scenario, no buffer is used on the parcels whose polygon after the buffer application is either empty or too thin to contain any pixel (1% of the parcels). In the −15 m scenario, for the parcels whose polygon after the −15 m buffer application is either empty or too thin to contain any pixel, the polygon from the −5 m buffer scenario is used (−5 m buffer for 7% of the parcels and no buffer for 1%). Notice that even when no buffer is applied, some parcels are too thin to contain the center of a SAR pixel, which implies that no signal can be extracted (0.007% of the parcels). Moreover, notice that, because of the "cascading" buffer application, the use of a buffer is more computationally demanding. Indeed, the extractions must be done for the buffer for all the parcels, then the extractions must be done for all smaller buffers for subsets of the parcels, since the number of pixels contained in a parcel is only known after the extraction. The percentage of parcels corresponding to each buffer is summarized in Table 2.

**Table 2.** Total number of parcels and percentage of parcels for which a buffer of 0 m, −5 m, and −15 m is applied for each buffer scenario.

| Scenario | N_tot | %_buf00 | %_buf05 | %_buf15 | No Data |
|---|---|---|---|---|---|
| Buffer 0 m | 211,875 | 100% | - | - | 0.007% |
| Buffer −5 m | 211,875 | 1% | 99 % | - | 0.007% |
| Buffer −15 m | 211,875 | 1% | 7% | 92% | 0.007% |

Parameter 4: training set

The set of all parcels is split in two disjoint halves, one being used to train the model and the other to score it. In order to improve the model quality, it is trained only on a subset of the training parcels, considered as having a good signal quality. For instance, small parcels are discarded since, because of the statistical uncertainties encountered in the SAR signal due to speckle, [45] recommends averaging the signal over a certain number of pixels. Therefore, a 'clean' training dataset is built by selecting only the training parcels which contain at least 200 pixels for all dates after applying the buffer. Notice that in preliminary exploratory studies, some other criteria were considered. For instance, oddly shaped parcels were discarded based on the value of the ratio perimeter over square root of the area. Moreover, the parcel homogeneity was quantified in terms of the per-field standard deviation of the SAR sigma nought. However, these criteria were not retained since they did not seem to lead to significant improvements. The clean dataset contains 64.5% of the parcels in the full training set, and it is used in all the scenarios, unless otherwise specified.

To assess the effect of using such a strict screening, a specific scenario is considered that uses the full training dataset to check if the score is then indeed lower.

On another vein, in some situations, only a very small training dataset might be available. In order to investigate the impact on the classification accuracy of a drastically limited training dataset, another scenario is considered where the model is trained using only 10% of the clean dataset. This leads to 3.2% of the full dataset and corresponds to about 6850 parcels. Such a calibration dataset is of the order of magnitude of a large field campaign feasible on the ground.

Parameter 5: the period of SAR acquisitions

This study compares different periods of the year for the time series used by the classification model. First, four periods are defined as starting in January and having different lengths: until May (5 months), June (6 months), July (7 months), or August (8 months). In this case, a shorter period would have the operational advantage of providing classification results earlier in the year, before the harvest time. In addition, 7 periods are defined as ending in August and beginning on different months: from February (7 months) to August (1 month). In this case, the operational advantage of a shorter period would be to deal with a smaller amount of data, which eases the procedure.

Scenarios

Table 3 gives a summary of all the scenarios that are considered, listing the corresponding choices made for each parameter.

**Table 3.** Values of the parameters for the different scenarios.

| | Orbits | Expl. Var. | Period | Buffer | Training Set | Dates per Orbit | Nfeatures | %cal | %val |
|---|---|---|---|---|---|---|---|---|---|
| 1 (reference) | 37 + 161 | all | Jan–Aug | −5 m | clean | 40 | 280 | 32% | 50% |
| 2 | 37 | all | Jan–Aug | −5 m | clean | 40 | 140 | 32% | 50% |
| 3 | 161 | all | Jan–Aug | −5 m | clean | 40 | 140 | 32% | 50% |
| 4 | 37 + 161 | all | Jan–Aug | 0 m | clean | 40 | 280 | 32% | 50% |
| 5 | 37 + 161 | all | Jan–Aug | −15 m | clean | 40 | 280 | 32% | 50% |
| 6 | 37 + 161 | VV + VH + VRAT | Jan–Aug | −5 m | clean | 40 | 240 | 32% | 50% |
| 7 | 37 + 161 | VV + VH + static | Jan–Aug | −5 m | clean | 40 | 188 | 32% | 50% |
| 8 | 37 + 161 | all | Jan–Aug | −5 m | full | 40 | 280 | 50% | 50% |
| 9 | 37 + 161 | all | Jan–Aug | −5 m | 10% clean | 40 | 280 | 3.20% | 50% |
| 10 | 37 + 161 | all | Jan–May | −5 m | clean | 24 | 184 | 32% | 50% |
| 11 | 37 + 161 | all | Jan–Jun | −5 m | clean | 29 | 214 | 32% | 50% |
| 12 | 37 + 161 | all | Jan–Jul | −5 m | clean | 34 | 244 | 32% | 50% |
| 13 | 37 + 161 | all | Feb–Aug | −5m | clean | 36 | 256 | 32% | 50% |
| 14 | 37 + 161 | all | Mar–Aug | −5m | clean | 31 | 226 | 32% | 50% |
| 15 | 37 + 161 | all | Apr–Aug | −5m | clean | 26 | 196 | 32% | 50% |
| 16 | 37 + 161 | all | May–Aug | −5 m | clean | 21 | 166 | 32% | 50% |
| 17 | 37 + 161 | all | Jun–Aug | −5 m | clean | 16 | 136 | 32% | 50% |
| 18 | 37 + 161 | all | Jul–Aug | −5 m | clean | 11 | 106 | 32% | 50% |
| 19 | 37 + 161 | all | Aug–Aug | −5 m | clean | 6 | 76 | 32% | 50% |

In the training set column, "full" means that the full calibration set is used, without any quality filtering, and "10% clean" means that only 10% of the clean parcels are used. The number of acquisition dates for each orbit is also given. Nfeatures is the resulting number of features (i.e., explanatory variables) attached to each parcel that will feed the classification model. For instance, in the reference scenario, for each orbit (37 and 161), there are 2 features for the local incidence angle, plus, for each time series (VV, VH, and VRAT), 40 SAR images and 6 temporal statistics (mean, min, max, std, var, and range). This gives a total of 2*(2 + 3*(40 + 6)) = 280 features. The last two columns give the relative size of the calibration and validation sets compared to the total number (211,875) of considered parcels in the study area.

#### 2.3.3. Classification Model

For our purpose, supervised machine learning is a natural choice since a lot of reference data are available from the farmers' declaration. Dealing with a large input dataset and a big number of features (but not big enough to require neural networks), a random forest (RF) classifier is chosen as the classification model. Moreover, previous experience carried out in the benchmarking of Support Vector Machine (SVM), decision trees, gradient boosted trees, and RF models in the context of crop classification [20] showed that RF performs generally better, even if SVM is expected to provide better classification results in the specific case of classes with few calibration samples.

The RF is implemented and trained using the python scikit-learn package. Because the goal of this study is not to optimize the machine learning model itself, the default hyper-parameters are kept. Exception is made on the number of trees, which is fixed to 250 in order to reduce the training time without significant performance drop.

Cross-validation is used to assess the model's performance (full details are given in Section 2.3.4). As previously mentioned, the set of considered parcels consists in those agricultural fields of the LPIS that are located under both orbits 37 and 161. When training a RF model, this set is randomly split in two disjoint equally sized subsets, a validation dataset and a training dataset. The splitting is stratified, that is, each crop group is equally represented in both datasets.

The training dataset is used to calibrate the RF model and is totally independent of the validation dataset.

The validation dataset is used to compute two overall performance scores for the RF: OAnum, the overall accuracy based on the number of parcels (percentage of well-classified parcels compared to the total number of parcels), and OAarea, the overall accuracy based on the area (percentage of well-classified hectares).

In addition, the F1-score of each crop group is used. In this multi-class classification context, it is defined as the harmonic mean of the group precision and recall (whose definitions are standard):

$$F1(group = g) = 2\frac{Precision(group = g)\ *\ Recall(group = g)}{Precision(group = g)\ +\ Recall(group = g)}. \quad (1)$$

For each classified parcel, a confidence level is also assigned to the predicted class. It corresponds to the percentage of trees in the RF that predict the given class.

Notice that the few parcels for which no signal can be extracted because they are too small (see Table 2) are considered as misclassified in the validation set. However, this has a negligeable effect on the scores since they represent 0.007% of the parcels (0.0001% in terms of area).

#### 2.3.4. Test of Statistical Significance

There are several sources of randomness in the building of the RF classification model corresponding to each scenario: the training/validation split and several random choices in the RF initialization and training. This leads to some randomness in their performance scores, and therefore, the best scenario for a given training/validation split might not be the same for another split. This is especially true given that, as will be seen in the next section, the scores are sometimes very close. A test of statistical significance is therefore conducted in order to assess whether the difference in score for two scenarios is significant or if it might just be due to a random fluctuation.

The idea is to define several pairs of training/validation datasets and, for each scenario, to build several RFs, one for each pair. This gives an insight on how strongly the score of each scenario fluctuates. However, naively estimating the variance by computing the standard deviation of the scores is not recommended because the latter are strongly correlated (typically, the training datasets of each pair overlap and are thus not independent (same for the validation datasets)). Using a very high number of pairs would lead to an underestimated variance. This is therefore a complex question, to which many answers are proposed in the literature, some of them being summarized and compared in [46]. Given the situation encountered in this paper, the 5 × 2 cross-validation F-test introduced in [47] is chosen—which is a slight modification of the popular 5 × 2 cross-validation *t*-test originally defined in [48]. Here is a quick overview of the procedure. As previously mentioned, the goal is to repeat the training/validation several times for each scenario to estimate the score fluctuations, while finding a right balance between having a high number of repetitions, lowering the overlapping between all the training datasets (same for the validation datasets), and keeping enough data in each set. First, the full dataset is randomly split in two equally sized stratified parts, and this is independently repeated 5 times. This leads to 5 pairs of 50–50% disjoint subsets of the full dataset. Then, for each scenario and each pair, two classification models are built: the first one using one part as the training set and the other part as the validation set, and the second one by switching the role of each subset. For each scenario, this gives thus 10 performance scores. When comparing two scenarios, a standard hypothesis test is conducted, taking as null hypothesis that they both lead to the same score. From the 20 scores, a number t is computed—its exact definition can be found in [47]—and the author argues that its statistic approximately follows a F-distribution, allowing to compute the corresponding $p$-value. If this $p$-value is higher than a chosen threshold $\alpha$, the null hypothesis that the scenario scores are equal can be rejected at the corresponding confidence level. Now, the fact that several scenarios are compared has to be taken into account in order to get the expected $p$-value for the multiple comparisons altogether. The Bonferroni correction is chosen, which suggests using $\alpha$ divided by the total number of comparisons as the $p$-value threshold of each individual comparison. In this study, $\alpha$ is fixed to 0.05.

Since the goal of this study is to evaluate the effect on the classification performance of each parameter separately, there is no need to compare all the scenarios together. Instead, in order to reduce the number of comparisons, a reference scenario is chosen, against which all the other ones are compared. It corresponds to the scenario with both orbits 37 and 161 considered, all the explanatory variables (VV, VH, and VRAT time series + static variables), the 8 months period, a −5 m buffer, and the clean training set.

As a final note, it should be reminded that all these statistical tests are based on approximate hypotheses and that the numbers should therefore not be blindly taken as an absolute truth.

## 3. Results

This section first includes a general overview of the results in terms of the overall accuracies. A deeper analysis is then given in terms of the F1-scores of each crop groups. Finally, the classification performance is analyzed considering the scenario's parameters described in the previous section.

### 3.1. General Results of the Classifications Using SAR Data

Table 4 summarizes the performance scores for each scenario. For each of them, the values of the parameters described in the previous section are given. The scenario 1 is the reference scenario mentioned earlier. The overall accuracies (OA) correspond to the mean of the OAs of each of the 10 models and are supplemented by their standard deviation. Two kinds of Overall Accuracies are computed: the OAarea which is equal to the percentage of well classified area and the OAnum which is equal to the percentage of the number of well classified fields. This gives two different points of view when comparing the scenarios. In the following, OAarea is generally used, unless when OAnum is more relevant.

**Table 4.** Results of the classifications using different SAR datasets, different sizes for the internal buffer applied to the parcels geometry, different sets of explanatory variables, different numbers of parcels in the training set, and different periods of SAR data acquisition. (The significance includes a Bonferroni correction.)

| Scenario | Orbits | Expl. Var. | Period | Buffer | Training Set | OAarea (%) | *p*-Value | Significance | OAnum (%) |
|---|---|---|---|---|---|---|---|---|---|
| 1 (reference) | 37 + 161 | All | Jan–Aug | −5 m | clean | 93.43 ± 0.06 | | | 88.4 ± 0.03 |
| 2 | 37 | All | Jan–Aug | −5 m | clean | 92.76 ± 0.07 | $5.89 \times 10^{-5}$ | TRUE | 88.03 ± 0.05 |
| 3 | 161 | All | Jan–Aug | −5 m | clean | 92.8 ± 0.07 | $4.86 \times 10^{-5}$ | TRUE | 87.66 ± 0.04 |
| 4 | 37 + 161 | All | Jan–Aug | 0 m | clean | 93.16 ± 0.06 | $7.49 \times 10^{-4}$ | TRUE | 88.52 ± 0.03 |
| 5 | 37 + 161 | All | Jan–Aug | −15 m | clean | 93.42 ± 0.07 | $6.56 \times 10^{-1}$ | FALSE | 87.56 ± 0.06 |
| 6 | 37 + 161 | VV + VH + VRAT | Jan–Aug | −5 m | clean | 93.46 ± 0.04 | $3.77 \times 10^{-1}$ | FALSE | 89.33 ± 0.04 |
| 7 | 37 + 161 | VV + VH + static | Jan–Aug | −5 m | clean | 92.97 ± 0.05 | $1.60 \times 10^{-4}$ | TRUE | 87.35 ± 0.05 |
| 8 | 37 + 161 | All | Jan–Aug | −5 m | full | 93.08 ± 0.09 | $4.41 \times 10^{-4}$ | TRUE | 89.24 ± 0.05 |
| 9 | 37 + 161 | All | Jan–Aug | −5 m | 10% clean | 91.65 ± 0.13 | $7.53 \times 10^{-6}$ | TRUE | 86.83 ± 0.14 |
| 10 | 37 + 161 | All | Jan–May | −5 m | clean | 88.2 ± 0.10 | $2.90 \times 10^{-8}$ | TRUE | 84.04 ± 0.08 |
| 11 | 37 + 161 | All | Jan–Jun | −5 m | clean | 91.52 ± 0.07 | $2.06 \times 10^{-6}$ | TRUE | 86.93 ± 0.06 |
| 12 | 37 + 161 | All | Jan–Jul | −5 m | clean | 93.16 ± 0.06 | $4.46 \times 10^{-5}$ | TRUE | 88.27 ± 0.06 |
| 13 | 37 + 161 | All | Feb–Aug | −5 m | clean | 93.48 ± 0.07 | $1.12 \times 10^{-1}$ | FALSE | 88.24 ± 0.05 |
| 14 | 37 + 161 | All | Mar–Aug | −5 m | clean | 93.54 ± 0.05 | $9.93 \times 10^{-3}$ | FALSE | 87.94 ± 0.06 |
| 15 | 37 + 161 | All | Apr–Aug | −5 m | clean | 93.53 ± 0.06 | $8.31 \times 10^{-2}$ | FALSE | 87.21 ± 0.12 |
| 16 | 37 + 161 | All | May–Aug | −5 m | clean | 93.32 ± 0.07 | $7.82 \times 10^{-2}$ | FALSE | 87.17 ± 0.08 |
| 17 | 37 + 161 | All | Jun–Aug | −5 m | clean | 92.56 ± 0.07 | $4.13 \times 10^{-5}$ | TRUE | 85.47 ± 0.05 |
| 18 | 37 + 161 | All | Jul–Aug | −5 m | clean | 90.76 ± 0.08 | $1.09 \times 10^{-6}$ | TRUE | 83.1 ± 0.09 |
| 19 | 37 + 161 | All | Aug–Aug | −5 m | clean | 84.34 ± 0.06 | $2.69 \times 10^{-9}$ | TRUE | 77.93 ± 0.05 |

A first key finding of this study shows that for most scenarios, the scores are very close, which is the motivation to investigate whether the score differences are statistically significant. In the tables, the $p$-value corresponds to the aforementioned hypothesis testing that the scenario score (OAarea) is different than the reference one. The significance column indicates whether that $p$-value is lower than the chosen Bonferroni corrected threshold, i.e., 0.05/18 (18 being the number of comparisons made).

OAarea is always higher than 90% except for the two worst scenarios. For the reference scenario, OAarea reaches 93.43% (both orbits are used, the longest period is considered, all the explanatory variables are included, the clean training set is used, and an internal buffer of −5 m is applied). This is the best score, together with the other six top scores as they are not statistically different from the reference's one. These scores correspond to the scenarios 5 (−15 m buffer and all variables) and 6 (−5 m buffer and all variables excepted static variables) and to the four scenarios whose period begins in February, March, April, and May.

This method reaches high performance compared to other studies. Indeed, in the optical domain, in [49], the authors reach an overall accuracy of 83% using a random forest classifier for 14 classes from Landsat surface reflectance in Yolo County, California. In [3], the authors discriminated the 5 main crops in three entire countries with overall accuracies higher than 80% using Sentinel-2 and Landsat time series. In the SAR domain, in [32], the authors classified 14 crop types (including grasslands) in Denmark (over 254 thousand hectares) from SAR Sentinel-1 data time series with an average pixel-base accuracy of 86%. In [22], the authors predicted 8 crop types with a maximum accuracy of 82% (OAnum) with a pixel-based random forest classifier and Sentinel-1 and Sentinel-2 dataset over Belgium's agricultural land that covers more than 1,300,000 hectares. In [39], the authors reach an overall accuracy of 94.6% using 2-d CNNs model for classifying land cover into 11 classes including 8 agricultural classes in Ukraine from LandSAT and Sentinel-1A images. However, regarding their high score, it must be noticed that this study discriminates less classes of crop types and does not focus only on agricultural classes. It includes the classes "water" and "forest", whose User and Producer Accuracies are very high (near 100). This pushes the overall accuracy higher than if only agricultural crop types were considered.

A first look at the numbers indicates that the four parameters that give some noticeable performance differences are the orbit, the period, the presence of the VRAT in the explanatory variables, and the training sample. Further comments will be made later in the "Impact of the scenarios' parameters" subsection.

### 3.2. F1-Scores of the Crop Groups

Table 5 shows the F1-scores (in terms of the number of parcels, not the area) for each crop group, sorted by their prevalence. The prevalence is both given as the percentage of the number of parcels of the group in the validation dataset (Num) and as the percentage of the area covered by the parcels of the group (Area). Recall that, for each scenario, 10 classification models are computed, corresponding to different pairs of training/validation datasets. The prevalence is (almost) the same for all of them, but the F1-score does vary. The F1-score given in the table is the mean of the F1-scores corresponding to each model. Some crop groups, although present in the validation dataset, are never predicted by some models, and their F1-scores are then set to 'NaN'. The column 'no grasslands' will be described later.

In the reference scenario, the 6 most represented crop groups have an F1-score higher than 84%. It is actually higher than 90%, except for "Maize". In that case, its recall is high, but its F1-score is lowered by its low precision. This is because a large proportion of mis-classified parcels are classified as Maize (this is also the case for the grasslands, but it does not impact their precision as much because of their higher prevalence). Nine crop groups get F1-scores higher than 84%, and they represent 89.5% of the parcels (91.6% of the area).

Table 5 shows that "sugar beet" has the highest F1-score, which is equal to 95.52%. Moreover, F1-scores tend to be higher for the most represented crop groups, always higher than or equal to 83.97% for the six most represented groups. For the groups with lower prevalence, there is no clear relation between the prevalence and the F1-score. It is interesting to notice that three crop groups with a low prevalence get a high F1-score: winter rape, spring pea, and chicory. Figure 2 shows the temporal evolution of the mean of the backscattering coefficient over all winter rape parcels (in red), in comparison with the same profile evolution for all the other agricultural parcels (in grey). This figure highlights a specific behavior of VH for the winter rape during the end of the spring. The winter rape is indeed different from the other winter crops at this period. The end of the spring corresponds to the end of the flowering, the steam elongation and branching, and the pod and seed (fruit) development and ripening of winter rape. The specific geometry of the winter rape with a high stem and a lot of branches at this period certainly explains the higher VH signal since this crop exhibits more non vertical elements compared to the other winter crops.

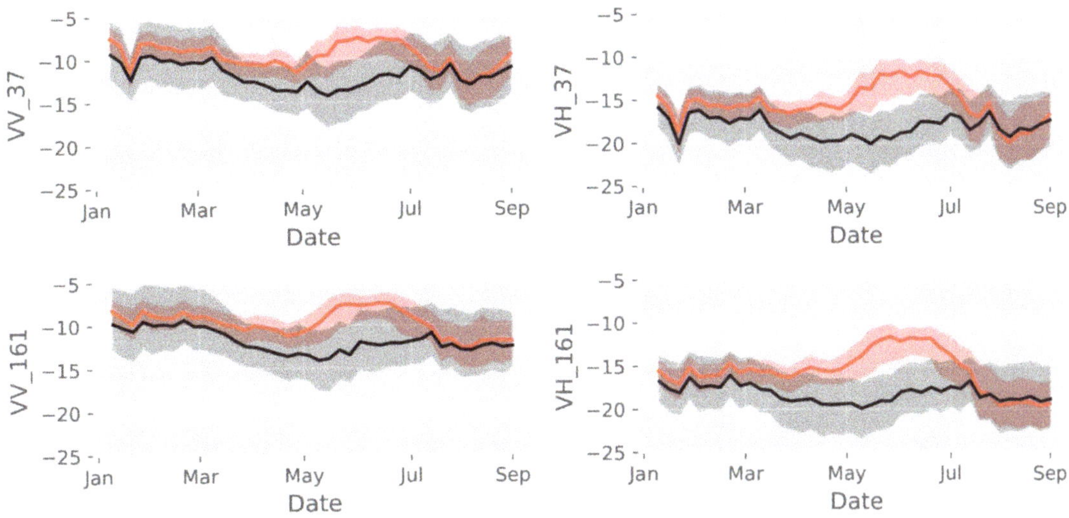


**Figure 2.** Evolution of the mean and interval between percentiles 5 and 95 of the backscattering coefficient for winter rape (in red) and all the crops (in grey).

Now let us look at some low F1-scores. The first three groups which get a low or no F1-score are "Alfalfa", "Fallows", and "Others". These groups are often classified as grasslands: 86%, 64%, and 55% of the parcels for "Alfalfa", "Fallows", and "Others", respectively. This can be explained by the fact that, although these groups are considered as distinct in the farmers' declarations, the reality on the field is very similar between these three groups. First, the group "Others" includes fodder crops (56% of the parcels of the group) that can be grasslands. Secondly, an alfalfa grassland can be claimed by the farmer either as alfalfa or as temporary grassland. Thirdly, the "Fallows" group includes herbaceous fallows that, on the field, are crop cover similar to grasslands, even if their management differs.

Regarding the bad score of the group "Others", it should also be noticed that it includes 13 different crop types, sometimes very dissimilar such as greenhouse vegetables, oleaginous or proteaginous crops. Therefore, that group does not really makes sense from an EO point of view. In light of the previous remark on grasslands, it is therefore interesting to notice that this bad score is in fact due to a strength of the classification model: its ability to recognize grasslands within a group of parcels of heterogeneous crop types.

**Table 5.** F1-score obtained per crop group when considering all crop groups, including grasslands (left), and when considering arable lands only (excluding grasslands) in the training and validation datasets (right).

| Crop Groups | Prevalence (%) | | F1-Scores (%) Scenario 1 | |
|---|---|---|---|---|
| | Num. | Area | All Groups | No Grasslands |
| Grassland | 55.7 | 46.8 | 94.51 | N– |
| Winter wheat | 11.3 | 17.1 | 90.3 | 90.66 |
| Maize | 9.9 | 8.5 | 83.97 | 88.49 |
| Potato | 3.8 | 5.9 | 93.24 | 93.97 |
| Sugar beet | 3.6 | 5.4 | 95.52 | 96.43 |
| Winter barley | 2.9 | 3.8 | 92.38 | 94.18 |
| Spelt | 2.1 | 2.2 | 49.13 | 51.19 |
| Others | 1.8 | 0.9 | 1.11 | 30.01 |
| Mixed cereals | 1.3 | 1.2 | 40.61 | 46.48 |
| Fallows | 1 | 0.2 | NaN | 0.16 |
| Winter rape | 0.8 | 1.2 | 92.6 | 95.99 |
| Spring pea | 0.8 | 1.6 | 91.24 | 91.52 |
| Alfalfa | 0.8 | 0.5 | 2 | 51.12 |
| Chicory | 0.7 | 1.2 | 87.44 | 87.43 |
| Spring oats | 0.7 | 0.5 | 61.34 | 67.02 |
| Permanent crop | 0.6 | 0.5 | 32.65 | 45.77 |
| Winter triticale | 0.4 | 0.5 | 39.7 | 42.84 |
| Spring barley | 0.4 | 0.4 | 66.23 | 70.29 |
| Spring wheat | 0.2 | 0.2 | 19.67 | 24.78 |
| Bean | 0.2 | 0.3 | 76.67 | 79.58 |
| Trefoil | 0.1 | 0.1 | NaN | NaN |
| Carrot | 0.1 | 0.2 | 59.6 | 59.82 |
| Bulb | 0.1 | 0.2 | 69.98 | 69.11 |
| Strawberry | 0.1 | 0.02 | NaN | 0 |
| Winter oats | 0.1 | 0.1 | 3.45 | 4.55 |
| Spring fava bean | 0.1 | 0.1 | 20.36 | 18.53 |
| Winter rye | 0.1 | 0.1 | 5.33 | 7.82 |
| Sorghum | 0.1 | 0.03 | NaN | NaN |
| Spring triticale | 0.04 | 0.04 | NaN | 4.17 |
| Angelic | 0.03 | 0.03 | 6.29 | 6.33 |
| Cabbage | 0.02 | 0.02 | NaN | NaN |
| Spinach | 0.02 | 0.04 | 23.77 | 27.99 |
| Courgette | 0.02 | 0.01 | NaN | NaN |
| Buckwheat | 0.02 | 0.01 | NaN | NaN |
| Winter fava bean | 0.01 | 0.01 | NaN | NaN |
| Sunflower | 0.01 | 0.005 | NaN | NaN |
| Flax | 0.01 | 0.02 | 22.73 | 23.45 |
| Spring rape | 0.01 | 0.01 | NaN | NaN |
| Lupin | 0.01 | 0.01 | NaN | NaN |
| Soybean | 0.01 | 0.01 | NaN | NaN |
| Tomato | 0.01 | 0.0002 | NaN | NaN |
| Winter pea | 0.01 | 0.01 | 26.45 | 26.2 |
| Quinoa | 0.01 | 0.01 | NaN | NaN |
| Butternut | 0.01 | 0.002 | NaN | NaN |
| Hemp | 0.01 | 0.004 | NaN | NaN |
| Spring rye | 0.004 | 0.005 | NaN | NaN |
| Parsley | 0.004 | 0.003 | NaN | NaN |
| Nettle | 0.003 | 0.003 | NaN | NaN |

Nevertheless, this illustrates a weakness of using the LPIS as a training set for crop classification. Indeed, the declaration of the farmers can be ambiguous for some crop types, which can confuse the random forest classifier.

Analysis excluding the grasslands

Since the high overall accuracies are driven by the large proportion of grasslands (more than half of the number of parcels), which get a high F1-score, the classification is also done for arable land only, excluding the grasslands. This represents 93,832 parcels, from which 50% are kept as a validation dataset, and 36% are used in the calibration dataset (half of the agricultural parcels excluding grasslands that contains more than 200 pixels after applying the buffer). The other parameters are the same than those used in the reference scenario 1. As for the other scenarios, 10 classification models are computed, corresponding to different random splits of training/validation datasets. OAarea, the percentage of well-classified area, equals 90.44%, which is 3% lower than for our reference scenario 1. Nevertheless, these results give globally better F1-scores for the arable land than the results obtained when the grasslands are included (see Table 5—right column), especially for the groups "alfalfa" and "other" (the latter being the 6th most prevalent crop group). Indeed, as was mentioned earlier, the parcels in these two groups are often grassland in reality, and when grasslands are included in the training sample, these parcels are often classed as grasslands, which is not a real error but an error due to the claim system that permits to a farmer to claim an alfalfa parcel as grassland or alfalfa. The F1-score of "permanent crop" also increases. In scenario 1, 50% of such parcels are classified as grasslands, which can be explained by the fact that fruit tree crops mainly represent this group (92% of the parcels). Hence, they can be mixed up with wooded grasslands. For the other crop groups, which are groups more different from the grasslands on the ground, the F1-score does not change drastically.

### 3.3. *Impact of the Scenarios' Parameters*

In this section, the classification results are further analyzed considering the variation of the five parameters described in the "Methodology" section: the SAR orbit, the explanatory variables set, the period of the time series used for signal extraction, the use of different internal buffers on parcel polygons before SAR signal extraction, and the quality and size of the training sample.

#### 3.3.1. Impact of the SAR Orbit on the Classification Performance

The effect of using different SAR orbits to compute the explanatory variables is described hereafter.

The considered orbits are orbit 161 and orbit 37. They differ by their hours of acquisition—which are around 5:30 p.m. and 6:00 a.m. (UTC)—and their orbit orientation: ascending and descending tracks respectively. The time of acquisition has a global impact on the signal because of the presence of dew during the morning that can enhance the signal [50]. The orbit track orientation also has an impact on the signal at the parcel level since the local incidence angle is different and the relative azimuth angle between the SAR beam and the plant rows differs as well (considering the crops cultivated in rows such as maize, potato and sugar beet).

The results first compare the classification performance corresponding to three different SAR datasets:

- SAR data acquired on both orbits 37 and 161 (reference scenario 1),
- SAR data acquired on orbit 37 (scenario 2),
- SAR data acquired on orbit 161 (scenario 3).

The results show that the higher number of explanatory variables given by combining the two orbits leads to a better overall accuracy (both in terms of area and of number of parcels), but the computing cost is doubled. When using only a single orbit, the score drops by less than 1%. This is a small difference, but the significance test shows that it is not due to random fluctuations. In the specific case of determining the crop type in the context of the CAP, the best accuracy is needed in order to reduce the number of fields visits. Indeed, for 100,000 parcels to monitor by a Member State, a drop of 1% of OAnum leads to 1000 additional misclassified parcels, increasing the required number of fields visits. However, a lower score might be perfectly adequate for other applications such as regional scale agricultural statistics or yield prediction for the main crops, where reducing the computational cost might be more important.

#### 3.3.2. Impact of the Set of Explanatory Variables on the Classification Performance

In all the scenarios, the VV and VH polarizations are used as explanatory variables. The effect of using three additional kinds of explanatory variables is here analyzed: the ratio VH/VV, local incidence angle per-field statistics, and temporal statistics.

In addition to the time series corresponding to the VV and VH polarizations, a third one corresponding to the ratio VH/VV, called VRAT, is added to the explanatory variables. In scenario 7, the ratio VH/VV (called VRAT) is excluded. This decreases the OAarea, by 0.46%, and the significance test shows that the difference is significative from a statistical point of view. Regarding OAnum, the performance drops by 1%. This shows that VRAT contains some useful information that the RF cannot extract solely from the values of VV and VH. However, depending on the application, the additional model complexity and computer processing time might not be worth the performance improvement.

The influence of including some static variables has also been studied. These static variables include, for each orbit, the per-field mean of the local incidence angle, together with its standard deviation. They also include some temporal statistics: the mean, the minimum, the maximum, the range, the standard deviation, and the variance are computed for each of the VV, VH, and VRAT time series and each orbit. This gives 2+36 additional variables per orbit. In order to assess whether these variables are useful, scenario 8 tests the performances of the classification without including them to the explanatory variables. The results show that removing them does not change the classification performance. Looking at the OAnum, one can even see that it even gets better for that scenario.

#### 3.3.3. Impact of the Buffer Size on the Classification Performance

As was explained earlier, applying an internal buffer to the parcel polygons before averaging the pixel values of the SAR signal might help reducing the signal noise. This section compares the use of three different buffer sizes (scenarios 1, 4, and 5): no buffer, −5 m, and −15 m.

Besides removing mixed pixels along the parcel borders, using a higher buffer size might remove noise introduced by different agricultural practices at the edges of the parcel. On the other hand, the application of a higher buffer size reduces the number of pixels considered to compute the per field mean, which might be in contradiction to the necessity of averaging over many pixels to increase the signal consistency regarding the speckle present in the SAR signal. Since the results show no difference between applying a buffer of −5 or −15 m, these two effects seem to either be negligible or to counterbalance themselves. A high buffer should be preferred in some specific cases: poor geometric quality of the polygons or misalignment of images.

Regarding the effect of applying a buffer at all, the results show a statistically significant difference between applying a −5 m buffer and using no buffer. However, because the improvement due to the buffer is very small (0.27% in terms of area), the added complexity might not be worth it, depending on the application.

#### 3.3.4. Impact of the Quality and Size of the Training Sample on the Classification Performance

One of the keys to building a good model is to have a good and big enough ground truth dataset for the training sample.

Comparison of the scenarios 1 and 8 focuses on the quality of the training dataset. Recall that it has been chosen to train the model using a 'clean' dataset, which contains only parcels containing at least 200 pixels, since they are considered to have a cleaner signal. Such parcels represent 64% of the parcels; thus, the training dataset contains 32% of the whole number of parcels. In order to assess whether such selection is useful, in scenario 8, no filtering is done, and the calibration dataset counts thus 50% of the whole number of parcels. The results show that the cleaning of the training set leads to slightly better results, increasing the OAarea by 0.35%. As this cleaning is easy to perform and also allows to reduce the processing time, it is thus advisable to apply it to the calibration dataset before training the model. However, it must be noticed that when considering OAnum, using the full calibration dataset leads to a better score. This is more deeply discussed in the "Discussion" section of this paper.

In another vein, to assess the impact of the size of the training dataset, scenario 9 uses a smaller training dataset. From the clean dataset used in scenario 1, only 10% of the parcels are used to train the model, which is about 6850 parcels, i.e., 3.2% of the whole number of parcels. When reducing the training sample size in such a drastic way, the OAarea drops by 1.78% only, from 93.43% to 91.65%.

The high OAarea obtained in each case presents very promising results for regions where poor ground data are available and field surveys are required. Moreover, considering that the declaration of the farmers can be ambiguous for certain crop types (for instance, the confusion between fodder, grasslands, and alfalfa which was mentioned earlier) or erroneous; a field visit of 6850 parcels could, at the Walloon Region scale, be considered to improve the training dataset quality.

#### 3.3.5. Impact of the Period of the SAR Time Series on the Classification Performance

One of the significant changes to be considered for the next CAP reform is to provide Near Real Time information to the farmers all along the year concerning their agricultural practices and changes observed by remote sensing images. To assess this issue, the evolution of crop identification quality over time is also analyzed by comparing scenarios 1, 10, 11, and 12. From the results of Table 6, one can see that the overall accuracies increase with the length of the period of the SAR time series, both in terms of area and number of parcels. To have a better insight of the impact of the period length on each crop group, Table 6 gives the F1-scores of each crop group for the 4 different periods. The F1-score of grasslands reaches more than 93.7% for all considered periods. Regarding the two most represented winter crops, which are harvested in July, they are classified with a F1-score of at least 0.86% already since the end of May, but the F1-scores increase by at least 3% if the considered period includes the whole winter crops season up to the harvest. Regarding the main spring crops, the F1-scores increase strongly between the first two considered periods and then increase more slightly. The spring crops are harvested in September or October. The present analysis shows that 5 of the 6 most represented crop groups can be recognized before their harvesting dates with a F1-score higher than 89%. The maize does not reach such high score and its F1-score equals 84% about 2 months before its harvest. As explained before, the lower F1-score of maize is due to the fact that a large proportion of the mis-classified parcels are classified as maize, which lowers its precision score. As before, some crop groups are never predicted by some models, in which case their F1-scores are set to 'NaN'.

Since three of the most represented crop groups (maize, potato, and sugar beet) are spring crops which are sown from March to May, the classification has also been tested for periods beginning later in the season. Indeed, the information derived from time series before their sowing date can bring noise for these crops as different field practices are encountered on these parcels before the sowing (presence of different types of catch crops or absence of catch crop for instance). Considering OAarea, the results in Table 7 do not show any significant score differences whether the classification starts in May or before. However, it should be noticed that from the point of view of OAnum, there appears to be a noticeable decrease in the performance. A higher decrease of the score is met if the three first months including March are excluded from the time series.

As before, the F1-scores of each crop group for the different periods are listed in Table 7. This gives a deeper insight on the performances in terms of the percentage of parcels. It also allows to see whether beginning later in the season might be beneficial for some crops by removing some noise coming from irrelevant months.

**Table 6.** Overall Accuracies and crop group F1-scores considering different periods for the SAR time series, beginning in January and ending in May up to August.

| | | Period of the Time Series | | | |
|---|---|---|---|---|---|
| | | Jan–May (Scenario 10) | Jan–June (Scenario 11) | Jan–July (Scenario 12) | Jan–August (Scenario 1) |
| OA_area (%) | | 88.2 | 91.52 | 93.16 | 93.43 |
| OA_num (%) | | 84.04 | 86.93 | 88.27 | 88.4 |
| **Crop Groups** | **Prevalence (%)** | **F1-Scores (%)** | | | |
| Grassland | 55.7 | 93.77 | 94.18 | 94.48 | 94.51 |
| Winter wheat | 11.3 | 87.52 | 89.01 | 90.34 | 90.3 |
| Maize (spring crop) | 9.9 | 74.48 | 81.28 | 84.36 | 83.97 |
| Potato (spring crop) | 3.8 | 71.28 | 90.1 | 93.08 | 93.24 |
| Sugar beet (spring crop) | 3.6 | 84.77 | 92.21 | 95.08 | 95.52 |
| Winter barley | 2.9 | 86.47 | 91.29 | 91.89 | 92.38 |
| Spelt | 2.1 | 32.95 | 42.88 | 50.05 | 49.13 |
| Others | 1.8 | 0.1 | 0.34 | 0.8 | 1.11 |
| Mixed cereals | 1.3 | 31.19 | 36.81 | 38.36 | 40.61 |
| Fallows | 1 | NaN | NaN | NaN | NaN |
| Winter rape | 0.8 | 77.49 | 87.9 | 87.33 | 92.6 |
| Spring pea | 0.8 | 62.63 | 86.28 | 91.12 | 91.24 |
| Alfalfa | 0.8 | 1.04 | 2.97 | 2.31 | 2 |
| Chicory | 0.7 | 58.14 | 67.99 | 83.32 | 87.44 |
| Spring oats | 0.7 | 45.19 | 52.28 | 59.57 | 61.34 |
| Permanent crop | 0.6 | 25.98 | 27.98 | 31.99 | 32.65 |
| Winter triticale | 0.4 | 10.16 | 34.09 | 40.54 | 39.7 |
| Spring barley | 0.4 | 36.99 | 58.33 | 66.39 | 66.23 |
| Spring wheat | 0.2 | 2.78 | 4.18 | 16.89 | 19.67 |
| Bean | 0.2 | 2.49 | 35.58 | 65.99 | 76.67 |
| Trefoil | 0.1 | NaN | NaN | NaN | NaN |
| Carrot | 0.1 | 8.56 | 37.14 | 54.24 | 59.6 |
| Bulb | 0.1 | 4.87 | 20.46 | 62.87 | 69.98 |
| Strawberry | 0.1 | 1.87 | NaN | NaN | NaN |
| Winter oats | 0.1 | NaN | NaN | 4.54 | 3.45 |
| Spring fava bean | 0.1 | 0 | 7.66 | 15.54 | 20.36 |
| Winter rye | 0.1 | 5.68 | 7.77 | 10.39 | 5.33 |
| Sorghum | 0.1 | NaN | NaN | NaN | NaN |
| Spring triticale | 0.04 | NaN | NaN | 4.17 | NaN |
| Angelic | 0.03 | 6.4 | NaN | NaN | 6.29 |
| Cabbage | 0.02 | NaN | NaN | NaN | NaN |
| Spinach | 0.02 | 3.17 | 21.33 | 26.18 | 23.77 |
| Courgette | 0.02 | NaN | NaN | NaN | NaN |
| Buckwheat | 0.02 | NaN | NaN | NaN | NaN |
| Winter fava bean | 0.01 | NaN | NaN | NaN | NaN |
| Sunflower | 0.01 | NaN | NaN | NaN | NaN |
| Flax | 0.01 | 17.42 | 14.86 | 20.54 | 22.73 |
| Spring rape | 0.01 | NaN | NaN | NaN | NaN |
| Lupin | 0.01 | NaN | NaN | NaN | NaN |
| Soybean | 0.01 | NaN | NaN | NaN | NaN |
| Tomato | 0.01 | NaN | NaN | NaN | NaN |
| Winter pea | 0.01 | 26.94 | 29.49 | 25.16 | 26.45 |
| Quinoa | 0.01 | NaN | NaN | NaN | NaN |
| Butternut | 0.01 | NaN | NaN | NaN | NaN |
| Hemp | 0.01 | NaN | NaN | NaN | NaN |
| Spring rye | 0.004 | NaN | NaN | 40 | NaN |
| Parsley | 0.004 | NaN | NaN | NaN | NaN |
| Nettle | 0.003 | NaN | NaN | NaN | NaN |

**Table 7.** Overall Accuracies and crop group F1-scores considering different periods for the SAR time series (beginning from January up to August 2020).

| | | Period of the Time Series | | | | | | | |
|---|---|---|---|---|---|---|---|---|---|
| | | **Jan–Aug** | **Feb–Aug** | **Mar–Aug** | **Apr–Aug** | **May–Aug** | **Jun–Aug** | **Jul–Aug** | **August** |
| **OA_area (%)** | | 93.43 | 93.48 | 93.54 | 93.53 | 93.32 | 92.56 | 90.76 | 84.34 |
| **OA_num (%)** | | 88.4 | 88.24 | 87.94 | 87.21 | 87.17 | 85.47 | 83.1 | 77.93 |
| **Crop Group** | **Prevalence (%)** | **F1-Scores (%)** | | | | | | | |
| Grassland | 55.7 | **94.51** | 94.37 | 94.16 | 93.57 | 93.57 | 92.43 | 91.44 | 89.78 |
| Winter wheat | 11.3 | **90.3** | 90.11 | 89.87 | 89.54 | 89.66 | 87.75 | 86.37 | 74.08 |
| Maize (spring crop) | 9.9 | **83.97** | 83.7 | 82.76 | 81.2 | 81.89 | 78.53 | 75.1 | 68.7 |
| Potato (spring crop) | 3.8 | 93.24 | 93.32 | 93.35 | 93.54 | **93.65** | 90.04 | 87.53 | 79.46 |
| Sugar beet (spring crop) | 3.6 | **95.52** | 95.47 | 95.47 | 95.3 | 93.53 | 92.08 | 93.41 | 90.36 |
| Winter barley | 2.9 | 92.38 | 92.7 | 92.84 | **92.85** | 92.02 | 85.11 | 69.23 | 48.27 |
| Spelt | 2.1 | 49.13 | 49.7 | 51.51 | **51.88** | 48.02 | 46.24 | 39.72 | 18.74 |
| Others | 1.8 | 1.11 | 1.23 | 1.31 | **1.47** | 1.14 | 0.97 | 0.66 | 0.09 |
| Mixed cereals | 1.3 | **40.61** | 39.41 | 38.31 | 35.68 | 34.93 | 31.36 | 18.81 | 4.56 |
| Fallows | 1 | NaN | NaN | NaN | 0 | NaN | NaN | 0 | NaN |
| Winter rape | 0.8 | 92.6 | 91.25 | 91.63 | 92.52 | 95.2 | **95.65** | 74.09 | 21.78 |
| Spring pea | 0.8 | 91.24 | 91.17 | 91.1 | **91.32** | 90.93 | 87.87 | 76.53 | 44.22 |
| Alfalfa | 0.8 | 2 | 2.67 | **3.19** | 3.07 | 3.11 | 2.15 | 0.94 | 1.73 |
| Chicory | 0.7 | 87.44 | **87.76** | 87.52 | 87.09 | 86 | 84.12 | 79.76 | 69.17 |
| Spring oats | 0.7 | 61.34 | 62.31 | **63.82** | 62.69 | 56.33 | 50.02 | 43.13 | 33.09 |
| Permanent crop | 0.6 | 32.65 | 33.12 | **33.17** | 32.36 | 32.84 | 26.88 | 17.2 | 7.27 |
| Winter triticale | 0.4 | 39.7 | 40.98 | 43.17 | 44.45 | **45.08** | 38.25 | 11.55 | 0.55 |
| Spring barley | 0.4 | 66.23 | 66.59 | 67.13 | **67.67** | 65.93 | 49.18 | 39.95 | 2.64 |
| Spring wheat | 0.2 | 19.67 | 23.09 | 29.78 | **32.94** | 29.33 | 1.7 | 2.96 | NaN |
| Bean | 0.2 | 76.67 | 77.97 | 78.75 | 79.11 | 80.15 | **80.87** | 75.29 | 54.14 |
| Trefoil | 0.1 | NaN | NaN | NaN | NaN | NaN | NaN | NaN | NaN |
| Carrot | 0.1 | 59.6 | **60.08** | 59.5 | 55.69 | 54.2 | 45.93 | 34.76 | 6.47 |
| Bulb | 0.1 | 69.98 | 70.21 | 70.59 | **71.65** | 69.39 | 68.04 | 69.44 | 29.94 |
| Strawberry | 0.1 | NaN | NaN | 0 | NaN | NaN | NaN | NaN | NaN |
| Winter oats | 0.1 | 3.45 | 3.03 | 6.68 | 10.95 | **13.16** | 3.21 | 2.33 | NaN |
| Spring fava bean | 0.1 | 20.36 | 19.74 | 22.44 | 28.69 | **29.4** | 17.63 | 3.15 | 1.43 |
| Winter rye | 0.1 | 5.33 | 7.76 | **11.1** | 8.58 | 10.93 | 4.69 | 2.9 | NaN |
| Sorghum | 0.1 | NaN | NaN | NaN | NaN | NaN | NaN | NaN | NaN |

**Table 7.** *Cont.*

| | | Period of the Time Series | | | | | | | |
|---|---|---|---|---|---|---|---|---|---|
| | | Jan–Aug | Feb–Aug | Mar–Aug | Apr–Aug | May–Aug | Jun–Aug | Jul–Aug | August |
| **OA_area (%)** | | 93.43 | 93.48 | 93.54 | 93.53 | 93.32 | 92.56 | 90.76 | 84.34 |
| **OA_num (%)** | | 88.4 | 88.24 | 87.94 | 87.21 | 87.17 | 85.47 | 83.1 | 77.93 |
| **Crop Group** | **Prevalence (%)** | **F1-Scores (%)** | | | | | | | |
| Spring triticale | 0.04 | NaN | 2.78 | 2.78 | 3.12 | **4.17** | NaN | NaN | NaN |
| Angelic | 0.03 | 6.29 | 6.32 | 6.32 | **7.86** | 7.08 | 7.08 | NaN | NaN |
| Cabbage | 0.02 | NaN | NaN | NaN | NaN | NaN | NaN | NaN | NaN |
| Spinach | 0.02 | 23.77 | 21.27 | 25.13 | **28.13** | 20.32 | 18.76 | 10.5 | 7.19 |
| Courgette | 0.02 | NaN | NaN | NaN | **10.53** | NaN | NaN | NaN | NaN |
| Buckwheat | 0.02 | NaN | NaN | NaN | NaN | NaN | NaN | NaN | NaN |
| Winter fava bean | 0.01 | NaN | NaN | NaN | NaN | NaN | NaN | NaN | NaN |
| Sunflower | 0.01 | NaN | NaN | NaN | NaN | NaN | NaN | NaN | NaN |
| Flax | 0.01 | 22.73 | 28.5 | 28.51 | 29.33 | **31.92** | 19.77 | 17.32 | NaN |
| Spring rape | 0.01 | NaN | NaN | NaN | NaN | NaN | NaN | NaN | NaN |
| Lupin | 0.01 | NaN | NaN | NaN | NaN | NaN | NaN | NaN | NaN |
| Soybean | 0.01 | NaN | NaN | NaN | NaN | NaN | NaN | NaN | NaN |
| Tomato | 0.01 | NaN | NaN | NaN | NaN | NaN | NaN | NaN | NaN |
| Winter pea | 0.01 | 26.45 | **30.9** | 22.33 | 23.61 | NaN | NaN | NaN | NaN |
| Quinoa | 0.01 | NaN | NaN | NaN | NaN | NaN | NaN | NaN | NaN |
| Butternut | 0.01 | NaN | NaN | NaN | NaN | NaN | NaN | NaN | NaN |
| Hemp | 0.01 | NaN | NaN | NaN | NaN | NaN | NaN | NaN | NaN |
| Spring rye | 0.004 | NaN | NaN | NaN | NaN | NaN | NaN | NaN | NaN |
| Parsley | 0.004 | NaN | NaN | NaN | NaN | NaN | NaN | NaN | NaN |
| Nettle | 0.003 | NaN | NaN | NaN | NaN | NaN | NaN | NaN | NaN |

The best F1-scores for each crop group are highlighted in bold black in Table 7. One could have expected that the period giving the best results for a given crop would depend on when the crop is in place. For grasslands and winter crops (the latter being mainly sewed in Autumn and harvested in July), this would correspond to the longest period, from January to August. For crops starting later in the season (like maize, potato and sugar beet which are sewed mainly in April and harvested in Autumn), one could think that the first months of the year would be irrelevant. However, the present results show that reality is more complex. Among the most represented crops that are in place during the beginning of the season, only grasslands and winter wheat get a higher score if the whole period is used. For the 2nd and 3rd most represented winter crops (winter barley and spelt), the best F1-scores are reached when the considered SAR time series begins in April. Regarding the spring crops, the conclusions are also diverse. Contrary to the previous hypothesis, maize gets a significantly better score when starting in January rather than in April or May. The January score is also better for sugar beet, although the scores are similar up to April (but lower starting from May). For potato, the best score is reached when starting in May, although longest periods show similar results.

Thus, the considered period may be adapted to the user's goal. Depending on whether the aim is to obtain an early map of the crop types, or to more accurately recognize crops appearing later in the season, a different ending month for the time series can be chosen. For some specific crops, using a times series starting later in the season can be beneficial. Subsequent classifications could thus be performed to first discriminate some crops and then the other ones.

## 4. Discussion

This section first addresses a major constraint linked to the use of SAR images: the presence of rain or snow that affects the SAR signal. The robustness of the classification regarding that aspect is discussed in Section 4.1.

Secondly, the discussion analyses how the classification accuracy depends on some characteristics of the parcels: their classification confidence level, their size, and their shape. The objective is two-fold. First, this allows to understand whether the classification is relevant on the whole range of that characteristic (for instance, is the classification accurate for oddly shaped parcels?). Furthermore, considering a threshold on those characteristics allows to identify a subset of the parcels for which the global classification accuracy is higher.

### 4.1. Impact of Rain and Snow on the SAR Signal

The SAR signal is not affected by the presence of clouds, but it is sensitive to rain and to the water drops on the vegetation. The presence of rain or water drops during the time acquisition of the SAR data used in the present study can be assessed thanks to ground truth meteorological data. The PAMESEB meteorological stations provide in situ precipitation measures recorded on an hourly basis at 30 locations in the WR (Source: CRA-W/Pameseb network—www.agromet.be—2 February 2021).

On the 80 selected SAR Sentinel-1 acquisitions on tracks 37 and 161 between 10 January 2020 and 02 September 2020, 52 images were acquired less than 2 h after or during a rainfall in at least one of the PAMESEB stations located in our study area. The average number of images from all Pameseb stations acquired just before or during a rainfall at the station is five. Thus, on average, the number of rain-free SAR images at a station is 75.

Comparatively, out of the 95 optical Sentinel-2 acquisitions over the T31UFR tile during the same period of eight months, only 24 acquisitions are useable (i.e., cloud-free) on average at the agricultural parcel level.

Moreover, the presence of snow also affects the SAR signal. On the 80 SAR Sentinel-1 acquisitions on tracks 37 and 161, 11 images were acquired when the presence of snow was recorded by the Royal Meteorological Institute of Belgium.

Despite the inclusion in the dataset of these images acquired when rain and snow occurred, the combination of the high number of acquisitions and the automatic selection of the most meaningful explanatory variables by the random forest seems to overcome the problem.

### *4.2. Evolution of the Classification Performance in Terms of the Classification Confidence Level*

To the class assigned to a parcel by the classification model is linked a confidence level of belonging to this class (percentage of the trees in the forest that predict the given class). Figure 3 shows the evolution of the overall accuracy in function of a threshold on the confidence level. For each value of the threshold, the corresponding OA is computed only on the parcels of the validation dataset which are classified with a confidence level higher than the given threshold. Recall that 10 models are trained for each scenario: the represented OA is, as before, the mean of the 10 corresponding overall accuracies. On the left graph, the OA is computed in terms of the area (OAarea), while on the right, in terms of the number of parcels (OAnum). The OA (black dots on the graph) increases with the confidence level threshold. However, this must be balanced by the fact that less parcels are classified since the number of parcels meeting the threshold criterion decreases. This is shown by the bars on the graph, which represent the proportion of parcels in the validation dataset which satisfy the confidence level threshold. For OAarea, this proportion is given as the percentage of the area covered by these parcels, and for OAnum, as the percentage of the number of parcels. These bars also represent the proportion of well- and mis-classified parcels (green and red bars, respectively).

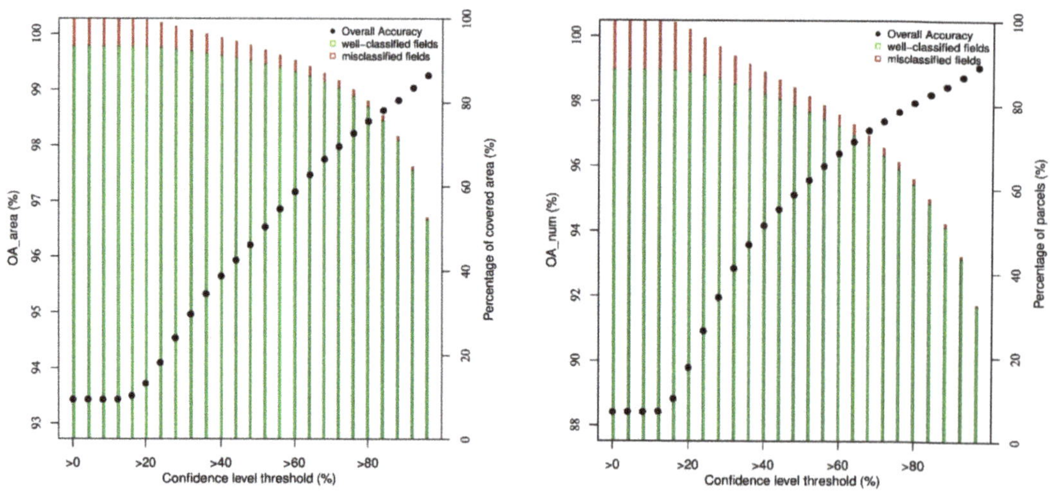


**Figure 3.** Evolution of the overall accuracy considering only the parcels in the validation dataset whose confidence level is above a given threshold (for scenario 1). (**Left**): OA in terms of the area. (**Right**): OA in terms of the number of parcels.

Considering a threshold on the confidence level allows to identify a subset of parcels on which the global accuracy is higher. For instance, considering only the subset of parcels classified with a confidence level higher than 92%, OAarea reaches 99.03%. However, the downside is that this subset covers only 64.7% of the total area of the validation dataset. This kind of analysis allows to choose the best compromise between the accuracy of the classification and the area covered by the classification. In this case, a good compromise is a threshold of 32% on the confidence level: it leads to an OAarea of 95% while still covering 97% of the area. In the case of OAnum, the decrease in the number of parcels is faster. If one is more concerned about keeping a high number of parcels instead of a high covered area, a threshold of 24% is therefore preferable: 95.7% of the parcels are classified and OAnum is

90.9%. Figure 3 also shows that the confidence level is linked to the size of the parcel as one can see that the parcels having a confidence level higher than 96% covers more than half of the area of the WR while it concerns only about 33% of the number of parcels.

In the framework of the CAP, a progressive classification process could be imagined over time. The classification could be done several times during the growing season and a parcel would be considered as "well-classified" as soon as its confidence level reaches a chosen threshold.

### 4.3. Evolution of the Classification Performance in Terms of the Parcels Size

In Table 4, the percentage of well classified area is always higher than the percentage of well-classified number of fields. This can be explained by a worse capacity of the classification model to classify small fields. Indeed, because of the statistical uncertainties in the SAR signal, the SAR speckle has a strong effect on the signal when considering a single pixel or an average over a small number of pixels. In order to analyze this constraint, Figure 4 shows the evolution of the overall accuracy when considering a threshold on the parcel size. For each value of the threshold, the corresponding overall accuracy is computed only on the parcels of the validation dataset whose area is higher than the threshold. This evolution of the OA is given for the reference scenario (black points) and the scenario 8 which includes all the parcels in the training dataset (black crosses). The bars on the graphs show, for the reference scenario, the proportion of the parcels in the validation dataset (in terms of the area on the left graph and of the number of parcels on the right graph) that satisfy the size threshold. The green and red bars represent the proportion of well- and mis-classified parcels, respectively.

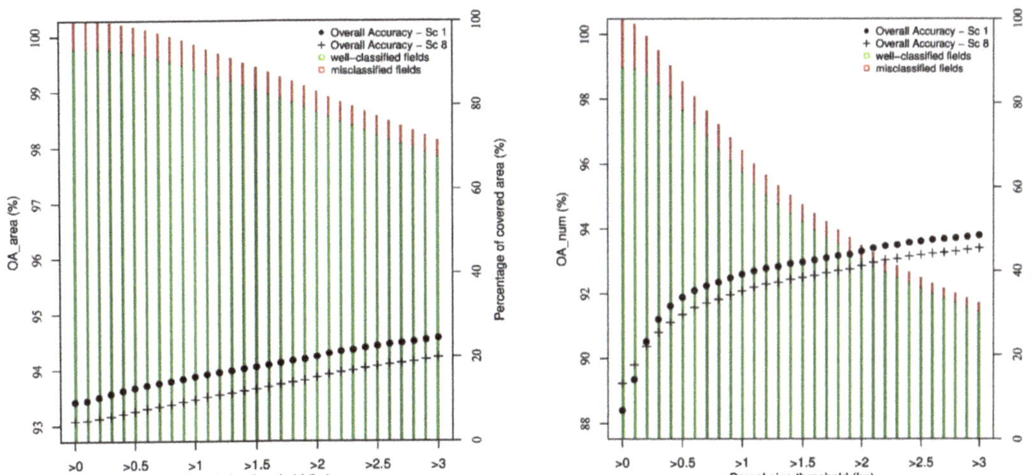


**Figure 4.** Evolution of the overall accuracy when considering a threshold on the parcel area. (**Left**): OA in terms of the area. (**Right**): OA in terms of the number of parcels.

The overall accuracy increases with the parcel size threshold, but at the same time, the number of fields for which a class is obtained decreases. When all the parcels in the validation dataset are included (as in Table 4), OAnum reaches 88.4%. At the other extreme, OAnum reaches 93.8% if only the parcels in the validation sample that are bigger than three hectares are considered, which represents 32.3% of the parcels. The steepest increase of OAnum happens in the range from 0 to 0.5 hectares. Over 1 ha, the increase is not yet significant. This clearly illustrates the difficulty of working at pixel level or with pixels averaging for a small number of pixels with SAR data due to the speckle.

Recall that in the reference scenario, only the 'clean' parcels (i.e., covering a minimum number of pixels) are included in the training dataset. Scenario 8 includes all the parcels in the training dataset, and it could be supposed that such scenario would be better at classifying small parcels (since the model is also trained with similar parcel sizes). To quantify this, the black crosses on Figure 4 show the evolution of OAnum for that scenario. One can see that, for small thresholds, the percentage of well classified parcels is indeed higher for scenario 8. However, for a threshold of 0.2 hectare or higher, scenario 1 is better. Notice that the same comparison can be made in terms of OAarea, and in that case, scenario 1 is always better.

Figure 5 shows the value of the F1-score of four different crop categories for five areas ranges. The crop categories are: the main spring crops (maize, potato, and sugar beet, representing 17.3% of the number of parcels and 20% of the area), the main winter crops (winter wheat and winter barley, representing 14.2% of the number of parcels and 21% of the area), the grasslands (55.7% of the parcels, 47% of the area), and the other crops (12.8% of the parcels, 12% of the area). For each of these categories, the F1-score is computed on the subsets of parcels whose size is contained in ranges of one hectare. Notice that we consider ranges here and not a threshold as before. The steepest increase of the F1-score is met for the main spring crops and for the other crops. For the main spring crops, the F1-score is significantly lower for parcels smaller than one hectare, which represent about 24% of the parcels in that category. Except for the other crops, beyond one hectare, the increase of the area has no significant impact on the F1-score anymore.

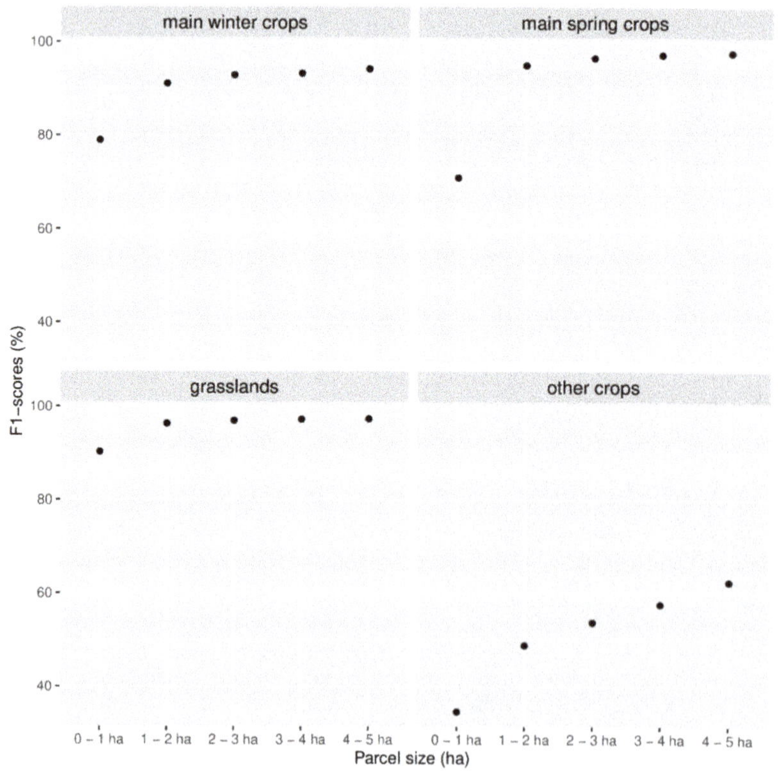


**Figure 5.** F1-scores of four crop categories computed on five subsets of parcels regrouped by their area.

In landscapes such as Belgian landscape where the agriculture is mainly intensive, the size of the parcel is not an excessively big challenge for working with Sentinel-1 data (the mean of the parcel areas is equal to three hectares). Nevertheless, in other regions, either the use of higher spatial resolution images or the aggregation of adjacent parcels considered as Feature of Interest (i.e., homogeneous areas covered by the same crop, uniformly managed) should be investigated. In [51], the authors demonstrated that the classification performance is much more dependent on the type of cropping systems (including the fields size) than on the classification method.

### 4.4. Evolution of the Classification Performance in Terms of the Parcels Shape

Other parcels can have very few numbers of pixels inside their polygons, despite having an honorable size: the oddly shaped parcels. For instance, a very long and thin parcel might have a big area but still contain very few pixels. A way to quantify this is to consider the shape index: the ratio between the perimeter of the parcel and the square root of its area. The smaller the shape index, the closer the parcel looks to a nice circle. In the LPIS, the shape index ranges from 3.7 to 57.1, and 99.9% of the parcels (99.98% of the area) have a shape index lower than 20. In order to analyze this constraint, Figure 6 shows the evolution of the overall accuracy in function of a threshold on the shape index. Only the parcels in the validation dataset whose shape index is under the given threshold are considered in the OA computation. On the left, the overall accuracy is computed in terms of the area of the parcels, and on the right, in terms of the number of parcels. As before, the bars represent the proportion of parcels in the validation dataset which meet the threshold criterion (in terms of the area on the left, and on the right, in terms of the number of parcels). The green and red bars correspond to the well- and mis-classified parcels, respectively.

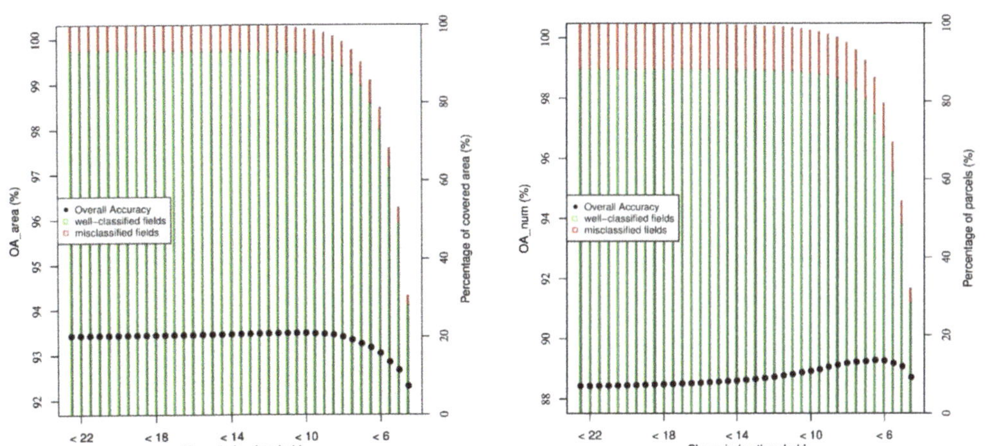


**Figure 6.** Evolution of the overall accuracy considering a threshold on the shape index. (**Left**): OA in terms of the area. (**Right**): OA in terms of the number of parcels.

When considering OAarea, the threshold does not lead to any significant score improvement. When considering OAnum, a slight score improvement can be observed, but it must be put in balance with the drop in the number of classified parcels. For instance, a threshold of 8 gives an overall accuracy of 89.2% on the subset of parcels meeting that threshold (instead of 88.4% when all parcels are considered), but this subset only counts 95% of the parcels.

## 5. Conclusions and Perspectives

In the particular context of the CAP crop monitoring, there is much interest in improving the performances of the crop classification, since it leads to less field visits, i.e., less work for the Paying Agencies.

Moreover, since the availability of optical images is often a problem in region with a frequent cloud cover, such as Belgium, Sentinel-1 SAR images give very promising results to identify the agricultural crop groups. Indeed, this study shows that the combination of the high number of acquisitions and the automatic selection of the most meaningful explanatory variables by the random forest seems to overcome the negative impact of rain and snow on the SAR signal. Using an object-level random forest classifier, 48 crop groups could be classified with an overall accuracy of 93.4% of well-classified area, which is a very nice result compared to the literature. The F1-scores of the six most represented crop groups are always higher than or equal to 84%.

Furthermore, one of the significant changes to be considered for the next CAP reform is the regular information to be provided to the farmers along the season about their agricultural practices. In this respect, the results show that five of the six most represented crop groups can be recognized before their harvesting dates with a F1-score higher than 89%.

Given the several kinds of information that can be extracted for one parcel from Sentinel-1 sensor, one of this study's aim is to define which are the most relevant explanatory variables that have to be used to improve the classification results quality while not unnecessarily increasing the model's complexity. The goal is to get the easiest and best performing model. The results show that the addition of the VH$\sigma$0/VV$\sigma$0 ratio in the explanatory variables and the use of information derived from two orbits (one ascending and one descending) increase the performances. The most relevant period of the time series to consider starts at the beginning of the spring crops growing season until the month after the harvest of the winter crops.

Other choices can improve the classification performances: the use of an internal buffer on the parcel polygons before signal extraction, the selection of the parcels containing more than 200 pixels for the signal extraction in the training sample.

In another vein, this study shows how the classification accuracy depends on some characteristics of the parcels. It is shown that small parcels are still difficult to classify. Indeed, the classification performance is significantly lower for parcels smaller than one hectare. On the other hand, considering a threshold on such a characteristic allows to identify a subset of parcels for which the global accuracy is higher. This kind of analysis helps to choose the best compromise between accuracy and the proportion of parcels that are still taken into account, depending on the user's need.

This study also addresses the main challenge of have good ground truth data to calibrate a model. Our results show that using a small training dataset (containing 3.2% of the total number of the fields, i.e., about 6850 parcels) leads to an overall accuracy of 91.65% of well-classified area. This is a promising result for regions where poor ground data are available but also in our region since it allows to improve the training dataset with a limited number of field visits.

Here are some perspectives to go beyond the present analysis.

In addition to the per-field mean of the SAR data, the intra-field heterogeneity might be relevant to the classification. The use of time series of, for instance, the intra-field standard deviation, should be investigated.

Moreover, other kinds of explanatory variables computed from the remote sensing data could be used (date of maximum/minimum backscattering coefficient, highest slope of the backscattering coefficient, period during which the backscattering coefficient is higher/lower than a threshold, etc.).

As mentioned earlier, subsequent classifications, using different time periods, could be performed in order to classify the crops one after the others, depending on when they are best recognized. Indeed, in the case of the Walloon crop calendar, a first classification could differentiate grasslands and orchards from other agricultural lands using a period covering autumn and winter since these crops would already be in place during that period. Then, winter crops could be separated from the spring and summer crops using the end of autumn, winter, and beginning of spring periods.

Beyond the improvement of the classification accuracy, the possibility of transferring the model calibration to subsequent crop seasons should be analyzed. On that respect, it should be mentioned that the analyses done in this paper have also been conducted on the 2019 data and that the results and conclusions are similar. However, this is different than using a model calibrated on one season and validated on the next growing season, which might be useful from an operational point of view.

**Author Contributions:** Conceptualization, P.D., E.B. and C.L.-D.; methodology, A.J. and E.B.; software, A.J. and E.B.; validation, A.J. and E.B.; formal analysis, A.J. and E.B.; investigation, E.B. and A.J.; resources, A.J. and E.B.; data curation, A.J. and E.B.; writing—original draft preparation, initial draft, E.B. and A.J.; review of the writing including results interpretation, A.J., E.B., P.D. and C.L.-D.; writing—review and editing, E.B. and A.J.; visualization, A.J. and E.B.; supervision, E.B.; project administration, E.B. and V.P.; funding acquisition, V.P. All authors have read and agreed to the published version of the manuscript.

**Funding:** This work was carried out within the framework of the SAGRIWASENT project founded by the Walloon region.

**Institutional Review Board Statement:** Not applicable.

**Data Availability Statement:** The SENTINEL data presented in this study are available freely through the COPERNICUS program of ESA. The parcel delineation data are available on request through https://geoportail.wallonie.be/ (13 July 2021). Meteorological data may be requested through https://agromet.be/ (2 February 2021). The exact details about crop groups may be requested from the authors.

**Acknowledgments:** We would like to thank the agents of the "Direction des Surfaces" of the "Service Public de Wallonie" (SPW) for their collaboration. We also would like to thank Emilie Mulot (CRA-W) for her contribution to the pre-processing of SAR data.

**Conflicts of Interest:** The authors declare no conflict of interest.

## References

1. Belgiu, M.; Csillik, O. Sentinel-2 Cropland Mapping Using Pixel-Based and Object-Based Time-Weighted Dynamic Time Warping Analysis. *Remote Sens. Environ.* **2018**, *204*, 509–523. [CrossRef]
2. Cai, Y.; Guan, K.; Peng, J.; Wang, S.; Seifert, C.; Wardlow, B.; Li, Z. A High-Performance and in-Season Classification System of Field-Level Crop Types Using Time-Series Landsat Data and a Machine Learning Approach. *Remote Sens. Environ.* **2018**, *210*, 35–47. [CrossRef]
3. Defourny, P.; Bontemps, S.; Bellemans, N.; Cara, C.; Dedieu, G.; Guzzonato, E.; Hagolle, O.; Inglada, J.; Nicola, L.; Rabaute, T.; et al. Near Real-Time Agriculture Monitoring at National Scale at Parcel Resolution: Performance Assessment of the Sen2-Agri Automated System in Various Cropping Systems around the World. *Remote Sens. Environ.* **2019**, *221*, 551–568. [CrossRef]
4. Hao, P.; Zhan, Y.; Wang, L.; Niu, Z.; Shakir, M. Feature Selection of Time Series MODIS Data for Early Crop Classification Using Random Forest: A Case Study in Kansas, USA. *Remote Sens.* **2015**, *7*, 5347–5369. [CrossRef]
5. Heupel, K.; Spengler, D.; Itzerott, S. A Progressive Crop-Type Classification Using Multitemporal Remote Sensing Data and Phenological Information. *PFG–J. Photogramm. Remote Sens. Geoinf. Sci.* **2018**, *86*, 53–69. [CrossRef]
6. Rahman, M.S.; LiPing, D.; Yu, E.; Chen, Z.; Mohiuddin, H. In-Season Major Crop-Type Identification for US Cropland from Landsat Images Using Crop-Rotation Pattern and Progressive Data Classification. *Agriculture* **2019**, *9*, 17. [CrossRef]
7. Xiong, J.; Thenkabail, P.S.; Gumma, M.K.; Teluguntla, P.; Poehnelt, J.; Congalton, R.G.; Yadav, K.; Thau, D. Automated Cropland Mapping of Continental Africa Using Google Earth Engine Cloud Computing. *ISPRS J. Photogramm. Remote Sens.* **2017**, *126*, 225–244. [CrossRef]
8. Friedl, M.A.; Sulla-Menashe, D.; Tan, B.; Schneider, A.; Ramankutty, N.; Sibley, A.; Huang, X. MODIS Collection 5 Global Land Cover: Algorithm Refinements and Characterization of New Datasets. *Remote Sens. Environ.* **2010**, *114*, 168–182. [CrossRef]
9. Radoux, J.; Lamarche, C.; Van Bogaert, E.; Bontemps, S.; Brockmann, C.; Defourny, P. Automated Training Sample Extraction for Global Land Cover Mapping. *Remote Sens.* **2014**, *6*, 3965–3987. [CrossRef]
10. Bellon De La Cruz, B.; Bégué, A.; Lo Seen, D.; Aparecido de Almeida, C.; Simoes, M. A Remote Sensing Approach for Regional-Scale Mapping of Agricultural Land-Use Systems Based on NDVI Time Series. *Remote Sens.* **2017**, *9*, 600. [CrossRef]

11. Matton, N.; Canto, G.; Waldner, F.; Valero, S.; Morin, D.; Inglada, J.; Arias, M.; Bontemps, S.; Koetz, B.; Defourny, P. An Automated Method for Annual Cropland Mapping along the Season for Various Globally-Distributed Agrosystems Using High Spatial and Temporal Resolution Time Series. *Remote Sens.* **2015**, *7*, 13208–13232. [CrossRef]
12. Wardlow, B.; Egbert, S. Large-Area Crop Mapping Using Time-Series MODIS 250 m NDVI Data: An Assessment for the U.S. Central Great Plains. *Remote Sens. Environ.* **2008**, *112*, 1096–1116. [CrossRef]
13. Zheng, B.; Myint, S.; Thenkabail, P.S.; Aggarwal, R. A Support Vector Machine to Identify Irrigated Crop Types Using Time-Series Landsat NDVI Data. *Int. J. Appl. Earth Obs. Geoinf.* **2015**, *34*, 103–112. [CrossRef]
14. Clark, M.L.; Aide, T.M.; Riner, G. Land Change for All Municipalities in Latin America and the Caribbean Assessed from 250-m MODIS Imagery (2001–2010). *Remote Sens. Environ.* **2012**, *126*, 84–103. [CrossRef]
15. Gebhardt, S.; Wehrmann, T.; Ruiz, M.A.M.; Maeda, P.; Bishop, J.; Schramm, M.; Kopeinig, R.; Cartus, O.; Kellndorfer, J.; Ressl, R.; et al. MAD-MEX: Automatic Wall-to-Wall Land Cover Monitoring for the Mexican REDD-MRV Program Using All Landsat Data. *Remote Sens.* **2014**, *6*, 3923–3943. [CrossRef]
16. Jia, K.; Liang, S.; Wei, X.; Yao, Y.; Su, Y.; Jiang, B.; Wang, X. Land Cover Classification of Landsat Data with Phenological Features Extracted from Time Series MODIS NDVI Data. *Remote Sens.* **2014**, *6*, 11518–11532. [CrossRef]
17. Gómez, C.; White, J.C.; Wulder, M.A. Optical Remotely Sensed Time Series Data for Land Cover Classification: A Review. *ISPRS J. Photogramm. Remote Sens.* **2016**, *116*, 55–72. [CrossRef]
18. Hermosilla, T.; Wulder, M.A.; White, J.C.; Coops, N.C.; Hobart, G.W. An Integrated Landsat Time Series Protocol for Change Detection and Generation of Annual Gap-Free Surface Reflectance Composites. *Remote Sens. Environ.* **2015**, *158*, 220–234. [CrossRef]
19. Weiss, D.J.; Atkinson, P.M.; Bhatt, S.; Mappin, B.; Hay, S.I.; Gething, P.W. An Effective Approach for Gap-Filling Continental Scale Remotely Sensed Time-Series. *ISPRS J. Photogramm. Remote Sens.* **2014**, *98*, 106–118. [CrossRef]
20. Inglada, J.; Arias, M.; Tardy, B.; Hagolle, O.; Valero, S.; Morin, D.; Dedieu, G.; Sepulcre-Canto, G.; Bontemps, S.; Defourny, P.; et al. Assessment of an Operational System for Crop Type Map Production Using High Temporal and Spatial Resolution Satellite Optical Imagery. *Remote Sens.* **2015**, *7*, 12356–12379. [CrossRef]
21. Swets, D.; Reed, B.C.; Rowland, J.; Marko, S.E. A Weighted Least-Squares Approach to Temporal NDVI Smoothing. In Proceedings of the From Image to Information: 1999 ASPRS Annual Conference, Portland, OR, USA, 17–21 May 1999.
22. Van Tricht, K.; Gobin, A.; Gilliams, S.; Piccard, I. Synergistic Use of Radar Sentinel-1 and Optical Sentinel-2 Imagery for Crop Mapping: A Case Study for Belgium. *Remote Sens.* **2018**, *10*, 1642. [CrossRef]
23. Wooding, M.G. *Satellite Radar in Agriculture*; ESA Publications Division ESTEC: Noordwijk, The Netherlands, 1995.
24. McNairn, H.; Shang, J.; Jiao, X.; Champagne, C. The Contribution of ALOS PALSAR Multipolarization and Polarimetric Data to Crop Classification. *IEEE Trans. Geosci. Remote Sens.* **2009**, *47*, 3981–3992. [CrossRef]
25. Silva, W.F.; Rudorff, B.F.T.; Formaggio, A.R.; Paradella, W.R.; Mura, J.C. Discrimination of Agricultural Crops in a Tropical Semi-Arid Region of Brazil Based on L-Band Polarimetric Airborne SAR Data. *ISPRS J. Photogramm. Remote Sens.* **2009**, *64*, 458–463. [CrossRef]
26. Dey, S.; Mandal, D.; Robertson, L.D.; Banerjee, B.; Kumar, V.; McNairn, H.; Bhattacharya, A.; Rao, Y.S. In-Season Crop Classification Using Elements of the Kennaugh Matrix Derived from Polarimetric RADARSAT-2 SAR Data. *Int. J. Appl. Earth Obs. Geoinf.* **2020**, *88*, 102059. [CrossRef]
27. Schulthess, U.; Kunze, M. *Crop Classification with Optical (RapidEye) and C-Band Radar Data (RADARSAT-2)*; AgriSAR 2009 Final Project Report (ESA Contract No. 22689/09/NL/FF/ef.); MDA: Richmond, BC, Canada, 2011; p. 39.
28. Bargiel, D. A New Method for Crop Classification Combining Time Series of Radar Images and Crop Phenology Information. *Remote Sens. Environ.* **2017**, *198*, 369–383. [CrossRef]
29. Boryan, C.; Yang, Z.; Haack, B. Evaluation of Sentinel-1 A C-Band Synthetic Aperture Radar for Citrus CROP Classification in Florida, United States. In Proceedings of the IGARSS 2018—2018 IEEE International Geoscience and Remote Sensing Symposium, Valencia, Spain, 22 July 2018; pp. 7369–7372.
30. Ndikumana, E.; Ho Tong Minh, D.; Baghdadi, N.; Courault, D.; Hossard, L. Deep Recurrent Neural Network for Agricultural Classification Using Multitemporal SAR Sentinel-1 for Camargue, France. *Remote Sens.* **2018**, *10*, 1217. [CrossRef]
31. Mullissa, A.G.; Persello, C.; Tolpekin, V.A. Fully Convolutional Networks for Multi-Temporal SAR Image Classification. In Proceedings of the IGARSS 2018—2018 IEEE International Geoscience and Remote Sensing Symposium, Valencia, Spain, 22–27 July 2018; pp. 3338–6635.
32. Teimouri, N.; Dyrmann, M.; Jørgensen, R. A Novel Spatio-Temporal FCN-LSTM Network for Recognizing Various Crop Types Using Multi-Temporal Radar Images. *Remote Sens.* **2019**, *11*, 990. [CrossRef]
33. Wei, S.; Zhang, H.; Wang, C.; Wang, Y.; Xu, L. Multi-Temporal SAR Data Large-Scale Crop Mapping Based on U-Net Model. *Remote Sens.* **2019**, *11*, 68. [CrossRef]
34. Arias, M.; Campo-Bescós, M.Á.; Álvarez-Mozos, J. Crop Classification Based on Temporal Signatures of Sentinel-1 Observations over Navarre Province, Spain. *Remote Sens.* **2020**, *12*, 278. [CrossRef]
35. Ban, Y. Synergy of Multitemporal ERS-1 SAR and Landsat TM Data for Classification of Agricultural Crops. *Can. J. Remote Sens.* **2003**, *29*, 518–526. [CrossRef]
36. Blaes, X.; Vanhalle, L.; Defourny, P. Efficiency of Crop Identification Based on Optical and SAR Image Time Series. *Remote Sens. Environ.* **2005**, *96*, 352–365. [CrossRef]

37. Zhao, W.; Qu, Y.; Chen, J.; Yuan, Z. Deeply Synergistic Optical and SAR Time Series for Crop Dynamic Monitoring. *Remote Sens. Environ.* **2020**, *247*, 111952. [CrossRef]
38. Orynbaikyzy, A.; Gessner, U.; Conrad, C. Crop Type Classification Using a Combination of Optical and Radar Remote Sensing Data: A Review. *Int. J. Remote Sens.* **2019**, *40*, 6553–6595. [CrossRef]
39. Kussul, N.; Lavreniuk, M.; Skakun, S.; Shelestov, A. Deep Learning Classification of Land Cover and Crop Types Using Remote Sensing Data. *IEEE Geosci. Remote Sens. Lett.* **2017**, *14*, 778–782. [CrossRef]
40. Veloso, A.; Mermoz, S.; Bouvet, A.; Le Toan, T.; Planells, M.; Dejoux, J.-F.; Ceschia, E. Understanding the Temporal Behavior of Crops Using Sentinel-1 and Sentinel-2-like Data for Agricultural Applications. *Remote Sens. Environ.* **2017**, *199*, 415–426. [CrossRef]
41. Blaes, X.; Defourny, P.; Wegmuller, U.; Della Vecchia, A.; Guerriero, L.; Ferrazzoli, P. C-Band Polarimetric Indexes for Maize Monitoring Based on a Validated Radiative Transfer Model. *IEEE Trans. Geosci. Remote Sens.* **2006**, *44*, 791–800. [CrossRef]
42. Della Vecchia, A.; Ferrazzoli, P.; Guerriero, L.; Ninivaggi, L.; Strozzi, T.; Wegmuller, U. Observing and Modeling Multifrequency Scattering of Maize During the Whole Growth Cycle. *IEEE Trans. Geosci. Remote Sens.* **2008**, *46*, 3709–3718. [CrossRef]
43. Lin, H.; Chen, J.; Pei, Z.; Zhang, S.; Hu, X. Monitoring Sugarcane Growth Using ENVISAT ASAR Data. *IEEE Trans. Geosci. Remote Sens.* **2009**, *47*, 2572–2580. [CrossRef]
44. Vajsova, B.; Fasbender, D.; Wirnhardt, C.; Lemajić, S.; Devos, W. Assessing Spatial Limits of Sentinel-2 Data on Arable Crops in the Context of Checks by Monitoring. *Remote Sens.* **2020**, *12*, 2195. [CrossRef]
45. Fellah, K.; Bally, P.; Besnus, Y.; Meyer, C.; Rast, M.; De Fraipont, P. *Impact of SAR Radiometric Resolution in Hydrological and Agro-Environemental Applications*; CNES: Toulouse, France, 1995.
46. Demšar, J. Statistical Comparisons of Classifiers over Multiple Data Sets. *J. Mach. Learn. Res.* **2006**, *7*, 1–30.
47. Alpaydm, E. Combined 5 × 2 Cv F Test for Comparing Supervised Classification Learning Algorithms. *Neural Comput.* **1999**, *11*, 1885–1892. [CrossRef] [PubMed]
48. Dietterich, T.G. Approximate Statistical Tests for Comparing Supervised Classification Learning Algorithms. *Neural Comput.* **1998**, *10*, 1895–1923. [CrossRef] [PubMed]
49. Zhong, L.; Hu, L.; Zhou, H. Deep Learning Based Multi-Temporal Crop Classification. *Remote Sens. Environ.* **2019**, *221*, 430–443. [CrossRef]
50. Auquière, E. *SAR Temporal Series Interpretation and Backscattering Modelling for Maize Growth Monitoring*; Presses Univ. de Louvain: Louvain-la-Neuve, Belgium, 2001.
51. Waldner, F.; De Abelleyra, D.; Verón, S.R.; Zhang, M.; Wu, B.; Plotnikov, D.; Bartalev, S.; Lavreniuk, M.; Skakun, S.; Kussul, N.; et al. Towards a Set of Agrosystem-Specific Cropland Mapping Methods to Address the Global Cropland Diversity. *Int. J. Remote Sens.* **2016**, *37*, 3196–3231. [CrossRef]

*remote sensing*

MDPI

*Article*

# Unsupervised Classification of Crop Growth Stages with Scattering Parameters from Dual-Pol Sentinel-1 SAR Data

**Subhadip Dey [1,*], Narayanarao Bhogapurapu [1], Saeid Homayouni [2], Avik Bhattacharya [1] and Heather McNairn [3]**

1 Microwave Remote Sensing Lab, Centre of Studies in Resources Engineering, Indian Institute of Technology Bombay, Mumbai 400076, India; bnarayanarao@iitb.ac.in (N.B.); avikb@csre.iitb.ac.in (A.B.)

2 Centre Eau Terre Environnement, INRS, 490 Couronne St, Quebec City, QC G1K 9A9, Canada; saeid.homayouni@inrs.ca

3 Ottawa Research and Development Centre, Agriculture and Agri-Food Canada, 1341 Baseline Road, Ottawa, ON K1A 0C5, Canada; heather.mcnairn@canada.ca

* Correspondence: sdey23@iitb.ac.in

**Citation:** Dey, S.; Bhogapurapu, N.; Homayouni, S.; Bhattacharya, A.; McNairn, H. Unsupervised Classification of Crop Growth Stages with Scattering Parameters from Dual-Pol Sentinel-1 SAR Data. *Remote Sens.* **2021**, *13*, 4412. https://doi.org/10.3390/rs13214412

Academic Editors: Olaniyi Ajadi and David McAlpin

Received: 18 September 2021
Accepted: 25 October 2021
Published: 2 November 2021



**Abstract:** Global crop mapping and monitoring requires high-resolution spatio-temporal information. In this regard, dual polarimetric Synthetic Aperture Radar (SAR) sensors provide high temporal and high spatial resolutions with large swath width. Generally, crop phenological development studies utilized SAR backscatter intensity-based descriptors. However, these descriptors are derived either from the covariance matrix elements or from the eigendecomposition. Therefore, this approach fails to utilize the complete polarization information of the scattered wave. In this study, we propose a target characterization parameter, $\theta_{xP}$ that utilizes the 2D Barakat degree of polarization and the elements of the covariance matrix. We also propose an unsupervised clustering scheme using $\theta_{xP}$ and the scattering entropy, $H_{xP}$. We utilize time-series Sentinel-1 data of canola and wheat fields over a Canadian test site to show the sensitivity of $\theta_{xP}$ to the development of crop morphology at different phenological stages. During the initial growth stages, $\theta_{xP}$ values are low due to the low vegetation density. In contrast, at advanced phenological stages, we observe decreased values of $\theta_{xP}$ due to the appearance of complex canopy structure. Similarly, the effectiveness of the unsupervised $H_{xP}/\theta_{xP}$ clustering plane is also evident from the temporal clustering plots. This innovative clustering framework is beneficial for the operational use of Sentinel-1 SAR data for agricultural applications.

**Keywords:** Sentinel-1; polarimetry; dual-pol; crop characterization; phenology; unsupervised classification

## 1. Introduction

Identification and monitoring of crop phenological stages are essential factors in agriculture for estimating crop production. In this context, Synthetic Aperture Radar (SAR) data have been extensively used, especially for crop classification [1,2], yield estimation, and biophysical parameter retrieval [3–5], formulation of vegetation indices [6,7] and descriptors [8]. Moreover, SAR sensors have high spatial resolution and all-weather imaging capability. The scattered SAR signal is significantly affected by the geometry of crop canopy, underlying soil roughness, and dielectric properties of both crop and soil. In addition to the acquisition wavelength, polarization and angle of incidence have a significant influence on the received information [9,10]. Over agricultural fields, crop biophysical parameters impact SAR backscatter, particularly at advanced vegetative stages. Soil roughness and moisture remain the governing factors for SAR backscatter response during initial growth periods [11].

In particular, Sentinel-1 dual-pol images provide opportunities for crop classification and biophysical retrieval-related applications. Sentinel-1 also provides large swaths and lower data volumes due to reduced polarimetric dimensionality [12]. In the literature, dual-pol intensity parameters are widely used for crop type identification [1,13–19]. The existing studies also extend to retrieval of crop biophysical parameters [4,20,21] and crop phenology monitoring [22,23].

For dual polarimetric HH-HV or VV-VH data, Cloude et al. [24] proposed an eigen-based decomposition technique. In this technique, the $2 \times 2$ covariance matrix is decomposed into two orthogonal eigenvectors that are further utilized to derive a target characterization parameter. This average scattering angle $\overline{\alpha}$ is obtained from the two orthogonal polarization states weighted by their respective pseudo probabilities obtained from the eigenvalues. The target scattering entropy $H_{\mathrm{xP}}$ is obtained from the pseudo probabilities. This target characterization parameter and $H_{\mathrm{xP}}$ are further utilized to propose an unsupervised clustering scheme. Similarly, Ainsworth et al. [25] proposed another characterization parameter, $\theta$, utilizing an eigen-based technique. The parameter is related to the cross, $\sigma^{\circ}_{\mathrm{XY}}$ and co-pol, $\sigma^{\circ}_{\mathrm{XX}}$ ratios. In addition, $\theta$ and $H_{\mathrm{xP}}$ are used to propose a clustering technique for scattering target identification. In their study, the clustering plane consists of eight zones that represent different scattering characteristics.

Sugimoto et al. [26] provided a comparison between polarimetric parameters obtained from dual and full polarimetric SAR data. They reported a high 2D correlation between the parameters derived from the full-pol $H/\alpha$ decomposition and the dual-pol (HH+VV) $H/\alpha$ decomposition. For the full polarimetric data, the analysis was performed using model-based three [27], and four [28] component decompositions. Several other descriptors were introduced for diverse applications: clustering [29], maritime applications [30,31], etc.

To monitor the growth of vegetation, different vegetation indices, such as the Radar Vegetation Index (RVI) [32], Dual-Pol SAR Vegetation Index (DPSVI) [33], and Dual-pol Radar Vegetation Index (DpRVI) [34], can be derived. In general, these descriptors capture the randomness from the complex crop canopy to describe their phenological stages. Apart from these, machine learning models have shown promising results for crop biophysical parameter estimation [22,35,36]. It was observed from these studies that the cross-pol ratio is the most important parameter for vegetation studies. Nasrallah et al. [37] utilized the Gaussian function on the time series SAR data to find the date of significant phenology stages for wheat. Wali et al. [38] analyzed the sensitivity of the temporal backscatter for rice crops using the line regression technique. Similarly, the interferometric phase for identifying wheat phenological stages is observed in a study by Schlund et al. [39].

Usually, polarimetric parameters have been suitably attributed to the physical properties of the crop canopy [40–42], and have therefore helped monitor crop phenology. Recently, Dey et al. [43,44] proposed a model-free target characterization parameter using full, compact and dual co-pol data [22]. This parameter utilizes the Barakat degree of polarization [45] and elements of the coherency matrix to characterize diverse target types. In their study, it was shown that this parameter enhanced the target characterization capability as compared to the $\overline{\alpha}$ parameter.

Motivated by this work, we derive a target characterization parameter for dual-pol SAR data in this study. This parameter jointly utilizes the 2D Barakat degree of polarization and the elements of the covariance matrix. Therefore, this parameter, $\theta_{\mathrm{xP}}$, captures the additional scattering information which the other existing descriptors might miss. Alongside this, we also propose an unsupervised clustering technique using $\theta_{\mathrm{xP}}$ and entropy, $H_{\mathrm{xP}}$ parameters. We utilize $\theta_{\mathrm{xP}}$ and the clustering scheme to analyze the temporal variation of canola and wheat over a Canadian test site derived from Sentinel-1 C-band SAR data.

## 2. Methodology

### 2.1. Study Area and Dataset

The test site in Canada is located in southern Manitoba (49°40′32″N, 97°59′57″S). This is one of the Canadian Joint Experiment for Crop Assessment and Monitoring (JECAM) test sites. Nominal field sizes range from 20–30 hectares to 50–60 hectares, with wheat, soybean, canola, and corn as major annual crops. The land under permanent grassland is $< 5\,\%$. The in-situ measurements were conducted during Soil Moisture Active Passive

Validation Experiment 2016 Manitoba (SMAPVEX16-MB) campaign. Figure 1 shows the location of the test site and the distribution of canola and wheat fields.

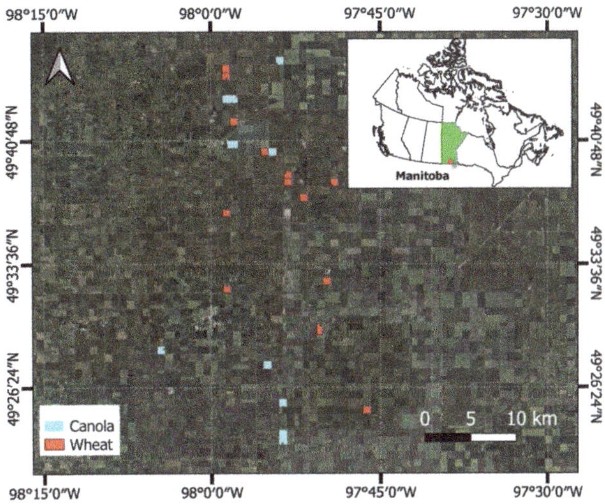


**Figure 1.** Map of the study area showing the locations of the canola and wheat fields over the Manitoba province of Canada.

The in-situ measurement campaign was carried out from June to July 2016 over 50 different fields. The measurement period mainly consisted of two temporal windows: 8 June to 22 June and 8 July to 22 July 2016. Within this temporal extent, most crops advanced from early to advanced phenological stages achieving peak biomass. The nominal size of each field is approximately 800 m × 800 m. Vegetation sampling, which includes biomass, Plant Area Index (PAI), and plant height, was performed in three points for each field. Crop biomass was collected via destructive sampling. A 0.5 m × 0.5 m square was placed over the canopy for canola and wheat. All above-ground biomass was collected by cutting all vegetation at the soil level within the square. The PAI was measured with hemispherical digital photos. In this technique, a camera with a fisheye lens captures photos of the crop canopy with the camera positioned at least 50 cm above or below the canopy. Details of the in-situ biophysical parameters (viz. PAI, Dry biomass and VWC) of Canola and Wheat are provided in Appendix A.3. One can find further details on the sampling schema in the SMPAVEX16-MB field report [46].

### 2.2. SAR Data Pre-Processing

We acquired the Terrain Observation with Progressive Scans SAR (TOPSAR) mode Level-1 Sentinel-1 Interferometric Wide (IW) swath Single Look Complex (SLC) data over the test site. To this SLC configuration, the swath length is ≈250 km and the spatial resolution is 5 m × 20 m. The whole IW swath is divided into three sub-swaths (IW1, IW2, and IW3) with nine bursts in the azimuth direction. Hence, prior to application, these images are pre-processed with standard correction steps [47].

The Sentinel-1 images are ingested in the SNAP 8.0 platform. Following this, the sub-swaths and bursts are selected based on the test site location, and the state vectors are updated with the Sentinel-1 precise orbit file. The images are then radiometrically calibrated. As we are interested in generating the complex covariance $\mathbf{C}_2$ matrix information, and hence, we need to extract both amplitude and phase values. Due to this reason, the calibration output is set to complex output values.

All Sentinel-1 images are then back-geocoded using the Sentinel-1 Back Geocoding operator. This operation essentially co-registers the temporal images with sub-pixel accuracy. We utilize the SRTM 1Sec Grid as the Digital Elevation Model (DEM). Subsequently, the Sentinel-1 TOPS deburst and merge operations are performed to produce a single SLC image. The stack of images is then clipped within the in-situ measurement location to reduce the data volume and increase the computation speed.

These image subsets are multi-looked with a factor of 4 in range and 1 in azimuth directions to generate a square pixel. The final pixel resolution is $\approx 15$ m. Finally, these multi-looked images are used to generate covariance elements images. A $5 \times 5$ boxcar filter is applied to further reduce the speckle information within the images. Next, the baseline information is deleted from the metadata, and the covariance images are exported in the PolSARpro format. Here, please note that the covariance elements have ensembled information that helps in applying the second-order statistics.

Then, we compute the target characterization parameter, $\theta_{\text{xP}}$ using the PolSARtools software in the QGIS platform [48]. Subsequently, we geocode the images with UTM projected coordinate system. We further analyzed the in-situ measurement locations and extracted target characterization parameters from the geocoded products. From several Sentinel-1 images acquired during the campaign, four dual-polarization (VV and VH) C-band Sentinel-1 Single Look Complex (SLC) data were selected for use in this study (Table 1). The selection of Sentinel-1 datasets was based on acquisition dates that were near coincident with in-situ measurement periods.

**Table 1.** Specification for Sentinel-1 data acquired for the Canadian test site.

| Acquisition Date | Beam Mode | Incidence Angle Range (Deg.) | Orbit | $az$ (m) $\times$ $rg$ (m) |
|---|---|---|---|---|
| 13 June 2016 | IW | 30.22–32.47 | Ascending | $15 \times 15$ |
| 7 July 2016 | IW | 30.22–32.47 | Ascending | $15 \times 15$ |
| 19 July 2016 | IW | 30.22–32.47 | Ascending | $15 \times 15$ |
| 24 August 2016 | IW | 30.22–32.47 | Ascending | $15 \times 15$ |

## 2.3. Target Characterization Parameter

The scattering vector for a dual-polarized SAR data are represented as, $k_d = [S_{XX}\ S_{XY}]^T$, where $X$ and $Y$ correspond to either horizontal (H) or vertical (V) polarization states. Using these scattering vectors, we define the $2 \times 2$ covariance matrix as

$$\mathbf{C}_2 = \langle k_d \cdot k_d^{*T} \rangle = \begin{bmatrix} \langle |S_{XX}|^2 \rangle & \langle S_{XX} S_{XY}^* \rangle \\ \langle S_{XY} S_{XX}^* \rangle & \langle |S_{XY}|^2 \rangle \end{bmatrix} \tag{1}$$

where $\langle \cdot \rangle$ denotes ensemble average, and $T$ denotes vector transpose. We define the elements of the $2 \times 2$ matrix as, $\mathbf{C}_{11} = \langle |S_{XX}|^2 \rangle, \mathbf{C}_{22} = \langle |S_{XY}|^2 \rangle$ and $\mathbf{C}_{12} = \langle S_{XX} S_{XY}^* \rangle$.

Similar to the conventional degree of polarization, the 2D Barakat degree of polarization [49] given in Equation (2) ($0 \leq m_{\text{xP}} \leq 1$) also characterizes the state of polarization (or purity) of an EM wave. For a completely polarized EM wave, $m_{\text{xP}} = 1$ and for a completely unpolarized EM wave, $m_{\text{xP}} = 0$. In between these two extreme cases, the EM wave is said to be partially polarized, $0 < m_{\text{xP}} < 1$:

$$m_{\text{xP}} = \sqrt{1 - \frac{4|\mathbf{C}_2|}{\left(\text{Tr}(\mathbf{C}_2)\right)^2}}, \tag{2}$$

where $|\cdot|$ is the determinant of a matrix and Tr is the trace of a matrix. Let us consider two auxiliary quantities defined as

$$\tan \eta_1 = \frac{C_{11}}{m_{\text{xP}}\,\text{Span}} \quad \text{and} \quad \tan \eta_2 = \frac{C_{22}}{m_{\text{xP}}\,\text{Span}}, \tag{3}$$

where $\eta_1$ and $\eta_2$ are two free variables, and Span $= C_{11} + C_{22}$. The quantity $\dfrac{C_{11}}{m_{\text{xP}}\,\text{Span}}$ indicates the co-polarized scattering power with respect to the total polarized power, while $\dfrac{C_{22}}{m_{\text{xP}}\,\text{Span}}$ indicates the cross-polarized scattering power with respect to the total polarized power. Furthermore, using a simple relationship, we obtain:

$$\begin{aligned}\tan\theta_{\text{xP}} &= \tan(\eta_1 - \eta_2)\\ &= \frac{m_{\text{xP}}\,\text{Span}\,(C_{11} - C_{22})}{C_{11}\,C_{22} + m_{\text{xP}}^2\,\text{Span}^2}.\end{aligned} \tag{4}$$

where $\theta_{\text{xP}} \in [-45^\circ, 45^\circ]$.

The eigen-decomposition of $\mathbf{C}_2$ can be expressed as

$$\mathbf{C}_2 = \mathbf{U}_2 \mathbf{\Sigma} \mathbf{U}_2^{-1}, \tag{5}$$

where $\mathbf{\Sigma}$ is a $2 \times 2$ diagonal matrix with non-negative real elements, $\lambda_1 \geq \lambda_2 \geq 0$, which are the eigenvalues of $\mathbf{C}_2$, and $\mathbf{U}_2$ is $2 \times 2$ unitary matrix, where $u_i$'s are the unit orthogonal eigenvectors. We define the pseudo probabilities, $p_i$, in terms of the eigenvalues as

$$p_i = \frac{\lambda_i}{\sum_{k=1}^{2} \lambda_k}, \tag{6}$$

which we then use to define the scattering entropy as

$$H_{\text{xP}} = -\sum_{k=1}^{2} p_k \log_2(p_k). \tag{7}$$

The degree of polarization $m_{\text{xP}}$ characterizes the degree of coherence of partially polarized waves. In contrast, the entropy $H_{\text{xP}}$ describes the degree of statistical disorder associated with partially polarized waves. The parameter $\theta_{\text{xP}}$ describes the scattering characteristics of a target. Therefore, in this work, these parameters provide information about the growth stage or phenology of crops.

### 2.4. $H_{\text{xP}}/\theta_{\text{xP}}$ Bound

The feasible regions in the $H_{\text{xP}}/\theta_{\text{xP}}$ plot is represented by the bounding curves, Curve I ($\mathbf{C}_{2(I)}$) and Curve II ($\mathbf{C}_{2(II)}$) as shown in Figure 2 and is given as

$$\mathbf{C}_{2(I)} = \frac{1}{1+k}\begin{bmatrix} 1 & 0 \\ 0 & k \end{bmatrix}, \quad \theta_{\text{xP}}^{\max} = \tan^{-1}\left(\frac{k^2 - 2k + 1}{k^2 - k + 1}\right) \tag{8}$$

$$\mathbf{C}_{2(II)} = \frac{1}{k+1}\begin{bmatrix} k & 0 \\ 0 & 1 \end{bmatrix}, \quad \theta_{\text{xP}}^{\min} = -\tan^{-1}\left(\frac{k^2 - 2k + 1}{k^2 - k + 1}\right) \tag{9}$$

where $k \in [0, 1]$ denotes the scattering amplitude ratio, and therefore, $\theta_{\text{xP}} \in [-45^\circ, 45^\circ]$. One can observe from Equation (4) that when:

- $m_{\text{xP}} = 0$ (i.e., when there exists no polarization structure in the scattered EM wave), then $\theta_{\text{xP}} = 0^\circ$ characterize random scattering from targets.
- $m_{\text{xP}} = 1$, and $\theta_{\text{xP}} = 45^\circ$, characterize coherent scattering from deterministic targets (i.e., trihedral or dihedral).
- $m_{\text{xP}} = 1$, and $\theta_{\text{xP}} = -45^\circ$, characterize cross-polarized scattering from complex targets.

Therefore, $\theta_{\text{xP}} \in [-45^\circ, 45^\circ]$ characterizes diverse scattering-type information in between these extreme cases, and hence it is suitable to characterize scattering-type information from various targets.

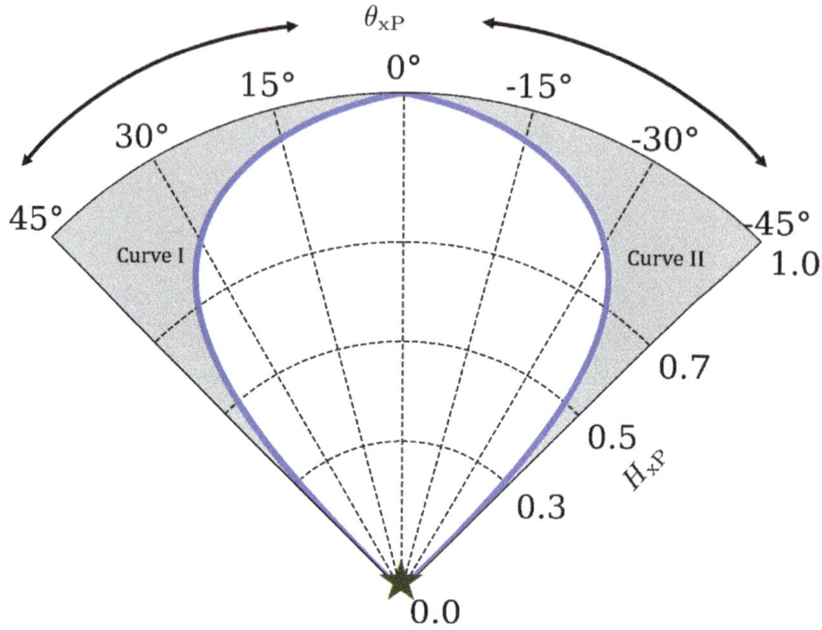


**Figure 2.** Feasible region of $H_{xP}/\theta_{xP}$ plane bounded by two curves (Curve I and Curve II). $\theta_{xP}$ is represented in the angular direction and $H_{xP}$ is represented in the radial direction. The non-feasible scattering region is shaded in gray.

*2.5. Example of Variation of $\widehat{\alpha}$ and $\theta_{xP}$*

In dual polarimetry, Cloude proposed a target characterization parameter, $\overline{\alpha} \in [0°, 90°]$ using the eigendecomposition approach [24]. In order to compare this with the proposed $\theta_{xP} \in [-45°, 45°]$, we have rescaled $\overline{\alpha}$ to $\widehat{\alpha} = 45° - \overline{\alpha}$, where both $\widehat{\alpha}$ and $\theta_{xP}$ vary from $-45°$ to 45°.

To show the efficacy of $\theta_{xP}$ over $\widehat{\alpha}$, we have considered three different land cover types, i.e., urban, oriented urban, and vegetation from the VV-VH data extracted from a full-pol C-band RADARSAT-2 data. The variations of $\theta_{xP}$ and $\widehat{\alpha}$ over these targets are shown in Figure 3. One can note from Figure 3 that the median value of $\widehat{\alpha}$ over orthogonal urban area is approximately 35°. In contrast, the median value of $\theta_{xP}$ is approximately 42°. The degree of polarization over the area is approximately 0.91. This high degree of polarization essentially confirms the nearly coherent scattering characteristics of the target. In this regard, $\theta_{xP}$ better characterizes the target to be coherent than $\widehat{\alpha}$.

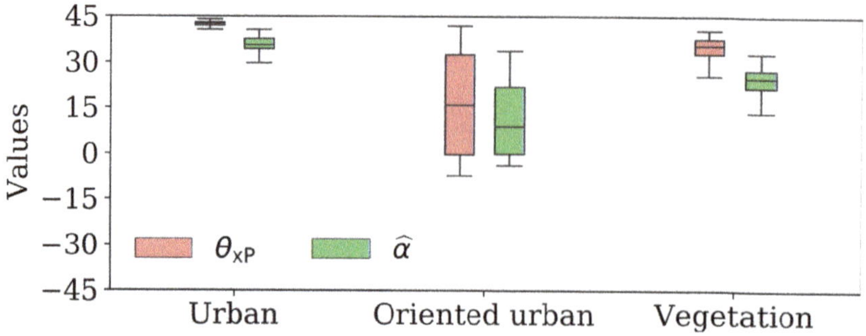


**Figure 3.** Variations of $\theta_{xP}$ and $\widehat{\alpha}$ for different scattering target from RADARSAT-2 data over San Francisco, CA, USA.

However, over oriented urban area, we observe a large variation of both $\theta_{xP}$ and $\widehat{\alpha}$. This large variation might be due to the diversity in the polarization randomness due to the orientation of buildings to the radar line of sight and orientation in the azimuth direction. Over vegetation surfaces, we observe a partially polarized scattered wave that consists of a major coherent part and a significant incoherent part. The coherent part might be due to the direct scattering from the nearly smooth soil surface, while the incoherent part is due to the scattering from the volume media. In this case, $\theta_{xP}$ also correctly characterizes the coherent contribution of the target during scattering as compared to $\widehat{\alpha}$. A $\widehat{\alpha} - \theta_{xP}$ plot for RADARSAT-2 data over San-Francisco in VV-VH mode is shown in Appendix A.1. Therefore, from these analyses, we can observe the enhancement of target scattering characterization using $\theta_{xP}$. Thus, $\theta_{xP}$ can be further utilized to characterize different phenological stages of diverse crop types.

### 2.6. Unsupervised Clustering Zones over Vegetative Surface

In this study, we are interested in observing the changes in the scattering mechanisms over different phenological stages of crops. One should note that, for natural targets, $C_{11} \geq C_{22}$ almost always, and hence, $\theta_{xP} \in [0°, 45°]$. Let us now characterize $m_{xP}$ and $\theta_{xP}$ for a few particular scattering scenarios:

- For a pure diffused target, $m_{xP} = 0$, implies, $\theta_{xP} = 0°$.
- For pure or point scatterer, $m_{xP} = 1$ and $\eta_2 = 0°$, implies $\theta_{xP} = 45°$.
- Infeasible scattering: $C_{11} = C_{22}$ and $m_{xP} = 1$.

Therefore, in terms of vegetation development, for a highly random vegetative structure, $m_{xP} \approx 0$ and $\theta_{xP} \approx 0°$. However, a slightly rough soil surface acts like a Bragg scatterer, and hence $m_{xP} \approx 1$ and $\theta_{xP} \approx 45°$.

The variations of $\theta_{xP}$ and $m_{xP}$ over bare field and vegetative field types are shown in Figure 4. It can be noted that $\theta_{xP}$ and $m_{xP}$ are highly sensitive with the canopy structure. Interestingly, we observed that the sensitivity is much higher for the highly erectophile crop structures such as canola, corn, and wheat. In contrast, the sensitivity is low for the soybean crop. This high sensitivity of $\theta_{xP}$ might be due to the volume decorrelation with an increased canopy structure.

In this study, we utilize the $\theta_{xP}$ and $H_{xP}$ parameters to propose a new unsupervised clustering scheme to describe the target scattering behavior. The overall clustering scheme consists of 12 zones: Z1, Z2, Z3, Z4, Z5, Z6, Z7, Z8, Z9, Z10, Z11, and Z12. These divisions of the $H_{xP}/\theta_{xP}$ plane are based on certain scattering symmetry assumptions. In particular, the scattered Stokes vector, $\vec{\mathbf{S}}_s$ is a function of the target property, represented in terms of the Kennaugh matrix, $\mathbf{K}$ and the transmitted Stokes vector, $\vec{\mathbf{S}}_t$. Moreover, the two Stokes vectors are related as $\vec{\mathbf{S}}_s = \mathbf{K}\,\vec{\mathbf{S}}_t$. Furthermore, one can note that $\vec{\mathbf{S}}_s$ is also a function of the scattering order ($n$) [50]. Thus, the definition of scattering entropy, $H$ [51,52] can be expressed in terms of $n$. Following this, we can observe that $H$ increases with increasing $n$. For example, $H = 0$ for $n = 0$; $H \approx 0.3$ for $n = 1$; $H \approx 0.5$ for $n = 2$, and, $H \approx 0.7$ for $n = 3$. However, for $n > 3$, change in $H$ is insignificant. Therefore, following this trend in the variation of the entropy, we divided the radial axis $H_{xP}$ into four sub-sections by considering the values: 0.3, 0.5, and 0.7. An illustration of the $H_{xP}/\theta_{xP}$ clustering plane with the zones is shown in Figure 5.

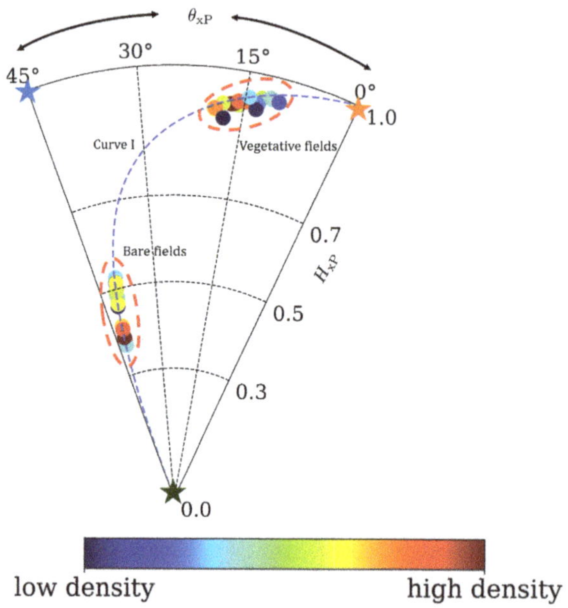


**Figure 4.** Variations of $\theta_{xP}$ and $H_{xP}$ for bare field conditions and fully developed crops. These measurements are derived over in-situ fields.

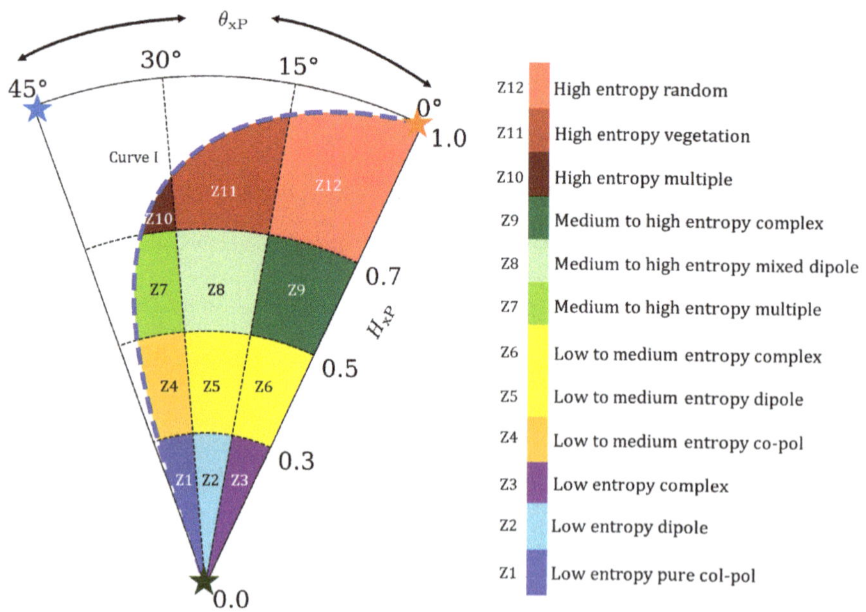


**Figure 5.** Twelve clustering zones in the $H_{xP}/\theta_{xP}$ plane.

## 3. Results and Discussion

In this section, we analyze the temporal dynamics of crops using the proposed dual-polarimetric descriptors. Furthermore, we utilize the proposed clustering framework to assess the phenological stages of the two crops, i.e., canola and wheat, from the C-band Sentinel-1 dual-pol SLC SAR data.

### 3.1. Canola

The spatio-temporal changes of $\theta_{xP}$ are shown in Figure 6. The temporal variations of $\theta_{xP}$ and $m_{xP}$ are shown in Figures 7 and 10. For temporal analysis, we considered 48 sampling points from three fields (Field No. 206, 208 and 224). We plot the variations of the proposed SAR descriptors with respect to the phenological changes of canola at different dates.

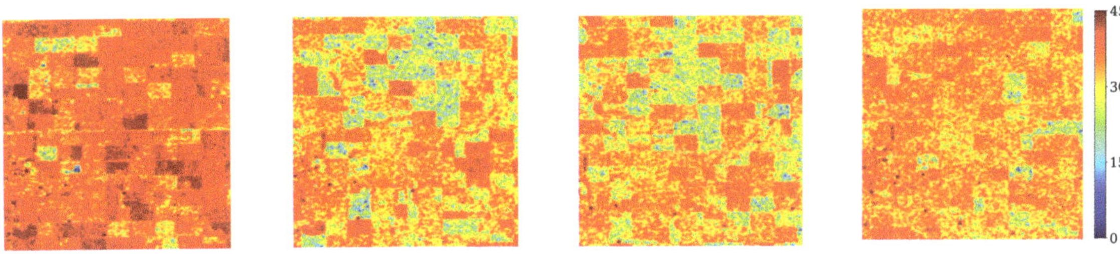


(a) 13 June (b) 7 July (c) 19 July (d) 24 August

**Figure 6.** Temporal variation of $\theta_{xP}$ over the study area. The dates are **(a)** 13 June, **(b)** 7 July, **(c)** 19 July and **(d)** 24 August.

According to the in-situ information, canola was sown at the end of May. Hence, during June, the phenological stage of canola was bounded within the early to mid vegetative stage. In this regard, it is worth mentioning that canola is a broad-leaf crop with comparable leaf size to the wavelength of C-band SAR ($\approx$5.6 cm). Therefore, due to the formation of dense rosette near the soil surface, the leaf structure considerably affects the SAR backscatter values. Thus, due to the high dynamic variations of crop morphology, canola is of particular interest in SAR research.

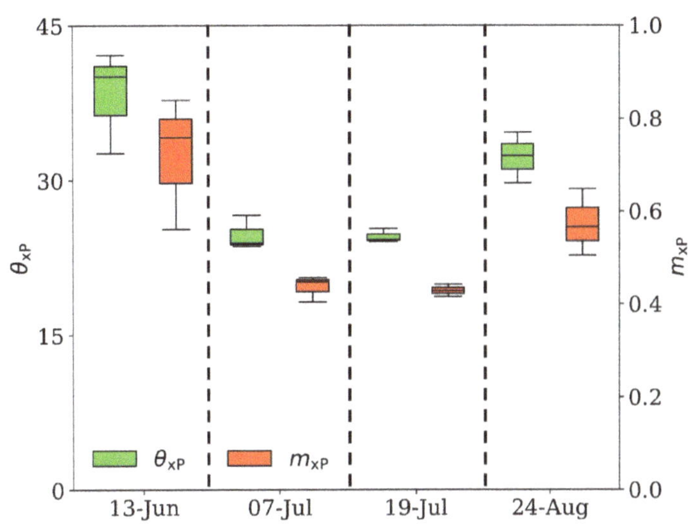


**Figure 7.** Temporal variation of $\theta_{xP}$ and $m_{xP}$ over the canola fields.

From Figure 7, we can observe that, on 13 June, the median value of $m_{xP} \approx 0.78$. This high value of $m_{xP}$ typically suggests high polarized scattered components from canola fields. This trend might be due to the low vegetation density in the fields. From the in-situ information, we observe that the canola was in the stem elongation stage during this time. Consequently, low biomass and low Plant Area Index (PAI) are evident at this period. Moreover, the vertical crop structure slightly attenuated the transmitted vertical

(V) polarization. Hence, we observe a median value of $\theta_{xP} \approx 40°$. Apart from this, the low depolarization in the scattered wave might be due to the scattering from leaves and canopy structures combined.

Furthermore, Figure 8 provides information about the different morphological characteristics of canola at this particular period. We observe that the points from these canola fields are mainly distributed in two separate zones: Z4 and Z7 in Figure 8a. This information typically infers that, during 13 June, the crop density in some canola fields was low. As a result, low entropy pure scattering was evident from those fields (Z4). In contrast, other fields had comparatively dense foliage structures, which increased scattering randomness increased (Z7). From in-situ measurement, we observe that fields 206 and 208 had higher PAI and biomass values than field 224. In field 224, most canola crops were at the leaf development stage. In addition, we observe a point in the Z1 region and points in the Z10 zone, indicating more advanced phenological stages. These variations from Z1 to Z10 reflect the differences in sowing dates for canola, and, consequently, differences in crop emergence and development among fields early in the season.

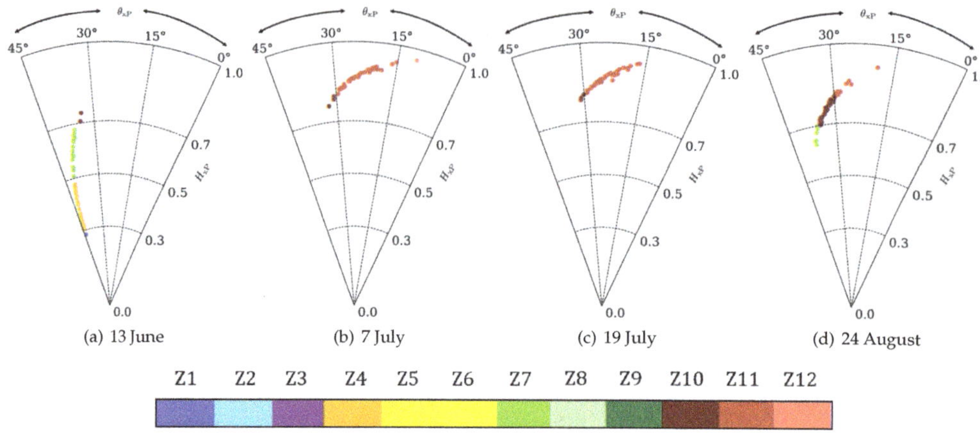


**Figure 8.** Temporal variation of clusters on $H_{xP}/\theta_{xP}$ plane over the canola fields. The dates are (**a**) 13 June, (**b**) 7 July, (**c**) 19 July, and (**d**) 24 August.

On 7 July, canola crops reached their flowering stage. During this time, we can observe a sudden drop in the $\theta_{xP}$ and $m_{xP}$ values. The median value of $m_{xP} \approx 0.44$. This sudden decrease in the scattered polarization structure is due to the development of flowers, stems, and branches. During this time, the high attenuation of the V-polarized transmitted wave also lowered the $\theta_{xP}$ value. Furthermore, the generation of high cross-polarization due to the complex crop canopy structure also aided in lowering $\theta_{xP}$ values. During this time, the median value of $\theta_{xP}$ is approximately 20°.

Similar changes in the clustering zones are evident from Figure 8b. We observe a shift of clustering zone from Z4 and Z7 to Z11. The formation of the dense cluster in Z11 indicates high randomness in the scattered wave due to the complex flower and branch structures. Moreover, during the flowering and early pod development stages, the leaf density drastically reduces. Canola crops drop their leaves during this period of pod and seed development. However, this dense cluster in Z11 also suggests that the number of flowers per plant was high, and the health of the crop was good. On the other hand, we observe some points in the comparatively low entropy region, Z10.

On 19 July, small changes in $\theta_{xP}$ and $m_{xP}$ are observed as compared to 7 July. This effect might be due to the existence of a similar complex canopy structure due to which the depolarization is evident in the scattered wave. During this early to the mid-pod development stage, ramified stems and seeds appear. In the early stage, seeds remain watery and translucent, while seeds reach maximum dry weight at the advanced stage. Therefore, depending upon this particular phenomenon, changes in the scattering mechanisms might be observed. However, during the satellite acquisition, most fields reached the early to mid-stage of pod development. As a result, the variance in $\theta_{\rm xP}$ is low during this time. The median value of $\theta_{\rm xP}$ is $\approx$18°, while the median of $m_{\rm xP}$ is approximately 0.4.

We also observe the accumulation of $H_{\rm xP}/\theta_{\rm xP}$ cluster in Z11 region in Figure 8c. During this time, the complex vegetation canopy structure increased the randomness in the scattered wave. Consequently, we observe a marginal increase in $H_{\rm xP}$ compared to 7 July. However, as stated earlier, most of the fields reached the early to mid-stage of pod development, and thus we observe the majority of the point cloud in the Z11 region.

Subsequently, on 24 August, we observe an increase in $\theta_{\rm xP}$ values. Similarly, the increase in $m_{\rm xP}$ suggests that the polarized component in the scattered wave has increased during this period. This increase in the values of $\theta_{\rm xP}$ and $m_{\rm xP}$ might be due to the changes in the canopy moisture content. However, during this period, high variance in the data is evident. During this period, the median value of $\theta_{\rm xP}$ is approximately 34°, while the median value of $m_{\rm xP}$ is approximately 0.58. In Figure 8d, we observe three distinct zones; Z7, Z10, and Z11. This phenomenon is due to the harvest of canola crops. The post-harvest residue might have generated marginal depolarization in the scattered wave. The percentages of pixels at each date are shown in Table 2.

**Table 2.** Temporal variation in the percentage of data points in each zone for different phenology stages of canola. The zones with high percentages of points at a particular phenology stage are highlighted in bold.

| Dates | Z1 | Z2 | Z3 | Z4 | Z5 | Z6 | Z7 | Z8 | Z9 | Z10 | Z11 | Z12 |
|---|---|---|---|---|---|---|---|---|---|---|---|---|
| 13 June | 2.0 | 0.0 | 0.0 | **58.0** | 0.0 | 0.0 | **35.4** | 0.0 | 0.0 | 4.6 | 0.0 | 0.0 |
| 7 July | 0.0 | 0.0 | 0.0 | 0.0 | 0.0 | 0.0 | 0.0 | 0.0 | 0.0 | 6.2 | **91.6** | 2.2 |
| 19 July | 0.0 | 0.0 | 0.0 | 0.0 | 0.0 | 0.0 | 0.0 | 0.0 | 0.0 | 6.2 | **93.8** | 0.0 |
| 24 August | 0.0 | 0.0 | 0.0 | 0.0 | 0.0 | 0.0 | 12.6 | 0.0 | 0.0 | **64.6** | **22.8** | 0.0 |

### 3.2. Wheat

In this section, we analyse the temporal variation of $\theta_{\rm xP}$ and $m_{\rm xP}$ over wheat fields. Wheat is an edible cereal grass with an erectophile canopy structure. Over the test site, sowing of wheat was performed during the beginning of May. This study has considered three wheat fields (Field No. 233, 220, 62) for the temporal evaluation. The temporal patterns of $\theta_{\rm xP}$ and $m_{\rm xP}$ are shown in Figure 10. Alongside this, we represent the dynamic behaviour of unsupervised clusters in Figure 9.

On 13 June, wheat crops reached the tillering stage. This stage starts with the appearance of tillers and ends with mature leaf sheaths. During this time, the crop density remains low. As a result, the scattered wave remains significantly polarized. It can be observed from Figure 10 that the median value of $m_{\rm xP}$ during this period is 0.78. This essentially indicates that approximately 78 % of the scattered wave is completely polarized, while approximately 22 % is depolarized. This marginal depolarization effect might be due to the canopy interaction. Alongside this, the median value of $\theta_{\rm xP}$ is $\approx$39°, indicating that the target is a closely coherent scatterer. However, the high standard deviations in both $\theta_{\rm xP}$ and $m_{\rm xP}$ represent the variations in the uneven phenological advancement within the fields due to variations in seeding dates.

The clustering zones on 13 June are shown in Figure 9a. Like canola, we can observe that the clusters are distributed in four distinct zones; Z4, Z7, Z10, and Z11. The existence of these distinct zones is due to the differences in the morphological characteristics of wheat crops. From in-situ measurements, it is observed that the plant density of Field No. 220 was low ($\approx$100 m$^{-2}$). In contrast, the plant density over Field Nos. 233 and 62 was $\approx$125 m$^{-2}$ and $\approx$190 m$^{-2}$, respectively. Similarly, high PAI and biomass are evident over these two fields. As a result, a high attenuation of the transmitted wave is observed over Field No. 233 and 62. Hence, we observe clouds of clusters at Z10 and Z11. On the other hand, the attenuation of the scattered wave is comparatively lower over Field No. 220. Hence, cluster formation at Z4 and Z7 is observed in Figure 9a.

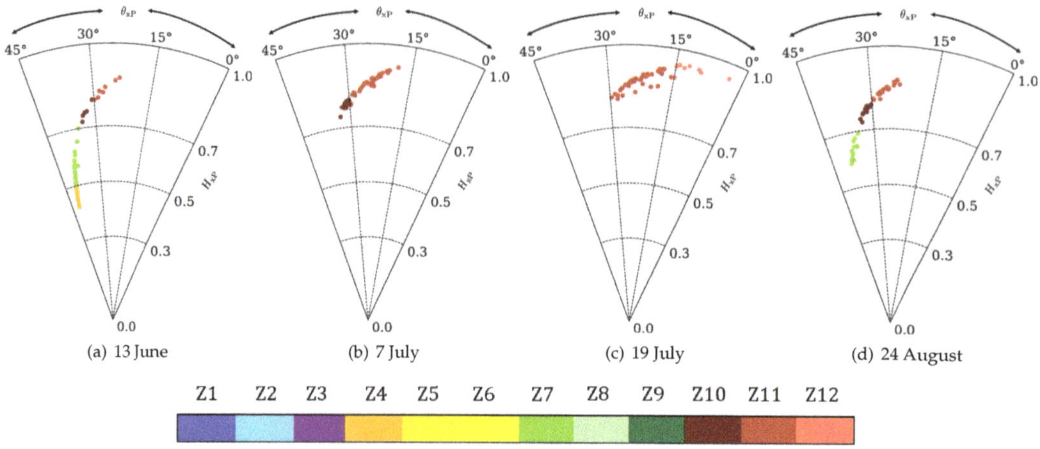


**Figure 9.** Temporal variation of clusters on $H_{xP}/\theta_{xP}$ plane over the wheat fields. The dates are (**a**) 13 June; (**b**) 7 July; (**c**) 19 July and (**d**) 24 August.

On 7 July, we observe a decrease in both $\theta_{xP}$ and $m_{xP}$ values due to the appearance of more branches and leaves. The median value of $\theta_{xP}$ is around 27°, while the median value of $m_{xP}$ is around 0.49. At this stage, wheat has advanced to the early flowering stage. At this stage, flowers appear on the upper portion of the canopy layer. As a result, the dense wheat structure and the flowers increase the multiple interactions of the EM wave. Hence, a high proportion of depolarized components in the scattered wave is evident during this period. Similar changes in the clusters are evident from Figure 9b. At this period, the clusters shift to Z11 from Z4 and Z7 due to the increased scattering entropy due to the randomness in the scattered wave by the targets. However, a cloud of clusters is seen at Z10, which might be due to some wheat crops having late phenological development.

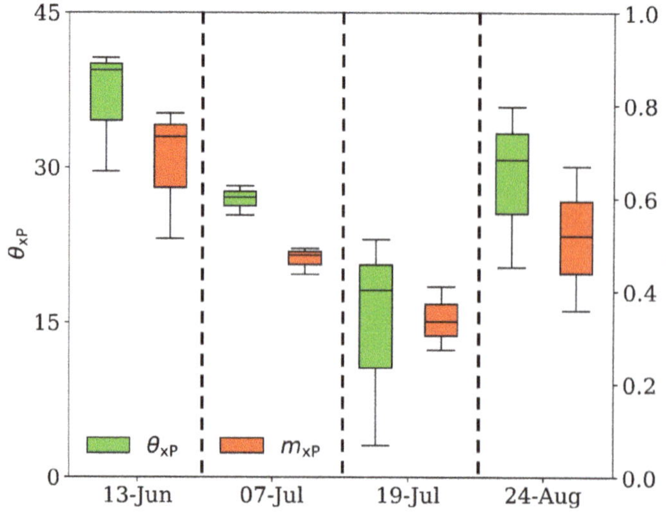


**Figure 10.** Temporal variation of $\theta_{xP}$ and $m_{xP}$ over the wheat fields.

On 19 July, wheat crops advanced to the early dough stage. At this phenological period, wheat grains appear, and it remains milky. Therefore, the dielectric property of grains affects the scattering mechanism. Moreover, the denser wheat canopy increases multiple scattering. As a result, the values of $m_{xP}$ reduce further as compared to 7 July. Similarly, the coherent component in the scattered EM wave also reduces. The median value of $m_{xP}$ is $\approx 0.37$. The median value of $\theta_{xP}$ is $\approx 20°$. The high variation in the values of $\theta_{xP}$ and $m_{xP}$ might be due to the randomly oriented wheat stems and heads. Similar scattering characteristics from the wheat during the dough stage are reported by Wu et al. [53]. In particular, during this period, the scattering phenomenon occurs primarily from two parts of the wheat canopy, i.e., thick upper canopy layer and relatively sparse lower canopy layer.

Further from Figure 9c, we can observe the formation of clusters in the Z11 region. However, we can also observe some points in Z12 due to additional complex scattering mechanisms from the wheat canopy subjected to partial lodging or related canopy anomalies. It is important to note that we observe much spread of points in the Z11 zone, unlike other phenological stages of canola and wheat. This suggests that, during this particular phenology period, the appearance of grains and their orientations might be responsible for breaking the reflection asymmetry condition during scattering from the wheat canopy. Moreover, as stated earlier, the complex canopy geometry, the spatial distribution of grains, and their random orientations increased the overall randomness in the scattered EM wave. As a result, for the majority of the points, we observe high entropy values.

During 24 August, the harvest of wheat started. At this stage, the interaction of EM waves mainly takes place with the soil surface and post-harvest residue. Consequently, a reverse trend is observed in $\theta_{xP}$ and $m_{xP}$ values. From Figure 10, we can observe that the median value of $m_{xP}$ is approximately 0.58, while the median value of $\theta_{xP}$ is approximately 34°. These values indicate the coherent component increase in the scattered EM wave from the wheat canopy. Additionally, Figure 9d shows three different clustering zones, Z7, Z10, and Z11, during this period. The percentages of temporal variations of clusters at different zones are shown in Table 3. In addition, the temporal variations of $\theta_{xP}$ and $\widehat{\alpha}$ are shown in Appendix A.2. One can observe that the temporal signature of these parameters differs for wheat and canola. Therefore, these descriptors can be efficiently utilized to distinguish wheat and canola fields using time series data. Furthermore, it might also be possible to distinguish these two crop fields from other land cover targets. The temporal variation of the clusters over the cultivation area is shown in Figure 11.

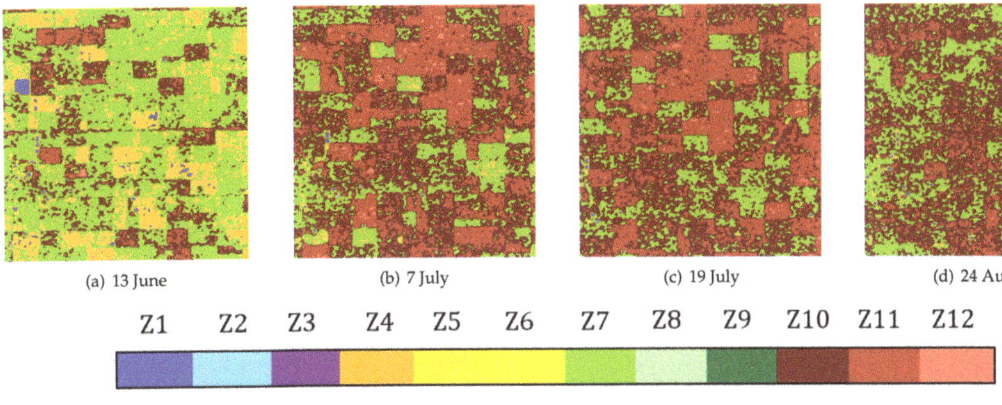


**Figure 11.** Images of temporal variation of clusters on $H_{xP}/\theta_{xP}$ plane over the study area. The dates are (**a**) 13 June; (**b**) 7 July; (**c**) 19 July and (**d**) 24 August.

**Table 3.** Temporal variation in the percentage of data points in each zone for different phenology stages of wheat. The zones with high percentages of points at a particular phenology stage are highlighted in bold.

| Dates | Z1 | Z2 | Z3 | Z4 | Z5 | Z6 | Z7 | Z8 | Z9 | Z10 | Z11 | Z12 |
|---|---|---|---|---|---|---|---|---|---|---|---|---|
| 13 June | 0.0 | 0.0 | 0.0 | 27.0 | 0.0 | 0.0 | **47.9** | 0.0 | 0.0 | 10.4 | 14.7 | 0.0 |
| 7 July | 0.0 | 0.0 | 0.0 | 0.0 | 0.0 | 0.0 | 0.0 | 0.0 | 0.0 | 18.8 | **81.2** | 0.0 |
| 19 July | 0.0 | 0.0 | 0.0 | 0.0 | 0.0 | 0.0 | 0.0 | 0.0 | 0.0 | 0.0 | **83.3** | 16.7 |
| 24 August | 0.0 | 0.0 | 0.0 | 0.0 | 0.0 | 0.0 | **31.3** | 0.0 | 0.0 | **29.1** | **39.6** | 0.0 |

## 4. Conclusions

In this study, we propose a new polarimetric target characterization parameter, $\theta_{\text{xP}} \in [-45°, 45°]$ for dual-pol Sentinel-1 Synthetic Aperture Radar (SAR) data. However, over natural surfaces, such as vegetation, $\theta_{\text{xP}} \in [0°, 45°]$, where $\theta_{\text{xP}} = 0°$ denotes totally incoherent targets and $\theta_{\text{xP}} = 45°$ denotes totally coherent targets. Utilizing the scattering entropy parameter, $H_{\text{xP}}$ along with $\theta_{\text{xP}}$, we proposed a novel unsupervised classification scheme for dual-pol SAR data. We further utilized $\theta_{\text{xP}}$ and the classification scheme to analyze the phenological development of canola and wheat over a Canadian test site. The results are promising and exhibit high sensitivity with the morphological changes of the crops at each phenological stage.

We observe a broad dynamic range of $\theta_{\text{xP}}$ from the temporal analysis of both canola and wheat starting from their early vegetative stage to maturity and harvest stage. For both crops, the variation of $\theta_{\text{xP}}$ within the desired phenology window is $\approx 40°$ to $\approx 15°$. High values of $\theta_{\text{xP}}$ during the early stages are due to the low crop density within the fields. In contrast, the low $\theta_{\text{xP}}$ values during the flowering stage are due to the complex canopy structure and additional appearance of flowers and grains. Moreover, we observe a trend reversal for both canola and wheat on 24 August. This trend reversal is due to the harvest of both crops during this period.

Within the scope of this study, we have characterized diverse crop phenological stages in terms of the physical scattering of the electromagnetic wave. The unsupervised classification scheme comprises twelve distinct zones, which represent these different physical scattering mechanisms. Thus, these clustering zones are beneficial for identifying the morphological status of the crop when a priori information is not available. Alongside this, the clustering scheme can also capture the difference in the growth stages simultaneously. Moreover, during the initial growth period, most of the points lay in the low entropy zones, while those points shifted towards high entropy zones at advanced phenological stages. This phenomenon is due to the changes in the canopy complexity with the advancement of the crop phenological stages. We already know that the dual-pol configuration does not have either the VV or HH polarization information. Hence, by utilizing $\theta_{\text{xP}}$, we are unable to uniquely characterize scattering-type information, i.e., whether it is an even-bounce or odd-bounce mechanism. In addition, as stated earlier, if the SAR signal gets highly saturated due to the morphological characteristics of targets, $\theta_{\text{xP}}$ might provide a similar saturation effect. This appearance is evident during the end of flowering to maturity stages of crops.

We can extend this study to several other crop types using different dual polarimetric SAR sensor configurations, such as HH-HV. The proposed parameter and the clustering framework will be valuable to analyze data from the upcoming dual-pol NASA-ISRO Synthetic Aperture Radar Mission (NISAR) and Copernicus Sentinel SAR missions for agricultural and environmental studies. The open-source code is available at: https://github.com/Subho07/dual_cross_pol_theta_clustering.

**Author Contributions:** Conceptualization, S.D., N.B., A.B. and S.H.; methodology, S.D. and A.B.; programming, S.D. and N.B.; validation, A.B., S.H. and H.M.; format analysis, all; writing, all; supervision, S.H., A.B. and H.M. All authors have read and agreed to the published version of the manuscript.

**Funding:** This research received no external funding.

**Institutional Review Board Statement:** Not applicable.

**Informed Consent Statement:** Not applicable.

**Data Availability Statement:** The Sentinel-1 SLC data were accessed from https://search.asf.alaska.edu/#/, accessed date: 16 July 2021.

| Date | Data ID |
|---|---|
| 13 June | S1A_IW_SLC__1SDV_20160613T001529_20160613T001556_011685_011E64_4083 |
| 7 July | S1A_IW_SLC__1SDV_20160707T001530_20160707T001557_012035_01298D_9FD9 |
| 19 July | S1A_IW_SLC__1SDV_20160719T001540_20160719T001604_012210_012F46_1DED |
| 24 August | S1A_IW_SLC__1SDV_20160824T001533_20160824T001600_012735_0140AB_4BC9 |

**Acknowledgments:** The authors would like to thank the European Space Agency (ESA) for providing Sentinel-1 images over the test site and the members of the SMAPVEX16 science team for providing ground truth information. The authors also acknowledge the GEO-AWS Earth Observation Cloud Credits Program, which supported the computation with Sentinel-1 on AWS cloud platform through the project: AWS4AgriSAR-Crop inventory mapping from SAR data on a cloud computing platform and formed the testbed for processing pipelines. N.B. and S.D. would like to acknowledge the support from the Ministry of Education (formerly the Ministry of Human Resource and Development-MHRD), Govt. of India, towards their doctoral research work.

**Conflicts of Interest:** The authors declare no conflict of interest.

## Appendix A

### *Appendix A.1. Relationship with Cloude's $\alpha_i$'s*

In this section, we derive the relationship of $\theta_{\text{xP}}$ with Cloude $\alpha_i$'s [54]. In this regard, please note that

$$\frac{C_{11}}{\text{Span}} = \sum_{i=1}^{2} p_i \cos^2 \alpha_i, \tag{A1}$$

where $(\alpha_i)_{i=1,2}$ are the individual scattering-type parameters obtained from Cloude target scattering vector [24]. Let us define

$$\tan \eta_1 = \frac{C_{11}}{m_{\text{xP}}\,\text{Span}} = \frac{\sum_{i=1}^{2} p_i \cos^2 \alpha_i}{\sqrt{1-\prod_{i=1}^{2} 2p_i}}, \tag{A2}$$

and

$$\tan \eta_2 = \frac{C_{22}}{m_{\text{xP}}\,\text{Span}} = \frac{1-\sum_{i=1}^{2} p_i \cos^2 \alpha_i}{\sqrt{1-\prod_{i=1}^{2} 2p_i}}. \tag{A3}$$

Therefore, using simple trigonometric formulation, we get

$$\tan(\eta_1 - \eta_2) = \frac{\dfrac{2\sum_{i=1}^{2} p_i \cos^2 \alpha_i - 1}{\sqrt{1-\prod_{i=1}^{2} 2p_i}}}{1 + \dfrac{\left(\sum_{i=1}^{2} p_i \cos^2 \alpha_i\right)\left(1-\sum_{i=1}^{2} p_i \cos^2 \alpha_i\right)}{1-\prod_{i=1}^{2} 2p_i}}, \tag{A4}$$

further simplifying the expression of $\theta_{xP}$ given in terms of $(p_i)_{i=1,2}$ and $(\alpha_i)_{i=1,2}$ as

$$\begin{aligned}\tan\theta_{xP} &= \frac{\text{Num}}{\text{Den}}, \text{ in which}\\ \text{Num} &= \left(2\sum_{i=1}^{2} p_i\cos^2\alpha_i - 1\right)\sqrt{1-\prod_{i=1}^{2} 2p_i}, \text{ and}\\ \text{Den} &= 1-\prod_{i=1}^{2} 2p_i +\\ &\left(\sum_{i=1}^{2} p_i\cos^2\alpha_i\right)\left(1-\prod_{i=1}^{2} p_i\cos^2\alpha_i\right).\end{aligned} \tag{A5}$$

Therefore, one can note that $\theta_{xP}$ is a function of $\sum_{i=1}^{2} p_i\cos^2\alpha_i$, and the Barakat degree of polarization, $\sqrt{1-\prod_{i=1}^{2} 2p_i}$. We have shown the $\widehat{\alpha} - \theta_{xP}$ scatter plot for VV-VH polarization modes in Figure A1. Here, $\widehat{\alpha} = 45° - \overline{\alpha}$

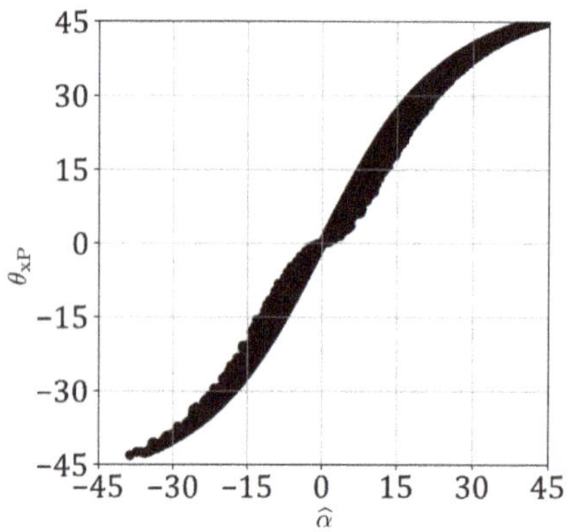


**Figure A1.** $\widehat{\alpha} - \theta_{xP}$ plot for RADARSAT-2 data over San Francisco in VV-VH mode.

In Figure A1, the deviation of data points from the 1:1 line in the plots can be observed in both VV-VH scatter plots. This deviation could be due to the presence of a partially polarized EM wave that is seized by $\theta_{xP}$ through the utilization of $m_{xP}$. On the contrary, $\widehat{\alpha}$ misses this information that is obtained by eigen-decomposition of the covariance matrix. One should note that for completely polarized ($m_{xP} = 1$) EM wave from coherent targets, both $\widehat{\alpha}$ and $\theta_{xP}$ will either be equal to $-45°$ or $45°$.

*Appendix A.2. Temporal Variations of $\theta_{xP}$ and $\widehat{\alpha}$*

**Table A1.** Temporal variations of $\theta_{xP}$ and $\widehat{\alpha}$ over canola and wheat at different dates.

| | | 13 June | 7 July | 19 July | 24 August |
|---|---|---|---|---|---|
| **Canola** | $\theta_{xP}$ | 40° | 20° | 18° | 34° |
| | $\widehat{\alpha}$ | 37° | 18° | 16° | 32° |
| **Wheat** | $\theta_{xP}$ | 39° | 27° | 20° | 34° |
| | $\widehat{\alpha}$ | 37° | 26° | 15° | 28° |

### Appendix A.3. Temporal Variations of In-Situ Measurements of Crops

**Table A2.** Temporal variations of Plant Area Index (PAI), dry biomass (kg $m^{-2}$), and Vegetation Water Content (VWC) (kg $m^{-2}$) for canola and wheat at different dates.

| | | 13 June | 7 July | 19 July | 24 August |
|---|---|---|---|---|---|
| **Canola** | Phenology | Leaf development | Flowering stage | Pod development | Maturity/ harvest |
| | PAI | $1.82 \pm 0.43$ | $4.02 \pm 0.62$ | $6.32 \pm 0.16$ | N/A |
| | dry biomass | $0.21 \pm 0.08$ | $0.43 \pm 0.04$ | $0.76 \pm 0.06$ | N/A |
| | VWC | $1.20 \pm 0.13$ | $5.82 \pm 0.32$ | $5.96 \pm 0.20$ | N/A |
| **Wheat** | Phenology | Tillering stage | Early flowering stage | Early dough stage | Maturity/ harvest |
| | PAI | $2.78 \pm 0.31$ | $5.92 \pm 0.22$ | $6.52 \pm 0.11$ | N/A |
| | dry biomass | $0.23 \pm 0.04$ | $0.57 \pm 0.02$ | $0.98 \pm 0.07$ | N/A |
| | VWC | $2.21 \pm 0.12$ | $5.74 \pm 0.24$ | $6.11 \pm 0.15$ | N/A |

## References


1. Mandal, D.; Kumar, V.; Bhattacharya, A.; Rao, Y.S.; Siqueira, P.; Bera, S. Sen4Rice: A processing chain for differentiating early and late transplanted rice using time-series Sentinel-1 SAR data with Google Earth Engine. *IEEE Geosci. Remote Sens. Lett.* **2018**, *15*, 1947–1951. [CrossRef]
2. Dey, S.; Mandal, D.; Robertson, L.D.; Banerjee, B.; Kumar, V.; McNairn, H.; Bhattacharya, A.; Rao, Y. In-season crop classification using elements of the Kennaugh matrix derived from polarimetric RADARSAT-2 SAR data. *Int. J. Appl. Earth Obs. Geoinf.* **2020**, *88*, 102059. [CrossRef]
3. Mandal, D.; Kumar, V.; McNairn, H.; Bhattacharya, A.; Rao, Y. Joint estimation of Plant Area Index (PAI) and wet biomass in wheat and soybean from C-band polarimetric SAR data. *Int. J. Appl. Earth Obs. Geoinf.* **2019**, *79*, 24–34. [CrossRef]
4. Mandal, D.; Kumar, V.; Lopez-Sanchez, J.M.; Bhattacharya, A.; McNairn, H.; Rao, Y. Crop biophysical parameter retrieval from Sentinel-1 SAR data with a multi-target inversion of Water Cloud Model. *Int. J. Remote Sens.* **2020**, *41*, 5503–5524. [CrossRef]
5. Mandal, D.; Hosseini, M.; McNairn, H.; Kumar, V.; Bhattacharya, A.; Rao, Y.; Mitchell, S.; Robertson, L.D.; Davidson, A.; Dabrowska-Zielinska, K. An investigation of inversion methodologies to retrieve the leaf area index of corn from C-band SAR data. *Int. J. Appl. Earth Obs. Geoinf.* **2019**, *82*, 101893. [CrossRef]
6. Mandal, D.; Ratha, D.; Bhattacharya, A.; Kumar, V.; McNairn, H.; Rao, Y.S.; Frery, A.C. A radar vegetation index for crop monitoring using compact polarimetric SAR data. *IEEE Trans. Geosci. Remote Sens.* **2020**, *58*, 6321–6335. [CrossRef]
7. Mandal, D.; Kumar, V.; Ratha, D.; Lopez-Sanchez, J.M.; Bhattacharya, A.; McNairn, H.; Rao, Y.; Ramana, K. Assessment of rice growth conditions in a semi-arid region of India using the Generalized Radar Vegetation Index derived from RADARSAT-2 polarimetric SAR data. *Remote Sens. Environ.* **2020**, *237*, 111561. [CrossRef]
8. Bhogapurapu, N.; Dey, S.; Bhattacharya, A.; Mandal, D.; Lopez-Sanchez, J.M.; McNairn, H.; López-Martínez, C.; Rao, Y.S. Dual-polarimetric descriptors from Sentinel-1 GRD SAR data for crop growth assessment. *ISPRS J. Photogramm. Remote Sens.* **2021**, *178*, 20–35. [CrossRef]
9. Ferrazzoli, P.; Paloscia, S.; Pampaloni, P.; Schiavon, G.; Solimini, D.; Coppo, P. Sensitivity of microwave measurements to vegetation biomass and soil moisture content: A case study. *IEEE Trans. Geosci. Remote Sens.* **1992**, *30*, 750–756. [CrossRef]
10. Davidson, M.W.; Le Toan, T.; Mattia, F.; Satalino, G.; Manninen, T.; Borgeaud, M. On the characterization of agricultural soil roughness for radar remote sensing studies. *IEEE Trans. Geosci. Remote Sens.* **2000**, *38*, 630–640. [CrossRef]
11. Wiseman, G.; McNairn, H.; Homayouni, S.; Shang, J. RADARSAT-2 polarimetric SAR response to crop biomass for agricultural production monitoring. *IEEE J. Sel. Top. Appl. Earth Obs. Remote Sens.* **2014**, *7*, 4461–4471. [CrossRef]
12. Lee, J.S.; Grunes, M.R.; Pottier, E. Quantitative comparison of classification capability: Fully polarimetric versus dual and single-polarization SAR. *IEEE Trans. Geosci. Remote Sens.* **2001**, *39*, 2343–2351.
13. Kussul, N.; Lemoine, G.; Gallego, F.J.; Skakun, S.V.; Lavreniuk, M.; Shelestov, A.Y. Parcel-based crop classification in Ukraine using Landsat-8 data and Sentinel-1A data. *IEEE J. Sel. Top. Appl. Earth Obs. Remote Sens.* **2016**, *9*, 2500–2508. [CrossRef]
14. Nguyen, D.B.; Gruber, A.; Wagner, W. Mapping rice extent and cropping scheme in the Mekong Delta using Sentinel-1A data. *Remote Sens. Lett.* **2016**, *7*, 1209–1218. [CrossRef]
15. Bargiel, D. A new method for crop classification combining time series of radar images and crop phenology information. *Remote Sens. Environ.* **2017**, *198*, 369–383. [CrossRef]
16. Van Tricht, K.; Gobin, A.; Gilliams, S.; Piccard, I. Synergistic use of radar Sentinel-1 and optical Sentinel-2 imagery for crop mapping: A case study for Belgium. *Remote Sens.* **2018**, *10*, 1642. [CrossRef]
17. Whelen, T.; Siqueira, P. Time-series classification of Sentinel-1 agricultural data over North Dakota. *Remote Sens. Lett.* **2018**, *9*, 411–420. [CrossRef]

18. Minasny, B.; Shah, R.M.; Che Soh, N.; Arif, C.; Indra Setiawan, B. Automated Near-Real-Time Mapping and Monitoring of Rice Extent, Cropping Patterns, and Growth Stages in Southeast Asia Using Sentinel-1 Time Series on a Google Earth Engine Platform. *Remote Sens.* **2019**, *11*, 1666.
19. Arias, M.; Campo-Bescós, M.Á.; Álvarez-Mozos, J. Crop Classification Based on Temporal Signatures of Sentinel-1 Observations over Navarre Province, Spain. *Remote Sens.* **2020**, *12*, 278. [CrossRef]
20. Bousbih, S.; Zribi, M.; Lili-Chabaane, Z.; Baghdadi, N.; El Hajj, M.; Gao, Q.; Mougenot, B. Potential of Sentinel-1 radar data for the assessment of soil and cereal cover parameters. *Sensors* **2017**, *17*, 2617. [CrossRef]
21. Kumar, P.; Prasad, R.; Gupta, D.; Mishra, V.; Vishwakarma, A.; Yadav, V.; Bala, R.; Choudhary, A.; Avtar, R. Estimation of winter wheat crop growth parameters using time series Sentinel-1A SAR data. *Geocarto Int.* **2018**, *33*, 942–956. [CrossRef]
22. Dey, S.; Bhogapurapu, N.; Bhattacharya, A.; Mandal, D.; Lopez-Sanchez, J.M.; McNairn, H.; Frery, A.C. Rice phenology mapping using novel target characterization parameters from polarimetric SAR data. *Int. J. Remote Sens.* **2021**, *42*, 5519–5543. [CrossRef]
23. Dey, S.; Bhattacharya, A.; Ratha, D.; Mandal, D.; McNairn, H.; Lopez-Sanchez, J.M.; Rao, Y. Novel clustering schemes for full and compact polarimetric SAR data: An application for rice phenology characterization. *ISPRS J. Photogramm. Remote Sens.* **2020**, *169*, 135–151. [CrossRef]
24. Cloude, S. The dual polarisation entropy/alpha decomposition. In Proceedings of the 3rd International Workshop on Science and Applications of SAR Polarimetry and Polarimetric Interferometry, Frascati, Italy, 22–26 January 2007; pp. 22–26.
25. Ainsworth, T.; Schuler, D.; Lee, J.S. Polarimetric SAR characterization of man-made structures in urban areas using normalized circular-pol correlation coefficients. *Remote Sens. Environ.* **2008**, *112*, 2876–2885. [CrossRef]
26. Sugimoto, M.; Ouchi, K.; Nakamura, Y. On the similarity between dual-and quad-eigenvalue analysis in SAR polarimetry. *Remote Sens. Lett.* **2013**, *4*, 956–964. [CrossRef]
27. Freeman, A.; Durden, S.L. A three-component scattering model for polarimetric SAR data. *IEEE Trans. Geosci. Remote Sens.* **1998**, *36*, 963–973. [CrossRef]
28. Yamaguchi, Y.; Moriyama, T.; Ishido, M.; Yamada, H. Four-component scattering model for polarimetric SAR image decomposition. *IEEE Trans. Geosci. Remote Sens.* **2005**, *43*, 1699–1706. [CrossRef]
29. Shan, Z.; Wang, C.; Zhang, H.; Chen, J. H-alpha decomposition and alternative parameters for dual Polarization SAR data. In Proceedings of the PIERS, Suzhou, China, 12–16 September 2011.
30. Xie, L.; Zhang, H.; Wang, C.; Wu, F.; Zhang, B.; Tang, Y. Maritime application using Hα decomposition in compact and dual-pol SAR. In Proceedings of the 2013 Asia-Pacific Conference on Synthetic Aperture Radar (APSAR), Tsukuba, Japan, 23–27 September 2013; pp. 563–566.
31. Sugimoto, M.; Ouchi, K.; Yang, C.S. On the eigenvalue analysis using HH-VV dual-polarization SAR data and its applications to monitoring of coastal oceans. In *Ocean Sensing and Monitoring V*; International Society for Optics and Photonics: Bellingham, WA, USA, 2013; Volume 8724, p. 87240G.
32. Trudel, M.; Charbonneau, F.; Leconte, R. Using RADARSAT-2 polarimetric and ENVISAT-ASAR dual-polarization data for estimating soil moisture over agricultural fields. *Can. J. Remote Sens.* **2012**, *38*, 514–527.
33. Periasamy, S. Significance of dual polarimetric synthetic aperture radar in biomass retrieval: An attempt on Sentinel-1. *Remote Sens. Environ.* **2018**, *217*, 537–549. [CrossRef]
34. Mandal, D.; Kumar, V.; Ratha, D.; Dey, S.; Bhattacharya, A.; Lopez-Sanchez, J.M.; McNairn, H.; Rao, Y.S. Dual polarimetric radar vegetation index for crop growth monitoring using sentinel-1 SAR data. *Remote Sens. Environ.* **2020**, *247*, 111954. [CrossRef]
35. Dey, S.; Chaudhuri, U.; Mandal, D.; Bhattacharya, A.; Banerjee, B.; McNairn, H. BiophyNet: A Regression Network for Joint Estimation of Plant Area Index and Wet Biomass From SAR Data. *IEEE Geosci. Remote Sens. Lett.* **2020**, *18*, 1701–1705. [CrossRef]
36. Dey, S.; Chaudhuri, U.; Bhogapurapu, N.R.; Lopez-Sanchez, J.M.; Banerjee, B.; Bhattacharya, A.; Mandal, D.; Rao, Y. Synergistic Use of TanDEM-X and Landsat-8 Data for Crop-type Classification and Monitoring. *IEEE J. Sel. Top. Appl. Earth Obs. Remote Sens.* **2021**, *14*, 8744–8760. [CrossRef]
37. Nasrallah, A.; Baghdadi, N.; El Hajj, M.; Darwish, T.; Belhouchette, H.; Faour, G.; Darwich, S.; Mhawej, M. Sentinel-1 data for winter wheat phenology monitoring and mapping. *Remote Sens.* **2019**, *11*, 2228. [CrossRef]
38. Wali, E.; Tasumi, M.; Moriyama, M. Combination of Linear Regression Lines to Understand the Response of Sentinel-1 Dual Polarization SAR Data with Crop Phenology—Case Study in Miyazaki, Japan. *Remote Sens.* **2020**, *12*, 189. [CrossRef]
39. Schlund, M.; Erasmi, S. Sentinel-1 time series data for monitoring the phenology of winter wheat. *Remote Sens. Environ.* **2020**, *246*, 111814. [CrossRef]
40. Lopez-Sanchez, J.M.; Cloude, S.R.; Ballester-Berman, J.D. Rice phenology monitoring by means of SAR polarimetry at X-band. *IEEE Trans. Geosci. Remote Sens.* **2011**, *50*, 2695–2709. [CrossRef]
41. Lopez-Sanchez, J.M.; Vicente-Guijalba, F.; Ballester-Berman, J.D.; Cloude, S.R. Polarimetric Response of Rice Fields at C-Band: Analysis and Phenology Retrieval. *IEEE Trans. Geosci. Remote Sens.* **2014**, *52*, 2977–2993. [CrossRef]
42. McNairn, H.; Jiao, X.; Pacheco, A.; Sinha, A.; Tan, W.; Li, Y. Estimating canola phenology using synthetic aperture radar. *Remote Sens. Environ.* **2018**, *219*, 196–205. [CrossRef]
43. Dey, S.; Bhattacharya, A.; Ratha, D.; Mandal, D.; Frery, A.C. Target Characterization and Scattering Power Decomposition for Full and Compact Polarimetric SAR Data. *IEEE Trans. Geosci. Remote Sens.* **2020**, *59*, 3981–3998. [CrossRef]
44. Dey, S.; Bhattacharya, A.; Frery, A.C.; López-Martínez, C.; Rao, Y.S. A Model-Free Four Component Scattering Power Decomposition for Polarimetric SAR Data. *IEEE J. Sel. Top. Appl. Earth Obs. Remote Sens.* **2021**, *14*, 3887–3902. [CrossRef]

45. Barakat, R. n-fold polarization measures and associated thermodynamic entropy of N partially coherent pencils of radiation. *Opt. Acta Int. J. Opt.* **1983**, *30*, 1171–1182. [CrossRef]
46. McNairn, H.; Tom, J.; Powers, J.; Bélair, J.; Berg, A.; Bullock, A.; Colliander, A.; Cosh, A.; Kim, M.; Ramata, S.; et al. Experimental Plan SMAP Validation Experiment 2016 in Manitoba, Canada (SMAPVEX16-MB). 2016. Available online: http://smapvex16-mb.espaceweb.usherbrooke.ca/ (accessed on 10 October 2021).
47. Mandal, D.; Vaka, D.S.; Bhogapurapu, N.R.; Vanama, V.; Kumar, V.; Rao, Y.S.; Bhattacharya, A. Sentinel-1 SLC preprocessing workflow for polarimetric applications: A generic practice for generating dual-pol covariance matrix elements in SNAP S-1 toolbox. *Preprints* **2019**. [CrossRef]
48. Bhogapurapu, N.; Dey, S.; Mandal, D.; Bhattacharya, A.; Rao, Y.S. PolSAR tools: A QGIS plugin for generating SAR descriptors. *J. Open Source Softw.* **2021**, *6*, 2970. [CrossRef]
49. Barakat, R. Degree of polarization and the principal idempotents of the coherency matrix. *Opt. Commun.* **1977**, *23*, 147–150. [CrossRef]
50. Cloude, S. *Polarisation: Applications in Remote Sensing*; Oxford University Press: Oxford, UK, 2009.
51. Brosseau, C. Polarization transfer and entropy transformation. *Optik* **1991**, *88*, 109–117.
52. Bicout, D.; Brosseau, C. Multiply scattered waves through a spatially random medium: Entropy production and depolarization. *J. Phys. I* **1992**, *2*, 2047–2063. [CrossRef]
53. Wu, L.k.; Moore, R.K.; Zoughi, R. Sources of scattering from vegetation canopies at 10 Ghz. *IEEE Trans. Geosci. Remote Sens.* **1985**, *5*, 737–745. [CrossRef]
54. Cloude, S.R.; Pottier, E. An entropy based classification scheme for land applications of polarimetric SAR. *IEEE Trans. Geosci. Remote Sens.* **1997**, *35*, 68–78. [CrossRef]

MDPI
St. Alban-Anlage 66
4052 Basel
Switzerland
Tel. +41 61 683 77 34
Fax +41 61 302 89 18
www.mdpi.com

*Remote Sensing* Editorial Office
E-mail: remotesensing@mdpi.com
www.mdpi.com/journal/remotesensing

www.ingramcontent.com/pod-product-compliance
Lightning Source LLC
Chambersburg PA
CBHW061438210926
51850CB00008B/1547
* 9 7 8 3 0 3 6 5 7 3 5 8 8 *